Journal of Applied Logics - IfCoLog Journal of Logics and their Applications

Volume 8, Number 1

February 2021

Disclaimer

Statements of fact and opinion in the articles in Journal of Applied Logics - IfCoLog Journal of Logics and their Applications (JALs-FLAP) are those of the respective authors and contributors and not of the JALs-FLAP. Neither College Publications nor the JALs-FLAP make any representation, express or implied, in respect of the accuracy of the material in this journal and cannot accept any legal responsibility or liability for any errors or omissions that may be made. The reader should make his/her own evaluation as to the appropriateness or otherwise of any experimental technique described.

© Individual authors and College Publications 2021
All rights reserved.

ISBN 978-1-84890-356-2
ISSN (E) 2631-9829
ISSN (P) 2631-9810

College Publications
Scientific Director: Dov Gabbay
Managing Director: Jane Spurr

http://www.collegepublications.co.uk

All rights reserved. No part of this publication may be reproduced, stored in a retrieval system or transmitted in any form, or by any means, electronic, mechanical, photocopying, recording or otherwise without prior permission, in writing, from the publisher.

Editorial Board

Editors-in-Chief
Dov M. Gabbay and Jörg Siekmann

Marcello D'Agostino
Natasha Alechina
Sandra Alves
Arnon Avron
Jan Broersen
Martin Caminada
Balder ten Cate
Agata Ciabttoni
Robin Cooper
Luis Farinas del Cerro
Esther David
Didier Dubois
PM Dung
David Fernandez Duque
Jan van Eijck
Marcelo Falappa
Amy Felty
Eduaro Fermé

Melvin Fitting
Michael Gabbay
Murdoch Gabbay
Thomas F. Gordon
Wesley H. Holliday
Sara Kalvala
Shalom Lappin
Beishui Liao
David Makinson
George Metcalfe
Claudia Nalon
Valeria de Paiva
Jeff Paris
David Pearce
Pavlos Peppas
Brigitte Pientka
Elaine Pimentel

Henri Prade
David Pym
Ruy de Queiroz
Ram Ramanujam
Chrtian Retoré
Ulrike Sattler
Jörg Siekmann
Jane Spurr
Kaile Su
Leon van der Torre
Yde Venema
Rineke Verbrugge
Heinrich Wansing
Jef Wijsen
John Woods
Michael Wooldridge
Anna Zamansky

Scope and Submissions

This journal considers submission in all areas of pure and applied logic, including:

pure logical systems
proof theory
constructive logic
categorical logic
modal and temporal logic
model theory
recursion theory
type theory
nominal theory
nonclassical logics
nonmonotonic logic
numerical and uncertainty reasoning
logic and AI
foundations of logic programming
belief change/revision
systems of knowledge and belief
logics and semantics of programming
specification and verification
agent theory
databases

dynamic logic
quantum logic
algebraic logic
logic and cognition
probabilistic logic
logic and networks
neuro-logical systems
complexity
argumentation theory
logic and computation
logic and language
logic engineering
knowledge-based systems
automated reasoning
knowledge representation
logic in hardware and VLSI
natural language
concurrent computation
planning

This journal will also consider papers on the application of logic in other subject areas: philosophy, cognitive science, physics etc. provided they have some formal content.

Submissions should be sent to Jane Spurr (jane@janespurr.net) as a pdf file, preferably compiled in LaTeX using the IFCoLog class file.

CONTENTS

ARTICLES

Guest Editor's Remarks 1
John Woods

Douglas Walton: The Early Years 3
John Woods

Douglas Walton and the Covid-19 crisis 31
Louise Cummings

Argumentation Schemes for Composition and Division Arguments:
A Critique of Walton's Account 53
Maurice A. Finocchiaro

Appealing to Ignorance? De-extinction and Accounts of a Fallacy 75
Trudy Govier

Turning the Tables: Up- and Downgrading of Evaluative Terms in Public
Controversies 89
Jan Albert van Laar and Erik C. W. Krabbe

Walton on Ethical Argumentation 115
James B. Freeman

Argument is Moral. Using Walton's Dialectical Tools to Evaluate Argumentation from a Moral Perspective 137
Katharina Stevens

Dialogue Types, Argumentation Schemes and Mathematical Practice: Douglas Walton and Mathematics 159
Andrew Aberdein

Walton on Argument, Arguments, and Argumentation 183
Harvey Siegel and John Biro

Aspects of Walton's Theory of Argumentation Schemes 195
Hans V. Hansen

Less Scheming, More Typing: Musings on the Waltonian Legacy in Argument Technologies 219
Fabio Paglieri

The Waltonian Foundations of Argument Technology 245
Chris Reed

Argument Schemes and Dialogue Protocols: Doug Walton's Legacy in Artificial Intelligence 263
Peter McBurney and Simon Parsons

Critical Questions to Argumentation Schemes in Statutory Interpretation . 291
Michał Araszkiewicz

Guest Editor's Remarks

John Woods
Department of Philosophy, University of British Columbia, 1866 Main Mall, Vancouver, BC, V6T 1Z1.
`john.woods@ubc.ca`

In the first week of the year just passed, Doug died in his sleep, without prior notice or expectation. His decease brought an end to an extraordinarily influential presence in informal logic, and did so at the subject's first mid-century from its nominal launch in 1970, initiated by seminal work by Hamblin and Harman. Doug's own pioneering work in that slipstream dates from 1972 and made its multipli-fruitful way for forty-eight more years. As the Church of England is said to be, argumentation theory is also said to be, and actually is, a big tent; and informal logic is one of its early founders, also itself something of a big tent. Hamblin was firmly of the view that fallacy theory is a fundamental part of logic. Harman was firmly of the view that the theory of inference is a fundamental part of epistemology. All his working life, Doug saw himself as a logician, albeit one with spiralling cross-disciplinary interests. His epistemological instincts were on the old-fashioned side — I mean old-fashioned in relation to his time. Sceptical of mentalist measures, and drawn to behaviourist ones, Doug saw pragmatics as the preferred theoretical context for his own work.

Neither lamentation nor hagiography, the purpose of this Special Issue is to mark informal logic's first fifty years and, by examination of Doug's formidable presence of it, help to assess the present state and future prospects of informal logic in all its manifestations. All the essays selected for the task were by invitation. Their very capable authors have been drawn from every age group of the subject's past half-century, many of whom have made formidable contributions of their own and several of whom are well-set on a course of like importance. To the extent possible, the contributions have been organized thematically in accordance with Doug's evolving interests. Coverage begins with the early years and Doug's transition from the philosophy of action to fallacy theory. It is not widely-enough known that Doug had an early interest in ethical matters. Indeed his first book is a contribution to biomedical ethics, written in the busy decade of the Woods-Walton Approach to fallacy theory. Doug's ethical interests continued, and are discussed in the text tranche of the issue's papers. There follow contributions on Doug's philosophy of argument more generally conceived of, and on the emphasis he places on its dialogical, dialectical, and structural features, and on his looming and soon to be dominant preoccupation with argument types, argumentation schemes and the importance of critical questions for their profitable implementation. At the heart of it all lies the question of the relation between argument and inference, and a years-long attachment to the importance of third-way reasoning. It is a matter of some interest (to me) that my own opinions of each of these matters lie athwart Doug's own. The issue closes with contributions on Doug's pioneering work in argument technology, the place where he found his most creative pulse. It is here that he achieved his greatest theoretical maturity by finding in AI and its applications to legal reasoning a productive home for his complex philosophy of argument-structure.

I warmly thank the authors below for their valuable and stimulating essays, and Dov Gabbay, JAL's founding and present editor, for the hospitality of a journal whose name so perfectly fits the character of Doug's work. As usual, the journal's managing editor, Jane Spurr, has been enormously helpful in translating most files into Latex, no doubt with the deep gratitude of all involved. Jane does the heavy lifting in London, and Carol Woods does the heavy lifting in Vancouver. Without Jane Dov would have to seek a career in real estate and, without Carol, I might have to give dentistry a shot. For refreshment of memory, I am especially grateful to Christopher Tindale and Karen Walton in Windsor.

Douglas Walton: The Early Years

John Woods
Department of Philosophy, University of British Columbia, 1866 Main Mall, Vancouver, BC, V6T 1Z1.
john.woods@ubc.ca

Abstract

In one of his last writings, Douglas Walton tells of his journey as a young modal logician to the heights of his achievements as a computational analyst of argument.[1] In it, he makes no mention of the work that had won him considerable early recognition in the years from 1972 to 1982. Of course, there was no space in that short 2019 note for exposure of all that had mattered in Doug's philosophical working life. But in not mentioning it there, it was made plain how thoroughly he had shred methodological assumptions of his early work. Since I myself was privy, indeed party, to them, it falls to me to try to give some sense of their later loss of purchase. In what follows, I cover the first two intervals of the early period, beginning with the doctoral years in Toronto, then moving to the ten years of his and my work on fallacy theory and related matters. In so doing, I will say more of myself than is fitting in a paper about Doug. But, as I was an eye-witness to the events I'm about to recount, I don't see how it can be avoided. For reasons that will also be obvious, in referring to him and to self, I shall forego the conventional use of surnames.

Part I: The doctoral years 1964-1972

In this first part of the paper, I would like to give some impression of what would have awaited Doug in 1964 when he arrived in Toronto for the PhD. I would also like to say something about the conditions in play when Doug and I made our first acquaintance in the one course he took from me there. I will conclude this part with remarks about the unusual circumstances under which Doug wrote his dissertation under my supervision.

For valuable comments and suggestions, my thanks to David Godden and Alirio Rosales. For help in jogging my memory, special thanks to Karen Walton and Chris Tindale.

[1]Douglas Walton, "My continuing journey from modal logic to computational logic", *Felsefe Arkivi*, 51 (2019), 321-331. *Felsefe Arkivi* is published by the Faculty of Philosophy at the University of Istanbul, and was co-founded in 1946 by Hans Reichenbach.

1 The Toronto philosophy department

In 1964, the University of Toronto's department of philosophy was in the throes of a great change from one of the English world's foremost places for the history of philosophy to yet another place in which, in a loosely convenient manner of speaking, analytic philosophy would aspire to prevail. In the course of this change, most of what would remain of Toronto's historical importance for philosophy would find safe harbour where it had always flourished, and still does, in the University's confederated St. Michael's College and the Pontifical Institute for Mediaeval Studies. I mention in passing that St. Mike's was the undergraduate collegiate home-base for our other lately lamented friend Bob Pinto.[2] Underlying all this strife of self-reimagining were the forces loosed on the Western democracies by the Soviet launch of Sputnik in 1957. This was the founding moment of the space-race between the powers of darkness and the now troubled places of freedom and light. In his inaugural address in January 1961, U. S. President Kennedy announced that before the decade had passed, the United States would place a man on the moon. In the aftermath of Sputnik and the president's inaugural address, Canadian universities began transitioning from their historic emphasis on the liberal arts and sciences to their present attachment to STEM. In 1959, except for a few examples of departments lodged in what used to be comfortable and amply-sized single-family houses on the westside of it, St. George Street was the western academic extremity of the University of Toronto. Three years later, the spanking new Sydney Smith Hall, now the largest academic building in the University, filled a square block bounded on the east by the aforementioned St. George Street, and not entirely willingly, the department of philosophy had been installed there from the main part of the campus. Three years after that, the philosophy department was evicted to a nonacademic building one block further west and three blocks to the south. It was the Superintendent's Building, ascending heavenwards in ten stories, three floors higher "than a buildin' oughter grow".[3] There was simply no room for the ever-expanding philosophy department that was convenient to where its classes were still being taught.

As can easily be imagined in these circumstances, for most years of that decade, academic appointment was a buyers' market, and there arose in Toronto and everywhere else an unhealthy gap between supply and demand. It hardly needs saying that Toronto's great philosophical make-over would be faced with the need for new people in sufficient numbers and quality to endow the emergingly analytical department with the sterling reputation of its former historical self. In my own nine years

---

[2]I wrote of Bob's sad decease in "In memoriam: Robert C. Pinto (1935-2019)" for *Argumentation*, 33 (2019), 459-463.

[3]"Everythin's up to date in Kansas City", from Rodgers & Hammerstein, "Oklahoma", 1943.

there, and Doug's five, Toronto's philosophy department found itself in this buyers' market. Though I say so myself, I don't doubt that my own unexpected presence in the department was a reflection of the then-current state of the market. (See section 2 below.)

At this same time, the University of Toronto had established satellite campuses just past the eastern and western boundaries of Metropolitan Toronto. Each new campus was built from scratch, and was amply staffed in a tight market. The surprising thing is that, every now and then, people of high quality would be appointed. Howard Sobel, a UCLA philosopher was appointed to what is now called U of T Scarborough (on the east of Metro), and Alasdair Urquhart at U of T Mississauga (on the west). Sobel would serve (at least for awhile) as one of Doug's dissertation advisors. Urquhart has gone on to have a distinguished career as a logician and computer scientist. I once asked Doug whether he had had any contact with Alasdair, and he said "regrettably not". Both Scarborough and Mississauga are undergraduate-only institutions, but anyone in those satellite campuses who was considered suited to teach a course on the graduate curriculum of the full-service campus U of T St. George in mid-town Toronto, was eligible for membership in the University's Faculty of Graduate Studies. The call, I think, was ultimately St. George's. Since all its graduate courses are taught on the St. George campus, it can be a bit tricky to make the trip in inclement weather. On the days when Sobel came into town, I saw a lot of him, partly because of his advisory role in Doug's dissertation work, but also because he and I had been on friendly terms as graduate students. I was interested to learn that Sobel had developed a fond respect for an undergraduate education in philosophy. He told me once that he thought Scarborough's *undergraduates* were better undergraduates than his St. George *graduate* students were graduate students. I pressed him on this, and he assured me that he himself felt well-nourished by an undergraduate-teaching diet. "Besides", he said "if not highly talented, graduate students are a nuisance." Well, they certainly are time-consuming and, if one's time is already fully-booked, one could understand the reluctance of someone as busy as Doug came to be.[4]

In my own undergraduate days in Toronto (1954-1958), the department was already graced by some people of high reputation in the analytic field. T. A. Goudge, an eminent Peirce scholar, published a pathway book in the philosophy of biology, for which to this day he lacks the credit due him. *The Ascent of Life* (1950) won the Governor-General's medal for nonfiction, and played a foundational role in weaning

---

[4]Of course, Sobel's was an over-small sample. Off-hand, I could easily name half a dozen of Toronto's graduate students from the era who have had very successful careers, some of whom have achieved considerable distinction. But I could see what Howard meant. Perhaps he had felt somewhat put upon by some of his UCLA graduate students.

the philosophy of biology from the ways in which philosophers had been modelling physics. William Dray had done the same for the philosophy of history with *Laws and Explanation in History* (1957), which effectively dethroned Hempel's convering law model of historical explanation. David Gauthier was a well-received and somewhat revisionist Hobbes scholar and a rapidly rising star whose reputation would be made with the publication of *Morals by Agreement* (1986). Gauthier was my first exposure, and Doug's too I am sure, to game-theoretic methods in philosophy. I must also mention another new arrival, J. M. O. Wheatley, a student of A. J. Ayer's, for his superb course in mathematical logic. I took it in my MA year of 1958-1959, and we were all encouraged to fathom the English version of *Grundzüge der Theoretische Logik* of Hilbert and Ackermann. Jim was a pioneer in the development of optative logics, yet another expansion of modal talk beyond the alethic properties of necessity and possibility. Jim was also the supervisor of my MA thesis on perceptual relativity, in 1959. Among the notables, Bas van Fraassen arrived in Toronto from Pittsburgh in the academic year 1969-1970, as had (or soon would) Jack Canfield from Cornell, Jack Stevenson from Manitoba, and Ronald Butler from Princeton. The two Jacks remained until retirement but, not long after his arrival, Bas began dividing his year between Toronto and the University of Southern California, and later would move full-time to Princeton. So the losses amounted to something — Dray and Gallop, to Trent, Gauthier to Pittsburgh, Ronald Butler to Waterloo and thence to the University of Kent, and now Bas.[5]

2 Doug's arrival in Toronto

Doug came to the University of Toronto in 1964 as a first-year PhD student, having achieved an Honours degree in philosophy from the University of Waterloo. This was two years after I myself arrived as a newly appointed member of faculty, following three years of residency in Ann Arbor as a doctoral student at the University of Michigan. My own arrival had been sudden. I had just finished my coursework-requirements and passed the rather tough week-long Preliminary Exams, and only then had been admitted to candidacy and had been free to start work on my doctoral dissertation. But when the Toronto call came, I was honoured to be restored to my *alma mater* and my country. I mention this to make the point that when Doug arrived at age twenty-two, I was not yet twenty-seven and, in matters of completed formal qualifications, I bettered him by a meagre one-year MA. Although Doug

---

[5]I would not like to leave the impression that these were the sole people of note in the intake years from 1962 onwards. Among those who come readily to mind are Hans Herzberger, Ronald de Sousa, Fred Wilson, Lloyd Gerson and Wayne Sumner.

was just starting his doctoral studies, I was only half way through my own. We were fellow questers for the PhD. Later, Bob Pinto would playfully observe that Doug and I were just a couple of young whipper-snappers on the make. In large measure, I wrote my doctoral dissertation by teaching it. Doug enrolled in my first graduate seminar, which focused on entailment and the paradoxes of strict implication (my dissertation title). It took account of the fact that, notwithstanding the considerable differences between them, each of C. I. Lewis' systems for modal propositional logic — S1, S2, S3, S4 and S5 — generates a theorem known as *ex falso quodlibet*, according to which any contradictory sentence sanctioned by a system strictly implies every sentence whatever expressible in the language of that system, including all their negations.[6] In their 1932 book *Symbolic Logic*, Lewis and his University of Michigan co-author C. H. Langford, advanced a conditional proof of *ex falso* from the assumption of a contradictory conjunction applying, in this order, the proof-rules ∧-elimination, ∨-introduction, ∧-elimination again, disjunctive syllogism, and conditionalization.[7] Since Lewis and Langford took these rules to be valid for natural languages, they concluded that *ex falso* records a logical fact about English: *ex falso* is a true entailment-statement of English. Although Lewis and Langford held that their proof was valid for strict implication, it was actually set out in a classical notation and its proof rules were also classically valid. In all our dealings with it in the graduate seminar, it could hardly be said that this was an exercise in modal logic.[8]

The greater part of the course was reserved for the consideration of relevant logics as the better theoretical working-up of the entailment relation. It wouldn't be far wrong to say that the course was an introduction to relevant logics of the sort ensuing from Anderson and Belnap, first at Yale and later at Pittsburgh, and originating in a joint paper in 1959 on a simple treatment of truth-functions, in which it is reported in a footnote that the disjunctive syllogism rule just *happens* to fail (!).[9] Soon to stir were similar — and later not so similar — logics from the likes of Richard and Val Routley, Bob Martin, and Graham Priest in Australia, and Stephen Read in

---

[6] The same is true of intuitionist systems. *Ex falso* fails in most relevant logics and all paraconsistent and dialetheic ones.

[7] C. I. Lewis and C. H. Langford, *Symbolic Logic*, New York: Appleton Century-Croft, 1932. Reissued in New York by Dover, 1959.

[8] It might be of some interest that Lewis and Langford weren't the originators of that *ex falso* claim and their proof of it was first advanced by Alexander Neckam in the year 1200.

[9] Alan Ross Anderson and Nuel D. Balnap, Jr., "A simple treatment of truth functions", *Journal of Symbolic Logic*, 24 (1959), 301-312. See also their *Entailment: The Logic of Relevance and Necessity*, volume 1, Princeton: Princeton University Press, 1975. Volume 2, with J. Michael Dunn as third editor, appeared in 1995.

Scotland; and battles royal ensued about the *bona fides* of the DS rule.[10] The First International Conference on Relevant Logics took place in St. Louis in 1974. I was there and, although we were a bit more than two years into our co-authorships, Doug was not. His own later work on relevance show clear signs of an awareness of the influence of the Anderson & Belnap *et al.* literature, but Doug made his way with relevance pretty much on his own.[11] It bears mention that *ex falso* follows directly from what, then and now, are the most widely received characterizations of the deductive consequence relation. Conceived of model-theoretically, Q is a consequence of $P_1, \ldots, P_n$ just in case there is no respect in which it is in any sense possible for the P_i to be true and Q not. Conceived of informationally. Q is a consequence of the P_i just in case, there is no information that Q could possibly carry that wouldn't also be carried by the P_i. *Ex falso* is immediate in the first case, and, once it is pointed out that contradictions carry all possible information, it is also immediate in the second. One of the things that attracted out attention at the time was that the Lewis-Langford proof was a supplementary consideration. Doug and I were puzzled that virtually all the critical discussions focused on the Lewis–Langford proof rules. At times we thought it was a feint, a distraction from ex falso's founding home, inhering in the consequence-relation itself.[12]

In some of our later joint work, Doug and I made use of an apparatus employed by Kripke in his 1965 semantical analysis of intuitionist modal logic. In 1978, Doug knew of this proof and so did I, and it was easy to see its usefulness for our purposes in "Arresting circles in formal dialogues".[13] But Doug did not learn of the metatheorem from me, nor did I from him. We each learned it from being around and breathing in the second half of the 1960s. Moreover, since our whole 1964 seminar ranged over the adequacy or otherwise of the Lewis and Langford inference-rules for a proof set out in *classical logic*, again it cannot be said that, on that occasion, the young modal logician learned his modal logic from me.[14]

---

[10] See, for example, Richard Routley, Val Plumwood, R. K. Meyer and Ross Brady, editors, *Relevant Logics and Their Rivals*, Atascadero: Ridgeview, 1984; and Graham Priest, Richard Routley and Jean Norman, editors *Paraconsistent Logic: Essays on the Consistent*, Munich: Philosophia Verlag, 1989.

[11] Walton, *Topical Relevance in Argumentation* (1982) in the last year of our co-authoring decade. Relevance in Argumentation appeared in 2004. Doug was also influenced by the relatedness logic of the American mathematician and logician Richard L. Epstein. It gained some traction in Doug's and my textbook, *Argument: The Logic of the Fallacies* (1982).

[12] A second reason for not favouring the relevantist turn was, and still is, the sheer difficulty of finding a plausible semantics . By "plausible semantics", I now mean one that's not more implausible than the already implausible ones of classical logics.

[13] *Journal of Philosophical Logic* (1978).

[14] I was asked to organize and teach Toronto's first dedicated course in modal logic in the academic year 1969-70. It was a third-year undergraduate course, hence not a course for PhD students.

I have no clear memory of the other people with whom Doug studied in the courses-segment of his doctoral programme, possibly Sobel and John Hunter. Also recently arrived, Hunter was the department's specialist on the later Wittgenstein, and was one of the jurors at the oral defence of Doug's dissertation. Doug would have missed van Fraassen and Canfield, but may have had some contact with Butler. However, there isn't the slightest doubt of Bill Dray's impact on Doug's later work. Dray had very insightful things to say about the role of empathy in hypothetical reasoning and historical explanation. Later Doug would adapt those insights for his analysis of abductive reasoning in a book of that same name (2004). Bill had read and made helpful comments on the typescript at various stages of its preparation. If there were any respect in which his time in Toronto helped permanently shape Doug's philosophical intelligence to marked advantage, this surely would have been it.

I would like to pause now to give my impressions of the man that Doug was when I knew him in Toronto and ever after. Doug has been an international star for many years. He is so well-known that there must be throngs of his readers who have never had the chance to speak to him or attend his lectures. If we were to judge him from the thickness of his books, and their quite extraordinary number and the speed of their appearance, we might see a grimly determined man of such ambition and reach as to have condemned himself to an unsmiling life-long *hermitage*. This is not the man who arrived in Toronto at age twenty-two and he is not the man who departed this world on January 3rd, 2020, or at any time in between. What Douglas Walton was in life is a bit of an enigma. At 22, he looked about 16, and he retained his youthful look all his life. He spoke quietly, without much inflection or cadence, and interacted with people pleasantly but without, we could say, emphatic conviction. Here was a pleasant and unassuming young man who wouldn't hurt a fly and probably wouldn't amount to anything all that much in life. True, there were hidden facets — exceptions, really. When you got to know him better, you'd find a fellow who enjoyed a night on the town, not in any kind of hell-raising way, but as a nice guy to drink some beer with, and an easy chap with whom to share a chuckle. And relax with: Doug was always relaxed. And after awhile, it would become apparent that here is a chap with a delicious and mischievous sense of humour. Doug had a gift for finding the perfect moniker was to affix to some or other third-party student or teacher — Smith was "Skull", Brown was "Dr. Frown", Jones was "The loan officer", and Robinson was "Professor Manic Panic". None of this involved any meanness, or sense of purpose beyond some discreet off-the-record amusement. Doug's funny-bone was readily accessible to stimulation, and you could not know

---

Besides, Doug had just then left Toronto for Winnipeg.

him for long without some ken of his readiness to laugh, not belly-laugh, but rather in quiet spasms of pink-faced wrinkled mirth. I suppose that those moments of cheer were the closest that Doug and I came to intimacy. I could be wrong, but in all my time with him I've never seen displays of it, and from what little I know of his other collaborations, they too were partnerships between professionals. It is said that the brilliant Gilbert and Sullivan, the geniuses of West End musical theatrical wit, could scarcely abide one another. Doug was never one for Gilbert and Sullivan hookups.

In the entailment seminar, Doug listened attentively, but rarely spoke. He let the quality of his work speak for itself, and the work was always well-executed and on time. In the privacy of an office-hour, where I saw a lot of him, he'd lighten up a bit and we'd have some fun. But you couldn't really know what made Doug tick. You wouldn't know, unless you were told it, that this apparition from a Charles Atlas held a Black Belt in Karate. You would never believe that here is a young fellow who's going to write sixty books about difficult and elusive subjects. And if you knew what's good for you, you wouldn't play high-stakes cards with him. With some people, what you see is what you get. With Douglas Walton, what you saw is what you wouldn't get. It is not that Doug was a furtive man. He was, rather, a man who kept his own counsel. Unlike many of his grad student contemporaries, Doug appears not to have been part of the recreational circles of the more sociable of the grad faculty. Something else I didn't know at the time is that Doug was a rock-solidly, utterly self-disciplined philosopher, with a capacity for work directly proportional to scale of his ambition. Hardly anyone publishes a book a year — large, well-researched, clearly written and interesting books — except by foregoance of discretionary time. Discretionary time is what philosophers have lots of, and it is the single biggest suppressor of innovative productivity.[15] The established career-path guide — a peer-reviewed paper in a respectable journal every other year and a research monograph out, or contracted, before tenure, then perhaps others every six to ten years or so — would have been anathema to Doug. Foregoance of discretionary time was not an option. He yielded nothing to the costs of undue postponement, still less the abandonment, of the routine pleasures of leisure, recreation and relaxation. Doug's professional life was a master class in life-balance time-management.

In the main, Doug's books were written to be accessible to the well-educated general reader, even when they dealt with technically difficult matters. Doug didn't write for the esoterically minded. He wrote in the manner in which he would present their subject-matters to the better upperclass undergraduates of the University of Winnipeg. One of the reasons for this is that Doug was more-aware than many oth-

---

[15] It would do some good for philosophers who think that they are too busy for words, to spend a week with associates in a major downtown law firm or with emergency room resident MDs. Better still, were they to have signed on for a postdoc with Doug.

ers of the challenges and opportunities presented by the interdisciplinary character of argumentation theory, ranging from formal to informal logic, to the functional linguistics of Amsterdam, to the historiography of scientific conflict, and on to the schools of speech communication, critical thinking and rhetoric and, soon to come, developments in legal reasoning and computer science. Doug was quick to see that writings of cross-disciplinary concern would likely attract a larger readership if they attained good levels of cross-disciplinary intelligibility. This also explains why he would have placed a number of his works with regional academic presses. It was not for their regionality. It was for their openness to other parts of the motley of argumentation studies, not excluding philosophical contributions, but not centred on them, as in the case of the U of T Press's Studies in Philosophy series.[16]

3 Dissertation and departure

I come now to the period during which Doug was writing his dissertation on action theory. I am unable to remember the precise date of his assignment to me, but I surmise that it would have been in the fall of 1967. By that time, Doug had been appointed Junior Fellow of Massey College which, under the guiding eye of former Governor-General Vincent Massey, was designed to be Canada's answer to Oxford's All Souls College, and whose first Master was the most High Table of Canadian literary toffs, Roberson Davies.[17] I had just returned from another stint in Ann Arbor, where the University of Michigan had hosted the Summer Institute of the

---

[16] Although it lies beyond the remit assigned by the guest editor of this Special Issue, I would like to add a few words about what it takes to write at the pace of a research monograph every year. Leaving aside the other calls on one's time — teaching, conference-going, and streams of scholarly articles — to produce a book a year, something like the following routines would have to be assiduously maintained — in Doug's case, every day of the year but Christmas. Let us begin in *medias res*. In any given year, one must write the book that will appear in the year to follow and concurrently see through the presses the book that was written the year before, and, in time for to appear in the present year. To produce a book of 300 pages, it would be necessary to write a page a day, leaving time for necessary revisions. But if one doesn't write clean copy as a matter of course, the pace will slacken dramatically, especially when one is also involved in all the fuss and bother of getting the book-in-waiting to press. To maintain this pace, year-in and year-out, without sacrificing recreational relaxation, one would have to be a veritable *homme d'affairs*, a person of business. For what one will have made of oneself is a one-person industry, of which one is the chief executive officer, the chief operating officer, the chief financial officer and its director of communications. For these conditions to be met, one must acquire, hold and pay for a support system, and to do that that, one must be a dab hand at grantsmanship. Doug was all that and more, and in his 77 years of life he published approximately sixteen more books than the also copious Bertrand Russell, who in 98 years, published only 42, albeit two of which ran to 3 volumes; so add four more.

[17] And, I think, the best Canadian novelist writing in English in the second half of the 20th century.

Linguistics Society of America, at which I taught a graduate course on what would in time become "Semantic kinds" (1973), as well as a logic course in the department of philosophy. Doug had invited me to Massey for an evening of chitchat with some Junior Fellows, and I remember the occasion because it was my first visit as his guest to Massey. I mention it here only for the certainty of its date, which confirms that I was by then Doug's PhD supervisor.[18] We were not then to know that Doug was two scant years away from his move to the University of Winnipeg in 1969. At that point Doug, was in the position I, too, was in after my 1962 passage from the University of Michigan to Toronto. He, as with me, would be writing his dissertation out of sight of its supervisor. One day in late April of 1965, I had received an urgent call from Ann Arbor. Julius Moravcsik was phoning to say that at the end of June, my supervisor Art Burks was off to India for a two-year stint sponsored by the U. S. Department of State.[19] If my dissertation could be completed and examined before his departure, I would be spared the nuisance, and delay, of re-assignment to the less than tender mercies of Irv Copi, who took pride in refusing to take anyone else's

---

[18] Here is another example of the Doug enigma. Judging from face-value appearances, the last thing that Doug Walton would have aspired to is fellowship in a place like Massey College, whose Master was an Old Vic-trained former actor and many of whose collegiate entertainments were so high-culture as to give a regular guy a nosebleed. When Doug applied, I wrote for him, and urged it upon the electors that there was a good deal more to the duck-tailed young Walton than meets the eye. He was elected and I was now his guest. It was put about in some quarters that Doug's sole interest in Massey was economic. Its location couldn't be bettered; the best kitchen on campus; well-designed comfortable bedrooms; and no bill to pay. Has this been the whole of Doug's rationale, it would have been an excellent example of practical reasoning. However, I didn't doubt it then and I'm sure of it now, that although Doug was in process of building a strong c. v., he also had both taste and capacity for the finer things of life. Not too long after, he was a photographer of professional quality. Doug and Karen visited Carol and me in the Victoria years, and two of our most treasured photos are his, one of me driving the old Volvo Canadian with Doug in the passenger seat, and the other of our son Michael, a young boy in his school uniform, gazing wistfully over the waters of his shoreline Victoria school, and shot unposed at an angle from behind that revealed enough of the young fellow's face to discern his state of mind. In short order, Doug would also develop an interest and talent for the appreciation of first editions of classical works of literature and letters.

[19] In all my time at Michigan and the time of writing my dissertation, Burks was heavily engaged, and concomitantly out of sight, in hush-hush work on AI in collaboration, it is said, with "Johnny" von Neumann. Two of my Michigan postdoc acquaintances and later-to-be Toronto colleagues were party to these developments. One was Will Crichton and the other John Slater, who would succeed Tom Goudge as department chair. Crichton had approached me about appointment prospects at Toronto, and I passed word of it without comment to Goudge. Later, Slater in turn would explore this same option with Crichton, who passed word of it to Goudge. I might add that, in the circumstances in which Burks was then in, I could not on the head of my sainted grandmother swear as fact that Art actually got around to reading *Entailment and the Paradoxes of Strict Implication*, possibly until the night before I was examined on it, when there'd have been only enough time to give it the once-over.

supervisory leavings. I would have to return to square one and negotiate with Irv an entirely new dissertation programme. So I lay on my tummy on our dining room floor and wrote the damn thing by hand, passing to up to a stoically typewriting Carol page after finished page. Happily the gods were kind. I defended the work in June and was awarded the degree that same year. But who could have predicted the fix that Doug would be in when *his* supervisor would decamp to Stanford for the academic year 1968-1969, and upon returning to Toronto, Doug would be in Winnipeg. Three years later I would head to Victoria, not on a two-year leave from Toronto, but having quit Toronto for good. How was Doug going to earn his Toronto degree when his supervisor has ceased employment there? In the end, thank goodness, it was all sorted out, but I don't quite remember how.

I think that we must say that Doug wrote his dissertation under the same sort of shifting circumstances as I had written my own, that is to say, with rather scant supervisory interference. It should be noticed, however, that during my time at Stanford, Doug became involved with the very able mathematical linguist Barron Brainerd in the department of mathematics. It is possible that Doug had taken a "cognates" course with him. I knew and liked Brainerd as a member of the University's newly created Linguistics Centre, where I served on the executive committee, and was struck by his intelligence. When I returned from Stanford in 1969, Doug's involvement with Brainerd continued and I encouraged it. It is possible, following Howard Sobel's own reservations about what Doug was writing, that Barron replaced him as advisor. I mention this development for two reasons. One is that Doug was not in the least cowed by mathematical technicalities. The other reason is that what started out as a dissertation on the logic of action sentences ended up as an essay in the philosophy of language — possibly under the influence of Brainerd's linguistics. In any case, as would later become apparent, Doug was drawn to areas of technical enquiry with good prospects for growth. It would help to keep this in mind for when we come a bit later to Doug's first exposure to work in computer science.

I never knew what induced Doug to seek pre-doctoral fulltime employment or what brought him to accept one at the University of Winnipeg, beyond the chance that Doug had taken early notice of the rather abrupt change in philosophy from a buyers' to a sellers' market. It would, in any event, have been a move that surprised fellow degree-seekers. Very few saw themselves as flourishing in an undergraduate-only department of philosophy. The University of Winnipeg was an example of the spurt of new universities in Canada; but, like the newly created University of Victoria, Winnipeg's new university had been preceded for some generations by the well-favoured United College, as had UVic been by the equally reputed Victoria College. The same sort of transition was underway in Lethbridge where its new university also emerged from the respected Lethbridge Community College. All three predecessors had had successful university-transfer units, in which the first two years of university credit at the University of Manitoba (1877), the University of British Columbia (1908) and the University of Alberta (1908) could be acquired. At the same time, a new university, Trent, was founded in Peterborough, also institutionally preceded in this same manner. Given their antecedents, there was reason to believe (entirely rightly), that a sound education at the undergraduate level could be had at these universities. It was also thought, again rightly, that without the complex encumbrances of the warpspeed multiversity, the new universities could attract faculty of high quality, and often large reputation, to the delights of a thriving arts and science undergraduate curriculum.

This is not idle conjecture. Bill Dray moved to Trent University as department chair. He was soon joined by his Toronto colleague David Gallop, highly reputed in Greek philosophy, and another marvellous teacher. In time, Trudy Govier would also accept an appointment there. This was when Doug embarked for Winnipeg and, unrelatedly, I had formed what would be an unbroken attachment to the idea of the four-year arts and science undergraduate university. It was, as we could say, all Dray's doing, possibly lightly preceded by Sobel's enthusiasm for undergraduate-only teaching. Here is why. Shortly before Bill's removal to Trent, there had arrived out of the blue an invitation from Trent for me to interview for the department's chairmanship (as it was then called). A bit earlier while I was at Stanford, John Hunter had unbiddenly floated the idea of me as the next chair of the Toronto department, and never particularly interested in them, would not have thought myself properly fit for such duties. That turned out to be a view shared by my Toronto colleagues and, as mentioned, John Slater would succeed Tom Goudge. But now another nomination had arrived, this time from a place whose cut of jib, little as I knew of it then, appealed to me. The new university's charter president was the estimable Tom Symons, and I was interviewed by the political scientist Denis Smith, who had himself been a member of Toronto's department of political economy. The interview made a lasting impression. Here was a young senior administrator at a start-up university who had no intention to forgo a vigorous academic career. Smith would soon publish important books about contemporary politics in Canada. Also evident was the skill with which he could make an undergraduate university appealing to people as smart as he himself clearly was (but not framing it that way). In the end, to my surprise and the utter shock of the Toronto department, it was Bill Dray whom Trent appointed. In short order I resolved to leave the University of Toronto in quest of a department like the one Bill now chaired and the one of which Doug was now a member. I started putting out feelers with no desire for administrative preferment, and UVic was the first to bite. In 1971, In academic, not geographic, terms I was in the place where Doug was. It remains to add that Winnipeg and Trent have remained faithful to their undergraduate missions, but UVic and Lethbridge have abandoned them. The defections were less a matter of faculty restlessness and more a matter of corporate, downtown and governmental intrusiveness. For seven years, I was president of the University of Lethbridge, for heaven's sake, but in the end could not entirely prevail against the forces opposite. At the close of presidential doings in 1986, I resolved to hold fast in the ranks lower down. By the time I reached then-mandatory retirement in 2002, the Lethbridge department of philosophy had adopted a "special-case" MA.[20]

When preparing his dissertation on action theory, Doug and I were in regular contact during the times we were both in Toronto. In the latter stages, with him in Winnipeg and me in Victoria, there was good information flow via Canada Post. I don't recall either then or later in our co-authoring decade, any contact by telephone. Doug wrote promptly and coherently at first-go; I mean by this that he wrote clean first-drafts, which proved a considerable advantage for someone who would go on to write so plentifully. What slowed things down at the dissertation stage were initial uncertainties about programme-design. Anyone writing about action in the latter 1960s and early 1970s would have ready sources to explore. One was the literature ensuing from Elizabeth Anscombe's *Intention* (1957), which launched her as a major figure in Anglo-American analytic philosophy, and following on, Anthony Kenny's *Action, Emotion and Will* (1963). A not unrelated source was philosophical theology, what with its interest in the causal powers of non-natural agents, especially in relation to the receptive susceptibilities of the natural order. Think here of the significance for mankind of God's omnipotence. The tie that bound Anscombe to theology was her conversion (and that of her equally talented husband Peter Geach) to Roman Catholicism and the influence of Wittgenstein in his more mystical moments. Kenny himself was a laicized Jesuit priest. Some of Doug's early papers

---

[20]When I was still at UVic, I was startled when the department received permission to mount a very modest MA programme, and I got to be the first chair of the department's Graduate Studies Committee. I surmise that I succumbed to appointment as graduate admissions officer to discourage any and all who would seek it. It turns out that this is exactly what I appear to have done when a young chap with a recent degree from Lakehead University came to see me. I have no memory of what was said at that meeting, and no visual memory of my visitor. It was only years later that it was made known to me that my visitor was the now-distinguished Hans Hansen. I had no reason to doubt my informant, for he was none other than Hansen himself. As we see from Hans' contribution to this special issue, Hans moved on to the University of Manitoba, in which Doug, at the University of Winnipeg, had acquired associate status and, as I believe, taught some courses there. Hans speaks of this as the beginning of a nourishing forty-year association with Doug. And wouldn't you know, I'm the guy who set these connections in motion. Hans finished his MA in Manitoba and moved on to a PhD at Wayne State, the killing fields of analytic philosophy.

were contributions to the agency-side of philosophical theology. See, for example "Putrill on power and evil", *International Journal for Philosophy of Religion* (1977), "Some theorems of Fitch on omnipotence", *Sophia* (1976), "Language, God and evil", *International Journal for Philosophy of Religion* (1975), and "St. Anselm on the verb 'to do' (facere)", *Proceedings of the Linguistic Circle of Manitoba and North Dakota* (1974). Perhaps the first published indication that Doug was approaching agency theory from the perspective of a modal analysis is to be found in "The modal auxiliary verb can: Some semantic problems", *Proceedings of the Linguistic Circle of Manitoba and North Dakota* (1971) and, a bit later, in "Can, determinism and modal logic", *The Modern Schoolman* (1975). I think it worth noting this early contact with linguistics.

A further source of interest was Donald Davidson's logic of the truth conditions of action-sentences, which won a large readership at the beginning and retains a substantial part of it to this day. See his "The logical form of action sentence" (1967) and "Semantics for natural language" (1970). Davidson's work was in the descendent class from Tarski and Quine, beginning with Tarski's "The concept of truth in formalized languages" (1933). Provoked by the nuisance caused to formalized languages by the Liar Paradox, Tarski took evasive measures against, as he mistakenly supposed, a like nuisance in natural languages. He adjusted the metatheory of first-order logic in such a way that, in suitably regimented form, the truth-predicate of English and all the other mother tongues of humanity would have one or other of two equally unattractive futures. One is that the English predicate "is true" is transfinitely ambiguous. The other is that the lexicon of English contains a transfinite number of pairwise extensionally inequivalent truth-predicates. In fact, as we see, this was not a two-option future for English. The options are equivalent. The one both implies and is implied by the other. However, Doug and I thought that Tarski's evasions were unneeded by natural languages. We both held that, in all its variations the Liar sentences collapse semantically for want of a reference, thereby dispossessing themselves of any truth-value. Later on, the gap between formal and natural languages was subject to measures for ameliorative shrinkage advanced in Quine's doctrine of linguistic regimentation. It was meant to give us some confidence that properties of select classes of regimented natural-language sentences and arguments can be validated by measures regulating their formal representations in first-order logic. Davidson, in turn, attempted to enlarge the formal-representability class of regimented sentences of English, by provisions for adverbalized sentences and action-expressing ones. The extent of Doug's awareness of these developments preceding our discussion of them in Toronto, I cannot recall. But at the time, the attendant Tarskian demolition of truth in natural language and, by extension also of validity, persuaded Doug and me that a Davidsonian semantics for action sentences was not the road for him to travel in his dissertation. There is a further difficulty with a Tarski semantics for truth and validity. In his 1933 paper Tarski also showed that no consistent first-order theory that interprets Robinson Arithmetic Q can define truth.[21] A corresponding limitation for logical validity is also provable. Had Doug and I paid this any mind, we certainly would have decided that it had no bearing on where Doug should be heading. Besides, had it arisen, I would have pointed out that in the logic of the subject's great founder, Aristotle's concept of validity was an undefined theoretical primitive. One good thing that was retained from the seminar on relevant logic is the crucial importance of the difference between truth-conditions on the relation of deductive consequence, and adequacy conditions on rules of deductive inference. We would emphasize the difference in our first joint paper, "On fallacies" (1972), only to have been beaten to the punch by Gilbert Harman's "Induction" in 1970.[22]

On Tuesday April 25th, 1972, Doug successfully defended *The Meaning of Can: A Study in the Philosophy of Language* at 15 Hart House Circle in Toronto. Afterwards Dr. Walton and I joined fellow juror John Hunter at the Hunter residence for cakes and ale. Had he been invited and available, even the High-Table Robertson Davies would have had a jolly fine outing.

Part II: The co-authoring years 1972-1982

It all started in 1970 with the Australian logician and computer scientist whose *Fallacies* appeared that year. I learned of *Fallacies* from my Michigan teacher Irv Copi. He had been sent an advance copy by the publisher and was favourably impressed by it. Irv told me of it when, in March 1969, we spent some time at a conference at Arizona State University, organized by Morris Starsky, another Michigan friend from PhD days. I pre-ordered the book at once. It is possible that it was I who had made Doug aware of *Fallacies*, but I have no clear memory of having done so. Here, too, he might have become aware of it just by being breathing and on hand at the dawn of the 1970s. In this book, Charles Hamblin exposed the sorry state into which the logic of the fallacies had fallen, and lamented the silliness of their treatment in leading introductory logic textbooks of the day. But the crisis he called attention to was not only, or even most importantly, the dreadful pedagogical state

---

[21] Robinson arithmetic is a finitely axiomatized part of Peano arithmetic minus the axiom schema for mathematical induction.

[22] Harman's precedence was drawn to my attention by my UVic colleague Eike-Henner Kluge. When Doug learned of it, he said that this is what you get when you leave good ideas on the table. It was a remark that presaged his life-long habit of writing good ideas down and not leaving them unpublished.

in which it floundered. The principal complaint and the source of his *cri de coeur* was that fallacy theory had lost its place in *logic*, the place of its birth at the very founding of the subject, and safely a part of it until mid-19th century, at which point the fallacies were faced with the prospect of losing their theoretical home. Shortly after, a common diagnosis of this fall from grace was, and still is, that logic came of age only in 1879 and the years closely following, when it suffered a much-deserved hostile takeover by mathematics. There is little reason to deny the takeover-charge, but much less reason to see how it would sever the tie to fallacies. Are we really to think that the workings of mathematical logic are error-free and innocent of all taint of fallacy? Another diagnosis then making the rounds was that fallacies are mainly informal blunders, whereas all blunders in mathematical logic give only formal offence. What this overlooks is that Aristotle, the originator of the very idea of fallacy, did not divide them into the formal and the informal. Several chapters of *Fallacies* provide valuable historical account of how fallacy theory initially arose and how it fared in various iterations in the centuries to follow. (An especially interesting chapter deals with the place of sophismata in the dialectical logic of ancient India.)

Some readers may have had occasion to read the Introduction to our 1989 book *Fallacies: Selected Papers 1972-1982*, originally published by Foris and reissued in 2007, with a Foreword by Dale Jacquette, by College Publications.[23] In it I recount how it came to pass that Doug and I had decided to collaborate on the fallacies in the aftershock of Hamblin's revelations and call to arms. It was on an inaugural Western Airlines flight to the Pacific Division meetings of the American Philosophical Association, on which champagne was complimentary and abundant; and, by golly, by the time of arrival Doug and John had determined to settle all this fallacies turbulence within the next two years. Our first effort, "On fallacies", appeared later that year in *Journal of Critical Analysis* (1972). We approached our task with two assumptions in mind, both compatible with Hamblin's thinking. One was that fallacy theory is the proper business of logic, and the other was that formal methods have a proper role there. Less expressly arrived at, but soon to be apparent, was that, given their structural differences, the fallacies are amenable to differing methods of logical analysis. In so thinking, however, we were not in the least ill-disposed to the emerging contrivances of informal logic when we saw occasion to put them to fruitful analytical uses. Concurrently we formed an abiding interest in the sources and value of the formal-informal logic wars. All and all, Doug and I were philosophical parasites. When we saw that some already worked out system of logic could be adapted to

---

[23]It fell to me to write the Introduction, though it appeared with Doug's approval. Nineteen eight-nine marked a considerable growth in Doug's thinking in directions that model a change of mind about what we had been up to in the interval from 1972 to 1982. He had less of a stake in our old ways of thinking, but he acceded to the book's publication as a matter of record.

fulfill our own theoretical ends, we would give it a try and hope for the best. Since pluralism in logical theory had become an enduring facet of it, a further advantage redounded to us. It was clear on inspection that the traditional list of fallacies — what would later be called the gang of eighteen — exhibited striking differences in make-up and style. They were sitting ducks for pluralized logical analysis. Not everyone liked this way of proceeding. Rob Grootendorst complained that in taking the Woods-Walton approach one commits the folly of selecting a different logic for each of the fallacies.[24] This offended Amsterdam's instincts for theoretical neatness. Our pragma-dialectical colleagues haven't betrayed that instinct. Pragma-dialectics hasn't put itself through a significant structural makeover in a half-century. It is the neatest theory of argument on the theoretical shelf.

One of *Fallacies* many virtues is the clarity with which it charts the historical changes in how fallacies have been conceived of, and the corresponding changes in the shifting extensions of those conceptions. When we compare Aristotle's concept of fallacy and his list of its instantiation with the concept and list on offer in, say, Mill's *Logic* (1943), it is easy to see that, in neither place, need fallacies be errors of argument, and certainly need not involve dialectical or dialogical malfeasance. In two of *Fallacies'* most influential chapters, Hamblin seeks to resurrect both these latter constraints as a condition on fallacy theory's reinstatement to the bar of logical theory (chapter 7 on argument; chapter 8 on formal dialectic).[25] I cannot over-emphasize the impact of chapter 8 on Doug. Before long, he would see formal dialectic as the true path to fallacy-theory's repatriation in modern formal logic. Since dialogue is the natural home of dialectical contestation, Doug also came to the view that fallacy-making was an intrinsically dialogical error. It was a view that he held fast with for the rest of his life, even after he had rethought his commitment to dialectic. But there is no doubting that the source of these long-standing commitments was Hamblin's chapter 8 on formal dialectic.

Had Hamblin's chapter 8 had the influence I think it had on him, Doug would have formed the view that logic is not only the study of the logical requirements for good arguments, but that arguments as such are intrinsically dialectical. Since dialogue is a natural home for face-to-face contestative argument, it is natural to assume that arguments are always or nearly always dialogical. Later, he would modify

---

[24] The moniker "the Woods-Walton approach" was coined by Frans van Eemeren, who in his 2001 book, *Crucial Concepts in Argumentation Theory*, Amsterdam: University of Amsterdam Press, writes that it is "the most continuous and extensive post-Hamblin contribution to the study of fallacies."

[25] We should also note Hamblin's reservations about whether the logical concept of argument stretches far enough to accommodate arguments of the inductive type. See *Fallacies* (p. 226).

the view in a joint paper with David Godden.[26] The later view is that theories of argument nearly always do but best when approached under the assumption that argument is a dialogical enterprise.

Underlying Doug's perspective is the conviction that drawing conclusions from premises (inferring) is just a (sometimes solo) mode of arguing. This, to my own way of thinking is assuredly false. Argument is a kind of case-making and inferring is a matter of belief-formation. In deductive contexts, logically cogent belief-formation would be subject to the constraints under which a case is validly made, but that wouldn't alter the fact that belief-formation is not the same as case-making. In our first joint work, Doug and I called attention to this distinction but, as I have thought for many years, Doug's own theoretical work hasn't given the distinction adequate heed. This wouldn't matter if Doug's work were exclusively related to case-making. In fact, nothing like that is remotely true. Some of the highlights of the Walton project on argumentation-schemes are directly connected to modes of inference — plausible presumptive, abductive, legal, and so on. That Doug should persist in the conviction that patterns of argument call the shot for this multi-sorted reasoning or belief-formation, commits the fallacy of mistaking the apple of case-making for the goose of belief-formation.

Bearing directly on this dialectical/ dialogical influence are misconceptions sewn by Hamblin about the place of *On Sophistical Refutations* in Aristotle's logical theory. This is something that Doug and I (and everyone else) missed back then, with a lasting impact on Doug's subsequent work and on argumentation theory in general. It is necessary, I think, to pause awhile to see what went wrong.[27]

There are two fundamental misconceptions embodied in what are still widely held opinions of the character, objectives and importance of Aristotle's logic:

1. *First misconception*: All the mature insights of Aristotle's logic ramify outwards from the core logic of syllogisms.

2. *Second misconception*: One of the principal virtues of Aristotle's logic of sophistical refutations is the insight that argument is intrinsically dialectical.

Both these claims are false. They are revealed to be so as an (inadvertent) byproduct of the pathbreaking work of John Corcoran and Timothy Smiley and later by

---

[26] Douglas Walton and David M. Godden, "Informal logic and the dialectical approach to argument", in Hans V. Hansen and Robert C. Pinto, editors *Reason Reclaimed: Essays in Honor of J. Anthony Blair and Ralph H. Johnson*, pages 3-17, Newport News, VA: Vale Press, 2007.

[27] I don't want to leave the impression that Hamblin intends the formal dialectic of chapter 8 as the be-all and end-all of fallacy theory's restoration to logic, or that it is the superior model for the logical analysis of argument. It is more correctly thought of as an attempt to bend active research programmes in these subjects to those same ends.

Corcoran again.[28] Hamblin could not have been aware of these developments in the run-up to 1970. Doug and I could have, but weren't. It was an innocent but regrettable oversight.

Aristotle conceived of logic as the metatheory of the deductive sciences. The great achievement of the *Prior Analytic* was the near-perfect and repairable proof of the semi-decidability of validity in any argument meeting the conditions that define syllogisms.[29] These constraints require syllogisms to be valid two-premiss arguments made up of categorical propositions. They further provide for a syllogism's nonmonotonicity, term-sharing relevance, and hyperconsistency,[30] and some fair approximation to an intuitionistic notion of deductive conclusions. The principal importance of syllogisms lies in the *proof-rules* that correspond to them. If an argument $\langle P_1, P_2, C \rangle$, is a syllogism then the proof-rule: $P_1, P_2 \vdash C$ is a truth-preserving rule that also preserves subject-matter and adds none not already present in the premisses. It is important to note that the semi-decidability proof is validated by a combination of the syllogistic rules and what Aristotle called the common rules, such as *modus ponens* and forms of indirect proof such as *reductio ad absurdum*. There is *no* syllogism to which this metatheoretic proof corresponds. Syllogisms are a tool (*organon*). They generate proof-rules in whose absence the semi-decidability proof fails, and which are themselves *insufficient* for its success.

The other great achievement of Aristotle's logic was the creation in *Posterior Analytics* of a demonstrative logic for the deductive sciences.[31] It charted the way

---

[28] John Corcoran, "The completeness of an ancient logic", *Journal of Symbolic Logic*, 37 (1972), 696-702, and a year later and independently, Timothy Smiley, "What is a syllogism?" *Journal of Philosophical Logic*, 2 (1973), 136-154. See also Corcoran, "Aristotle's demonstrative logic", *History and Philosophy of Logic*, 30 (2009).

[29] In modern logic, a property F is semi-decidable just in case it is algorithmically possible to spot F in any arbitrarily selected object that has it, and to do so in finite time. Aristotle's search device is not strictly algorithmic, but also infallible and much, much faster.

[30] That is to say that all lines must be self-consistent and pairwise consistent as well.

[31] See again Corcoran, (2009). I should quickly add that that Corcoran greatly dislikes this account of the syllogistic constraints. But what he misses is that they are constraints on syllogisms only and that none of the *Analytic*'s proofs is a syllogism in the sense just defined. Applying the syllogistic constraints as necessary conditions for the adequacy of the metalogical proofs would be the disaster Corcoran mistakenly thinks they are for syllogisms. So one thing that we should not let stand is the idea that the metalogical proofs are chains whose links are triples of propositions regulated by the syllogistic rules alone, and connected to one another by term-sharing between the conclusion of a predecessor link and a premiss of the link that comes after. This is not something to be going on and on about here. Suffice it to say that Corcoran's difficulty arises from translations into English of *sulligismos*. In the Smith translation which Cordcoran favoured, the Greek is translated as "deduction" and is treated as a count noun. It is possible that Aristotle treats *sulligismos* ambiguously. In *Soph. Def.* it is plain that Aristotle attributes to syllogisms the characteristics that have drawn Corcoran's ire. It is also plain to see that, in his treatment of them

in which every truth of a deductive science — minus the axioms themselves — would recognizably repose in the demonstrative closure of the axioms or first principles, and would do so in a way that produced a knowledge of their truth for anyone able to follow the proof. For Aristotle, an axiom or first principle of a science is "true, primary, immediate, better known than, prior to, and causative of the conclusion. ...They must be primary and *indemonstrable*..." (*Post. An.* A 71^b 21-ff; emphasis mine)[32] I might note in passing that Aristotle's demonstrative logic bears an altogether striking architectural resemblance to Frege's *Grundgesetze* (1893, 1903), though differing very substantially on the operational mechanics respectively within. In Aristotle's case, the proof-rules of his demonstrative logic are not only truth-preserving; they are also *subject-matter preserving, new-content avoiding* and also *theorem-generating*. Consider, for example, the axiom of Peano arithmetic that 1 is a natural number. It is easy to see that the implied statement that 1 is a natural number or Nice is nice in November is a statement which, although true, is not a number-theoretic truth and not a theorem of Peano arithmetic. Aristotle is able to evade this embarrassment by driving the demonstrative proofs of his axiom system with a mixture of the common rules and the syllogistic rules, the use of which latter guarantees enough content-sharing to be subject-matter preserving, new-content avoiding, and theorem-generating. The knowledge-producing feature of demonstrative proofs derive from the requirement that the axioms be known to be true even though they are unprovably so. (It is a very old-fashioned notion of axiom, yet Frege, of all people, shared it.) Subject-matter preservation guides the reader of the proof from unprovable truth to recognizably truth-preserving disclosures, of the same subject-matter as that of the original axioms.

In book A of *Posterior Analytics*, Aristotle spots a chink in his axiomatic armour. He acknowledges that a proposition's axiomaticity is not a self-disclosing fact about it. Knowing of a proposition that it is a first principle is something that has to be laboured after. If anyone is to know them, it must surely be the experts in the science in question. It is important to emphasize that Aristotle's *Analytics* are not themselves laid out axiomatically. Aristotle identifies no proposition as an axiom

---

in the *Analytics, sulligismos* are not encumbered thus. Either Aristotle's carelessness led him into contradiction or it led him to leave the ambiguity without express notice. When Aristotle says that every demonstration is a syllogism in *Post. An.* he cannot as Corcoran rightly observes mean "syllogism" as described in *Soph. Ref.* But to conclude that the *sulligismos* of *Soph. Ref.* don't have the characteristics they lack in *Post. An.* is to override the documentary provisions of the former for the sake of safeguarding the proofs of the latter from sweeping failure. Ambiguity is the better route to take. If it were up to me, I'd leave the *Soph. Ref. sulligismos* untranslated and translate the proofs of the *Analytics* as demonstrations.

[32]Loeb Classical Library translated by Hugh Tredennick, Cambridge, MA: Harvard University Press, 1960.

of *logic*. The *Analytics* are metatheoretic essays about deductive science, and *Post. An.* gives to expert geometers, arithmeticians, theoretical physicists, and the like instructions about how to organize themselves axiomatically. We can say that a science is one of the deductive kind just in case it is finitely axiomatizable. For if it is, all its nonaxiomatic truths are guaranteed to lie in the demonstrative closure of its first principles. There is no need to determine of science's deductive status by reference to its subject-matter. Aristotle was a natural scientist of considerable range. It is certain that he would have had expertise enough in geometry to have been aware of its first principles. He does not, however, appear in *Post. An.* As a scientific expert. He appears as the founding presence in the philosophy of science. Let D be a deductive science. Tracking down first principles is the business of D's own experts. Aristotle's role is to fashion the manner in which this is done in any of the sciences in question. This is where *Sophistical Refutations* come into play.[33]

Sophistical Refutations is a study of dialectical argument, or more carefully, about a somewhat stylized version of cross-examination argument in the common law (a further interest which Doug and I independently came to share).[34] In *Prior Analytics* II 20, Aristotle defines the concept of refutation (*elenchus*):

> "If what is laid down is contrary to the conclusion, a refutation must take place; for a refutation is *syllogismos* which establishes the contradictory (*sullogismos antiphareôs*)."

Readers of this special issue of JAL will be familiar with the ins-and-outs of Aristotle's provisions for refutation-arguments. There are two parties, the questioner (Q) and the answerer (A). A comes forward with a thesis T. T is A's answer to a dialectical problem (question) — e.g. "Is everything that is virtue teachable?" It now falls to Q to put to A dialectical propositions (cross-examination questions). T itself is an *endoxon*, that is, an opinion universally held, or widely held, or held by the wise (the experts). The purpose of Q's questions is to induce A to make a contradictory defence of his thesis T. Questions must be asked, one by one, and admit of complete answers by Yes or No reply, and the propositional content of

---

[33] Scholars are divided about the order of appearance with respect to *Soph. Def.* and the *Analytics*. There is good reason to see *Post. An.* as succeeding *Pr. An.* (as their respective titles would suggest), but Soph. Def.'s arrival is less easily determined. On one finding, *Topics* and *Soph. Def.* antecede the *Analytics*. This would make sense if we were ranking according to theoretical power and sophistication. On a contrary finding, Aristotle had already written the *Analytics* before *Soph. Def.* appears. This would make sense if we ranked appearances on the basis of their centrality to the chief theoretical purposes of Aristotle's logical theory. For what concerns us here, however, we need not press the matter further.

[34] Hamblin speaks approvingly about a theory of argument that seeks guidance from legal argument, as did Toulmin twelve years later. Toulmin doesn't appear in the *Fallacies*' bibliography.

Q's questions must be categorical propositions.[35] Each of A's answers furnishes Q with a proposition eligible for selection as a premiss of a refutation-argument that Q might be able to press against A. If, drawing upon this resource, Q is able to construct a syllogism whose conclusion is the contradictory of A's thesis T, then Q will have made a refutation-argument against A. Using premisses that are solely sourced from A's own answers, Q will have induced him to make a contradictory defence of T. Of course, the refutation doesn't falsity T; it falsifies the set made up of the argument's premises and the original thesis T. The refutation therefore is only a proof against the man (ad hominem). In Aristotle's logic ad hominem arguments are not sophistical refutations, or fallacies either.[36]

What we have here is a heavily stylized model of how *ad hominem* arguments go in real life. In Aristotle's treatment, fallacies are introduced as sophistical refutations. There are two ways in which a refutation-argument can be sophistical. It can be an argument against A (in relation to his T) which is erroneously taken to be a syllogism, or it can be a syllogism whose conclusion contradicts something other than T. Although initially introduced as refutation-errors, Aristotle provides ample reason to think that the notion of sophistical refutation generalizes to the less contextually constrained notion of *fallacy*, which is the error of confusing a non-syllogism with a syllogism no matter the context. When one examines Aristotle's own examples of fallacies, it is apparent that these too are errors that can arise in plenty of contexts in which there is no call on the notion of syllogism. Think here of the many questions fallacy.[37] Equally, it is easily seen that some fallacies can be committed in ways external to the constraints of dialogue, still less dialectic. Think here of the non-cause as cause fallacy.[38]

Come back now to the question of how our grasp of a science's first principles or axioms to be *grasped*. Let D be a branch of theoretic physics. It is easy to see

---

[35] There is little doubt that in the Analytics, syllogisms must be made of a categorical propositions only. Some scholars think that in *Topics* and *Sophistical Refutations*, that constraint is waved. See here Enrico Berti "Objections to Aristotle's defence of the principle of non-contradiction", in E. Fibara, editor, *Logic, History*, pages 97-108, Berlin: de Gruyter, 2014; p. 100. One can see the sense in this suggestion. "*Sullogismos*" is a theoretical artefact of Aristotle's metalogic. In *Top.* and *Soph. Ref.*, however, they help model arguments of the kind that human beings engage in in real life, and normally beings like us don't, in our contestations, limit ourselves to categorical propositions.

[36] *Metaphysics*, 1062a 2-3. See also *Soph. Ref.* 22 178b, and *Top.* 161a 21, among many other references of like provision.

[37] John Woods, "SE 176a 10-12: Many questions for Julius Moravcsik", in Dagfinn Follesdal and John Woods, editors, *Logos and Language: Essays in Honour of Julius Moravcsik*, pages 211-220, London: College Publications, 2009.

[38] John Woods and Hans V. Hansen, "The subtleties of Aristotle on non-cause as cause", *Logique et Analyse*, 176 (2004), 395-415.

that an outsider might be unable to form any notion of its first principles. Even the fledgling physicist will have to be told them by his supervisors. But at no point in the evolution of D will its first principles be discernible at first presentation. First principleship can only be an emergent property of an active science in progress. In time, the recognition of D's first principles will coincide with their universal expert acceptance in D. The trouble is that, at any point, new information could destabilize this consensus and the experts would have to set about finding the means to restabilize. Aristotle is clear about this. As candidates for consideration arise, recovery lies in the adroit employment of *ad hominem* aggression by D's leading experts. If a candidate p withstands all attempts to reduce its holders to inconsistent defences of it, and none other fares better, then p wins its spurs, but only *provisionally*, lest any "dialectical objections" might arise and "further qualification...might [have to] be added" (*Met.* 1005b 19-23). In which case, "the exceptions will have to be agreed upon." (1008a 10-11) In sum, a proposition's first principleship is one that survives *ad hominem* attack, and retains that status until further attack overturns it. It only remains to say something of the *dialectical* characteristic's of grasping first principlehood, apart from those lent it by *ad hominem* contestation.

At *Topics* 101b 2, Aristotle writes that no science can verify its own foundations, and that this is a task which "belongs more properly to dialectic." At *Top.* A 12, he says that dialectic comprehends two types of reasoning, demonstration and induction (*epagôgê*); and moreover it is *epagôgê* that provides the foundations of the sciences (*Post. An.* 76a 38). Not only can a deductive science not foundationalize itself, the epagogic character of its foundations lends them only defeasible security.[39] As we now see, the foundations of a deductive science are thricewise dialectical. (1) They are propositions held by the top people in the field (the wise). (2) They are current survivors of *ad hominem* aggression. (3) They render the science that rests on them epagogic as opposed to demonstrative support. Any subsequent characterization of logic as intrinsically of the dialectical in sense (2) will look to Aristotle for corroboration in vain. Any description of logic as an intrinsically *dialogical* enterprise will have gone equally astray. I write this in the I hope not forlorn expectation that readers will excuse the length of it for the light it sheds on Hamblin's conception of argument and logic alike, and the enormous and tenacious grip it has had on informal logic and argumentation theory this past half-century, and, of course, on

---

[39] Further details can be found in my "What did Frege take Russell to have proved?" *Synthese*, DOI 10.1007/s11229-019-02324-4. Published online: 22 July 2019.

Doug's own work.[40,41]

I come back now to another of Hamblin's influences on Doug's work, especially I would say, since the early 2000s and onwards. Hamblin had done pioneering work in computer science in that late 1950s which, at the time, was in much the same place that mathematical linguistics was in, each of them technologically innovative enterprises with large potential for growth. Mathematical linguistics would appeal to any logician who was fully seized of the deeply mathematical character of logic's formal language and also of the fact that all its semantic and proof-theoretic properties are defined over structures from logic's language. Computer science in turn arose from mathematics and was set out, for foundational purposes, on a first-order logical platform. (Ray Reiter once told me, in the latter 1990s I think, that he couldn't make the long list of the appointments' committee of any computer science department anywhere.) At the first opportunity,[42] Doug visited Hamblin in Australia, and he remained in touch with him for the rest of Hamblin's life. In 2017 Hamblin's book *Linguistics and the Parts of the Mind: How to Build a Machine Worth Talking To*, which appeared posthumously with Cambridge Scholars Publishing of Newcastle-upon-Tyne. Doug wrote the Introduction, and I reviewed the book for the *Australasian Journal of Philosophy*.[43] One of the appeals of computer science

---

[40] For a small sample, see his *Logical-Dialogue Games and Fallacies*, Lanham: University Press of America, 1984; *Question-Reply Argumentation*, New York: Greenwood, 1989; Walton and Krabbe, *Commitment in Dialogue*, Albany: SUNY Press, 1995; *Ad Hominem Arguments*, Tuscaloosa: University of Alabama Press, 1998; *The New Dialectic: Conversational Contexts of Argument*, Toronto: University of Toronto Press, 1998; *One-Sided Arguments: A Dialectical Analysis of Bias*, Albany: SUNY Press, 1999; Walton, Reed and Macagno, *Argumentation Schemes; and Methods of Argumentation*, Cambridge: Cambridge University Press, 2013. Especially illuminating is Doug's joint paper (2007) with David Godden. Also valuable is the authors' recognition of the foundational importance of Nicholas Rescher's work, notably in dialectics and plausible reasoning. In their concluding section, Doug and David write, "Yet, we emphasize that our understanding of the nature, purpose workings and success of argument is deeply enriched by adopting a dialogic perspective whenever possible". (p. 17)

[41] The torrential forces of dialectic/game-theoretic/interrogative/dialogical frameworks in fallacy-theory and argumentation theory more broadly will be known to readers of this piece. To give their sundry influences due recognition, it will suffice here to call the role as the names occur to me: von Neumann and Morgenstern, Lorenzen, Harsanyi, Hamblin, Lorenzen and Lorenz and Barth & Krabbe. The young founders of pragma-dialectics were members of Else Barth's study group in Amsterdam. The group included van Eemeren, Grootendorst, Krabbe, van Benthem, Veltman and others. Else was in the descendent chain from Lorenz and Lorenzen but not, I think, Hamblin. Hintikka, in turn, also independently, was a principal purveyor of the heresy of Aristotle's attribution of argument's intrinsic dialecticality. Hintikka was also among the first of this group to embrace game-theoretic measures for logic. Meanwhile, Hamblin's dialectical sources had deep mediaeval roots. (Doug did some profitable mediaeval digging for some of our joint work).

[42] I think that this would have been on his first sabbatical leave from Winnipeg.

[43] *Australasian Journal of Philosophy*, 97 (2019); published online on 21 January, 2019.

for scholars of Doug's persuasion is that even if, *in ordo essendi*, logical *reasoning* takes priority over talking it out, computers don't yet have any close access to the cognitive doings of unvoiced reasonings. To make it worthwhile for us to *talk* to them, the best they can do is to equip themselves for machine-human conversation. Much of our daily conversational practice gives voice to differences of opinion, some of which are structured in adversarial case-making ways. Software engineers have no expertise in how to program these types of exchange; and it only stands to reason that they would turn to the experts. For the last twenty years of his life, Doug and his AI colleagues have been fulfilling this need.

Meanwhile, the unvoiced solo-reasoning crowd, of which I myself am a committed member, must bend its every effort to plumb the intersices of solo reasoning with the aid of the other branches of cognitive science. As I now see it, an essential component of our cognitive prosperity is a subject's background information, which is made up of common knowledge in the cognitive communities of which one is an active member, the provisions of memory and, such as they may be, the innate or learned provisions of instinct. It is easy to see that any subject's background information-count at any given time is several powers' larger than what can then and there be called to mind. It is also easy to see *that* (but not *how*) background considerations rise to the surface of awareness as the need for them arises. In its stored state, background information is tacit and implicit. Perhaps not as easily seen is that masses of background information will always be causally efficacious in reasonings even in the front of the mind. Neither need it be the case that all (or even most) of that information would have had a prior presence in the front of one's mind. The implicit and tacit is an unruly entanglement for people of my epistemological leanings, but I venture to say that it is not yet clear that it lies within the means of software engineers to computerize.[44]

The joint work of Doug and John arose in the climate of this dialectical heresy, and as I look back on it now, it is a relief to see the extent to which we avoided outright capitulation to it at the time. True, some of what we wrote did deal with dialectical and dialogical matters. "Arresting circles in formal dialogues", appeared in the *Journal of Philosophical Logic* in 1978,[45] and "Question-begging and cumulativeness in dialectical games" came out in 1982 with *Noûs*. In some cases, a paper would give some consideration of dialectical considerations, but they would not be the dominating focus. "Argumentum ad verecundiam" (*Philosophy and Rhetoric*, 1974) is a case in point. It reserves a page and a half for dialectical matters and directs the other fifteen and a half elsewhere. The same is true of "The fallacy of

---

[44] See, for example, Brian Cantwell Smith, *The Promise of Artificial Intelligence: Reckoning and Judgment*, Cambridge, MA: MIT Press, 2019.

[45] It is here that we invoked the aforementioned Kripke semantics for intuitionist modal logic.

ad ignorantiam (*Dialectica*, 1978): one and a half pages on dialectical matters, and nine and a half otherwise oriented. In still further instances, dialectical matters make no express footfall. "Composition and division" (*Studia Logica*, 1977) is free of dialectical/dialogical considerations, as are "Petitio principii" (*Synthese*, 1975) and "Post hoc, ergo propter hoc" (*Review of Metaphysics*, 1977). To the best of my recollection, Doug and I had formed no express intention to shape our joint work on the model of the Hamblin approach to argument and fallacy. I think that we may have known instinctively that fidelity to Hamblin's ways would in various respects discomply with the action plan on which we had implicitly converged at the outset of our work.

All of our co-authored work was written at a distance. One of us would write a first (and usually incomplete) draft and mail it to the other. The draft would be revised and usually extended, and returned by post. The process would continue until one of us would submit the paper to a journal. In the first instance, Doug was the initiator and, in due course, the one to seek editorial consideration. "On fallacies" appeared in the *Journal of Critical Analysis* in 1972, and its appearance prompted another UVic colleague, Danny Daniels, to remark, "Look, if you guys are trying to reinstall the study of fallacies in logical theory, why wouldn't you look for journals that publish papers in logical theory and, the more highly regarded, the better?" *Exactement, mon cher Danny*! We decided that the default position was to send our submissions to top journals. In this, we were met with unexpected success; not once did we receive a rejection letter. The two journals of record in today's argumentation-theory community, Windsor's *Informal Logic* and Amsterdam's *Argumentation* came on-stream too late for the Woods-Walton contributions to make footfall there. A natural outcome of our self-regulating production-cycle is that if one party stops sending initial drafts to the other, and the other does the same, the partnership dissolves with the implied consent of each. This is what happened to the Woods-Walton project. At some point c. 1982, we stopped sending papers to one another and started sending them to journals. I have no robust understanding of what brought this about. But I note Doug's drift towards a more pragma-dialectical way of proceeding and my disinclination to do likewise. It is also possible that a good deal of our non-WW work at the time was both off-topic and solo-authored. What is more, solo writing is lots faster than the co-authored variety.

Indeed it would be natural to think that, in having made the decision to clean up the fallacies mess, Doug and I have given pause to our other work. This was never our intention, and it didn't come to pass. In the ten years between 1972 and 1982, Doug published forty-two mainly solo papers on matters outside our fallacies project, against twenty-five on-project papers with me. In this same period, Doug published three books off-project, and two on-project, one of which was solo-authored. In this same decade, John published four off-project books, and one on-project one with Doug. John published sixteen off-project papers, in addition to the twenty-five with Doug. Doug's off-project books include two on the biomedical ethics, an edited book on philosophical theology, and another edited book on action theory. Another book, solo-authored, was on relevance, an on-project subject. Of John's off-project books, one was on the logic of fiction, another was an introduction to mathematical logic, and one was on the logic of engineered death. A co-edited book of the state of the humanities also appeared. The one on-project book was our co-authored textbook, *Argument: The Logic of the Fallacies*, a kind of sign-off at the end of our ten-year stint. It is plain to see in our off-project productivity, that overall, Doug was publishing much more copiously than John. Bob Pinto may have been on to something in his observation of two young guys on the make but, as was clear in 1982, Doug was making out at a rate that would soon leave John eating his dust. In 1982, we had left the logic of the fallacies in an improved condition, but we fell well short of leaving it at the desired levels of reflective equilibrium.

In his foreword to the second edition of *Fallacies: Selected Papers 1972-1982*, Dale Jacquette, writes,

> "Woods and Walton disavow any title to presenting a theory of fallacies in the full and proper sense of the word in this book. This is not just modesty, let alone false modesty, but a sound recognition that there is more solid work to do in meeting this requirement, and Woods and Walton quite reasonably have not set themselves so lofty a goal."[46]

A page later, Jacquette returns to this point:

> "What I especially appreciate and admire about the Woods-Walton fallacies collection is precisely its lack of an over-arching theory. Rather than prematurely fitting their discussions of particular fallacies to the requirements of a favorite theory, stretching, bending and lopping off parts to fit a Procrustean bed that analyzes the concepts by means of distorting simplifications, Woods and Walton treat every fallacy in its own terms and on its own merits." (p. viii)

Perhaps it has fallen to Dale Jacquette to have hit on the head the reason for the W-W subsidence. One W headed off in quest of theories, and the other W demurred, in favour of more empirically sensitive reflections on the behavioural data which theories are asked to account for. In time, Doug would head off to computational

---

[46] Second edition, volume 7 in Studies in Logic, London: College Publications, 2007; p. vii.

logic, and John would veer off to a naturalized logic for truth-preserving inference.[47] But, beyond doubt, 1972-1982 was a lovely ride.

---

[47] John Woods, *Errors of Reasoning: Naturalizing the Logic of Inference*, volume 45 of Studies in Logic, London: College Publications, 2013; reprinted with corrections 2014.

Douglas Walton and the Covid-19 crisis

Louise Cummings
*International Research Centre for the Advancement of Health Communication
(IRCAHC)
The Hong Kong Polytechnic University, Hong Kong SAR, China*
`louise.cummings@polyu.edu.hk`

Abstract

As we reflect on the work of Douglas Walton, I want to encourage readers of this journal to look beyond the usual applications of logic and consider the domains of medicine and health. It is testimony to the intellectual breadth of Walton's ideas in argumentation theory and fallacies that his work should find a home in medical and health disciplines, particularly epidemiology and public health. In this paper, I examine three areas of Walton's theoretical approach to argument and fallacies that I have found most beneficial to my work on reasoning in public health. First, Walton's collaboration with John Woods resulted in a new, rigorous program of fallacy research. Integral to this new approach to the fallacies was the characterization of non-fallacious variants of most of the major informal fallacies. Second, Walton advocated for a third category of presumptive argument to sit alongside deduction and induction, with plausibility as the standard of rational evaluation. Many so-called informal fallacies, he contended, are rationally warranted presumptive arguments in the practically oriented contexts in which they are advanced. Third, Walton argued that presumptive arguments like the argument from expert opinion can be scrutinized using critical questions during systematic reasoning. They may also bypass critical questions and facilitate a quick leap to a conclusion based on one or two explicit premises during heuristic reasoning. Each of these three areas in Walton's work is discussed in the context of medicine and health, with illustration provided by the current Covid-19 pandemic.

1 Introduction

Douglas Walton passed away suddenly on 3rd January 2020. The day before he died, a colleague in my academic department in Hong Kong emailed me to say a novel virus had emerged in Wuhan, China. No doubt guided by her experience of the SARS outbreak in 2003, an episode that left an indelible impression on the memory of all citizens of Hong Kong, my colleague warned me to pay attention to hand hygiene, especially before eating and when returning indoors. I immediately thanked her for getting in touch and went online to see if I could read anything further about this new virus. The *South China Morning Post* carried the story. It reported that health officials were linking early cases of a viral pneumonia to a seafood or 'wet' market in Wuhan, a central Chinese city some four hours by high-speed rail from Hong Kong. The market had been closed and the World Health Organization had been informed of the outbreak. I phoned my parents in the UK about the outbreak as I knew this story would be of interest to my mother. As I spoke to her, I felt hopeful that the health authorities in China would be able to bring the outbreak under control quite swiftly and before there would be significant loss of life and transmission of the virus outside of Wuhan. My assessment could not have been more wrong [18].

Douglas Walton did not live long enough to witness the horror of what was to become the Covid-19 pandemic. Some might say, fortunately so. But the purely coincidental timing of his death with the start of this global health emergency has forced me to think about what his applied approach to logic would make of some of the responses to this viral pandemic. Individuals, communities and governments have reacted in ways that seem to characterize human responses to crises — goodwill and determination are expressed by all parties, but actions are invariably confused and delayed, often with devastating consequences. Douglas Walton never set out in his work to address public health responses to global pandemics. Artificial intelligence and legal reasoning were much more likely to excite him. But Doug was very much concerned with the many practical applications of argumentation. As Covid-19 moves with frightening speed around the world, leaving a trail of death and hardship in its wake, it is undoubtedly the case that there is no more pressing application of logic and argumentation right now than to the domain of public health. If Doug were alive today and were witnessing the human loss and substantial economic damage caused by this viral pandemic, I am sure he would be loath to disagree. It is because of the depth and scope of his work that I have been able to apply his ideas to problems in medicine and health. It is in recognition of his substantial contribution to argumentation theory and beyond that I write a paper for this special issue of the *Journal of Applied Logics*.

There are three areas of Walton's work that I want to explore in this paper. They concern ideas that have been influential in my own research on reasoning in medicine and health. The first area is Walton's early work with John Woods on the fallacies. The recognition that the fallacies are not only an area of inquiry worthy of serious study but also that non-fallacious variants of these arguments are part of our daily discourse was a springboard for my thinking about the UK's public health response to the emergence of bovine spongiform encephalopathy (BSE) in British cattle. The second area is Walton's views on the standard of reasoning and argument that is best suited to deliberations in the practical sphere. For Walton, that standard is plausible argument. Plausible reasoning makes it possible to address a wide range of thinking in medicine and health that does not conform to deduction and induction but that is no less worthy in consequence. The third area is Walton's later work on the fallacies as heuristics. If we take plausible reasoning and its practical contexts seriously, we must engage with the demands that these contexts place on reasoners. They include the need to conduct reasoning based on the best available evidence, according to time constraints, and with maximally effective use of a reasoner's cognitive resources. These are the hallmarks of heuristics, as Walton rightly acknowledged. I will conclude by arguing that the view of fallacies as heuristics in reasoning finds one of its most important applications in medicine and health.

2 Rethinking the fallacies

Writing in 1970, Charles Hamblin is rightly credited with launching the modern study of the fallacies [21]. Hamblin's frustration with the 'standard treatment' of the fallacies, which he described as 'debased, worn-out and dogmatic a treatment as could be imagined' (p. 12), ushered in a new, systematic approach to the study of fallacies. However, in an important respect Hamblin's treatment of the fallacies was also tradition-bound (one of the criticisms he levelled at the standard treatment). For even as he criticised the approach to fallacies in introductory logic textbooks, he still subscribed to the view that these arguments were errors of reasoning that should be prohibited. Indeed, his formal dialectic was designed to do just that, with fallacies such as *petitio principii* (begging the question) effectively outlawed by means of dialectical rules [17]. Hamblin saw the need for a more systematic approach to the study of fallacies without also seeing the need to overturn the long-held view that arguments like *petitio principii* are inherently fallacious. In failing to challenge this assumption of generations of logicians and philosophers before him, Hamblin's approach, although bold, did not go far enough. It took early work by Douglas Walton and John Woods to force a re-examination of the logical merits of the fallacies and put the analysis of these arguments on a truly promising path.

And so there began a transformative episode in the history of the fallacies. Like Hamblin, Woods and Walton were unambiguous about the inadequacies of the standard treatment, describing it as an 'embarrassment' that was 'bereft of theory' and laden with 'hackneyed examples' [50, p. 133]. They advocated a broadening of the scope of philosophical logic to accommodate the dialectical and epistemic frameworks needed to capture the logical weaknesses in informal fallacies [51]. However, alongside these frameworks, a more benign view of the fallacies was beginning to take shape for these theorists. Dialogical and other frameworks struggled to prohibit the types of dialectical sequences that give rise to many of the most common fallacies. If these frameworks could not prohibit circles in argument, for example, then maybe the logical conclusion to draw is that such circles are not so fallacious after all. This is how Walton captured the direction that his thinking was taking in relation to *petitio principii*, a direction that put him at odds with the traditional view of this argument as a logical fallacy [34]:

> "[I]n the Hintikka games, like the Hamblin and Rescher games of dialogue, it remains unclear whether arguing in a circle is wrong (vicious, fallacious). Or if it is a wrong type of move or strategy in argument, it remains unclear why, or exactly when, if ever, it is wrong. The most reasonable conclusion generally seems to be that circular argumentation may be quite permissible in dialogue, for it appears to violate no general rule of reasonable dialogue, nor would it seem to frustrate the objectives or strategies of good dialogue" (pp. 267-268).

Woods and Walton would go on to characterize non-fallacious variants of most of the major informal fallacies, including circular or question-begging argument, the argument from ignorance, slippery slope, *ad baculum*, appeal to popular opinion, and *ad verecundiam* [34, 35, 36, 37, 38, 39, 40, 41, 42, 45, 46, 47, 48, 49]. Fallacy theorists identified reasonable forms of these arguments in fields like law, economics, palaeontology, and ethics, not to mention more mundane contexts of everyday reasoning. I could see the explanatory potential of this new approach to fallacies for understanding reasoning in epidemiology and public health and proceeded to analyse a range of informal fallacies in these contexts. This included the argument from ignorance [2, 6, 7, 10, 15], analogical argument [7, 3, 5, 11, 12], appeal to authority [10, 13, 14], circular argument [12, 9, 16], *ad baculum* or fear appeal [8], slippery slope argument [17], and *post hoc ergo propter hoc* [17], as well as two fallacies not included in the standard list [4]. The contexts of these analyses were the BSE crisis in the UK, the emergence of HIV/AIDS, and issues as wide-ranging as the prescription opioid epidemic, human genetic engineering, and microbial resistance.

To illustrate how informal fallacies may be used non-fallaciously in health contexts, we can do no better than turn to the current Covid-19 situation. In textbox (A) below, Dr van Kerkhove, an infectious disease epidemiologist at the World Health Organization, is describing the current state of development of serological tests for the detection of Covid-19 antibodies. She uses a *no evidence* statement. Clearly, her aim is to try and warn countries that are looking to these tests as a means of establishing immunity that currently available serological tests cannot be used for this purpose. Dr van Kerkhove's *no evidence* statement functions as the premise in an argument from ignorance:

Textbox (A)

Source	Dr Maria van Kerkhove, infectious disease epidemiologist, World Health Organization, 18 April 2020
"There are a lot of countries that are suggesting using rapid diagnostic serological tests to be able to capture what they think will be a measure of immunity. Right now, we have **no evidence** that the use of a serological test can show that an individual has immunity or is protected from reinfection."	

Argument from ignorance
There is *no evidence* that current serological tests can establish an individual's immunity.
Current serological tests *cannot* establish an individual's immunity.

Is this argument from ignorance rationally warranted? The answer to this question depends on the satisfaction of two conditions: a closed-world assumption and an exhaustive search criterion. For the closed-world assumption to be satisfied [27], Dr van Kerkhove and her colleagues at WHO would need to know what serological tests are currently available, if these tests can accurately measure antibodies present in blood serum, and if these antibodies can confer immunity on an individual. These three areas constitute the knowledge base on Covid-19 that must be present for the closed-world assumption to hold in this case.

The World Health Organization almost certainly had this knowledge of serological tests at its disposal. With its expertise in diagnostics, WHO would be aware of the serological tests that are available for Covid-19 antibody testing. Dr van Kerkhove and her colleagues also knew that these tests can accurately measure the presence of Covid-19 antibodies in blood serum. She remarked: "These antibody tests will be able to measure that level of seroprevalence — that level of antibodies — but that does not mean that somebody with antibodies, means that they are immune." But as the remainder of Dr van Kerkhove's statement indicates, the presence of antibodies does not necessarily establish that an individual has immunity to Covid-19.[1] In fact, this is not a question that a serological test alone can even

---

[1] Dr van Kerkhove of the World Health Organization was not alone in urging caution about antibody tests and what they can tell us about a person's immunity to Covid-19 (re)infection. On 14 April 2020, journalist Jennifer Smith in the Mail Online reported Carlos del Rio, Executive Associate Dean of the Emory School of Medicine in Georgia, as saying: "Just because you have antibodies

address. It requires knowledge of the antibody status of large numbers of people as well as epidemiological and clinical evidence of Covid-19 (re)infection rates in people with and without antibodies. To the extent that the presence of antibodies alone cannot address the issue of immunity, it is *a fortiori* the case that serological tests cannot establish an individual's immunity.

Based on these considerations, the closed-world assumption is fulfilled for the above argument from ignorance. The knowledge that is needed to assess serological tests is in the Covid-19 knowledge base. But we are not done. For there is a second condition that must be fulfilled, and that is an exhaustive search of the Covid-19 knowledge base. It is important to emphasize that this condition relates only to the knowledge base at a certain point in time, namely, when Dr van Kerkhove produced her statements about serological testing. Clearly, as more research is conducted into Covid-19, we can expect the knowledge base on this virus to expand considerably. But this expanded knowledge base can play no part in the rational evaluation of Dr van Kerkhove's ignorance argument. Are there grounds for claiming that Dr van Kerkhove and her colleagues had exhaustively searched the Covid-19 knowledge base that existed on 18 April 2020? Once again, we can answer this question in the affirmative. As an infectious disease epidemiologist for WHO, Dr van Kerkhove could be expected to have studied in some detail research findings about Covid-19 that would have amassed by 18 April 2020. This includes what was known about Covid-19 serological tests by this stage. With this second condition also fulfilled, we can reasonably assert that Dr van Kerkhove used a non-fallacious argument from ignorance when discussing Covid-19 serological testing.

The type of argument analysis that has allowed us to conclude that Dr van Kerkhove used a non-fallacious argument from ignorance might appear unremarkable to present-day fallacy theorists. But it would have been a marked departure from the analysis undertaken in the standard treatment of the fallacies and conducted by certain contemporaries of Woods and Walton. (Robinson, for example, steadfastly rejected arguments based on ignorance [28].) It was only possible because theorists like Walton saw the potential of analysing arguments in the actual contexts in which they were used. This forced a re-examination of the *standards* used to evaluate argument, with a new focus on presumptive and plausible models of argument. It is to this second aspect of Walton's work that I now turn.

---

doesn't mean you have immunity" [29]. Kelly Wroblewski, the Director of Infectious Diseases at the Association of Public Health Laboratories, also remarked of antibody tests: "Everybody is being optimistic you have some sort of sustained immunity for at least the ensuing months to a year. But it is still somewhat an assumption." [29]

3 A new type of argument

To make progress in understanding the fallacies, Walton recognized that how we evaluate reasoning and argument had to undergo a profound change. Deductive reasoning with its certain and known propositions (premises) providing deductive warrant for a claim (conclusion) seemed strangely at odds with the real-life contexts in which we all engage in reasoning. In these contexts, arguers are constrained by the evidence that *is* available to them and not by evidence that *might* be available to them in an ideal world. Arguers are also constrained by temporal considerations that preclude extended deliberation of an issue and that draw a process of reasoning to a close often before claims can be exhaustively debated and tested. We are much more likely to accept claims on a tentative basis and reject them should contrary evidence emerge rather than suspend judgement until such times as we have deductively certain or inductively probable claims within our grasp. Against a deductive or inductive standard of argument, many perfectly reasonable arguments, including so-called fallacies like the argument from ignorance, can appear flawed and not worthy of acceptance. But rather than dismiss these presumptive and plausible arguments, Walton urged us to take issue with the narrow conception of rationality that leads us to view them as inadequate [43]:

> "We are so accustomed to the basing of our notion of rationality on knowledge and belief, we tend to automatically dismiss plausibility as "subjective", and therefore of no worth as evidence of the kind required to rationally support a conclusion. The modern conventional wisdom is used to thinking of rationality as change of belief or knowledge guided by deductive reasoning and inductive probability. This modern way of thinking finds the notion of plausibility alien or even unintelligible, as an aspect of rational thinking" (p. 151).

Presumptive and plausible arguments are closely connected with actions and decisions in the practical sphere. Practically situated reasoners must often make decisions in advance of investigations during which evidence is gathered. In a public health context, the need for action and decisions can be particularly pressing. Decisions to impose, extend or lift lockdowns to prevent Covid-19 transmission, for example, are currently taxing the best public health authorities in the world. Governments and scientists charged with making these decisions must do so tentatively in the absence of complete evidence, whilst being aware that any delay could have disastrous health and economic consequences. Decision-making in the practical sphere cannot await claims arrived at by deductive and inductive reasoning. Presumptive reasoning can warrant actions and give decision-makers some foothold on an issue or problem, with the promise that if things proceed well, claims can grow in epistemic stature.

To get a sense of the epistemic terrain occupied by presumptive argument, we return to the Covid-19 pandemic. In response to growing public concerns that ibuprofen may be exacerbating Covid-19 infection, the Department of Health and Social Care in the UK reported the conclusion of an expert working group, the Commission on Human Medicines (CHM), on the matter. This expert body concluded that, at the present time, there was insufficient evidence that ibuprofen and other non-steroidal anti-inflammatory drugs posed risks in terms of Covid-19 infection. The argument put forward by the Department of Health and Social Care took the form of an argument from authority:

Textbox (B)

Source	Department of Health and Social Care, UK, 14 April 2020
\multicolumn{2}{l}{"The Commission on Human Medicines (CHM) Expert Working Group on coronavirus (COVID-19) has concluded that there is currently insufficient evidence to establish a link between use of ibuprofen, or other non-steroidal anti-inflammatory drugs (NSAIDs), and susceptibility to contracting COVID-19 or the worsening of its symptoms."}

Argument from authority
The CHM has expertise in human medicines and their health effects.
The CHM asserts that there is currently insufficient evidence to associate ibuprofen with worsening of Covid-19 symptoms.
Therefore, it is true that ibuprofen cannot currently be associated with worsening of Covid-19 symptoms.

The argument from authority is another of the so-called informal fallacies. But if, like Walton advises, we look beyond a notion of rationality founded on deduction and induction, we can begin to see its rational merits. The conclusion of this presumptive argument is a tentative claim based on two premises. Walton captures the premises and conclusion of the argument from expert opinion in the argumentation scheme in Figure 1 [41]. Implicit in the first premise is the assumption that the Commission on Human Medicines also has expertise in the Covid-19 health effects of ibuprofen. Although ibuprofen is a well-known drug that has been used for many years to treat inflammation and pain, not even an expert body like the Commission on Human Medicines could reasonably claim to *know* its effects on a recently emergent virus like Covid-19. But while *knowledge* of these effects was not possible in the early months of the Covid-19 pandemic, it is not true to say that members of the Commission on

Human Medicines could make no claim whatsoever about the Covid-19 health effects of ibuprofen. However, such claims as they did make had to be advanced tentatively and in the knowledge that they might need to be revised as new evidence emerged about the virus. This tentative commitment towards the Commission's conclusion was further signalled through its use of the word 'currently'. By means of this wording, the Commission is explicitly indicating that its present assessment of the Covid-19 health risks of ibuprofen may be shown to be incorrect and may need to be revised at a later point in time.

Major Premise:	Source E is an expert in subject domain S containing proposition A
Minor Premise:	E asserts that proposition A (in domain S) is true (false).
Conclusion:	A may plausibly be taken to be true (false)

Figure 1: Argumentation scheme for argument from expert opinion [41, p. 210].

For Walton, the argument from authority or expert opinion can shift the weight of presumption in a dialogue in favour of accepting the proposition or claim advanced by an expert, be that an individual or, as in this case, a body like the Commission on Human Medicines. But that presumption only holds for as long as the individual or body with expertise can respond satisfactorily to critical questions [36]. These questions are designed to interrogate an authority's expertise along several parameters such as an expert's competence and personal integrity. They also challenge us to consider if the expert's area of expertise is relevant to the question-at-issue, in this case the Covid-19 health effects of ibuprofen. If the expert can adequately respond to these questions, then there remains a presumption in favour of the truth of the proposition or claim advanced by the expert. If a critical question cannot be satisfactorily addressed, then the presumption in favour of the expert's claim must be retracted. For example, if it were to be discovered that several members of the Commission on Human Medicines were in receipt of undeclared payments from the UK's largest pharmaceutical manufacturer of ibuprofen, then it is difficult to imagine how the Commission could address a critical question about its personal integrity and independence. In that case, we would expect the presumption in favour of the safety of using ibuprofen during Covid-19 infection to lapse and to return to the side of those who would question its safety.

Quite apart from its typical characterization as a fallacy, the argument from authority may be a rationally warranted presumptive argument in certain contexts of use. But the argument from authority or expert opinion is not unique in this regard. For every informal fallacy may be a rationally warranted presumptive argument when assessed in the practical contexts in which they are advanced. These contexts are characterised by a practical imperative to come to judgement on an issue — also to take actions and to make decisions — often in the absence of knowledge and before extensive evidence has been obtained. Early in the Covid-19 pandemic, the Department of Health and Social Care in the UK also sought to address safety concerns of people who take certain high blood pressure medications. Two groups of these drugs — angiotensin converting enzyme inhibitors and angiotensin receptor blockers — were raising safety concerns as it was thought that they may exacerbate Covid-19 infection. The Department of Health and Social Care issued guidance on the matter on 27 March 2020. Its guidance took the form of an argument from ignorance:

Textbox (C)

Source	Department of Health and Social Care, UK, 27 March 2020

"If you are taking angiotensin converting enzyme inhibitors (ACE inhibitors or ACE-i) or angiotensin receptor blockers (ARBs) to treat high blood pressure, it is vitally important you continue your usual treatment.

Whilst some media reports and publications have suggested that treatment with ACE-I or ARBs might worsen COVID-19 infection, there is **no evidence** from clinical or epidemiological studies to support this.

We recognise the concern the COVID-19 outbreak is causing, and we are working closely with the Commission on Human Medicines and other regulatory bodies to ensure we can respond with further advice on this issue, **should any new data emerge**.

It is vital that anyone currently taking these medicines to treat their medical condition, continues to do so." (Bold and underlining added)

Argument from ignorance
There is *no evidence* that ACE-I and ARBs worsen Covid-19 infection.
Therefore, ACE-I and ARBs do not worsen Covid-19 infection.

Like the argument from authority before it, this argument from ignorance is a rationally warranted presumptive argument. Its conclusion, that certain groups of blood pressure medications do not worsen Covid-19 infection, is a tentative claim based on the minimal evidence base on Covid-19 that existed in March 2020. But a tentative claim can still be rationally warranted and have some traction within our deliberations. It is not the final statement on a matter but the first statement that can be revised as new evidence emerges. That the Department of Health and Social Care intended their conclusion about these medications to be just that — a tentative claim based on limited evidence that may later need to be revised — is suggested by their remarks that they would amend their advice *should any new data emerge*. Early clinical and epidemiological studies, combined with advice from expert advisory groups like the Commission on Human Medicines, succeeded in shifting the presumption in argument in favour of the claim that certain blood pressure medications do *not* worsen Covid-19 infection. But if a critical question can expose a weakness in these studies or the expert advice received, then this claim must be withdrawn and the presumption in favour of the safety of these drugs no longer holds. Imagine, for example, it was discovered that clinical and epidemiological studies only examined people on these medications who recovered from Covid-19 infection. Then the presumption in support of the safety of these drugs would have to be retracted to reflect the new evidential situation at hand.

4 Walton on heuristics

With Walton's emphasis on practical reasoning and plausible standards, it was not entirely surprising that he should turn his attention to heuristics [44]. Heuristics already had a long-established presence in the cognitive scientific literature by the time Walton began to discuss them in argumentation theory. In the now classic investigations of Tversky and Kahneman [32], it was found that people use heuristics to simplify probabilistic information: 'Many decisions are based on beliefs concerning the likelihood of uncertain events [...] people rely on a limited number of heuristic principles which reduce the complex tasks of assessing probabilities and predicting values to simpler judgmental operations' [32, p. 1124]. Although these heuristics can lead subjects to the correct answer, they may also be a source of error or bias in how people assess probabilities. One such error is known as the gambler's fallacy, the belief that random processes self-correct: 'if [a random] sequence has strayed from the population proportion, a corrective bias in the other direction is expected. This has been called the gambler's fallacy' [33, p. 193]. More recently, theorists have challenged the idea that heuristics are associated with error. It has been shown that simple heuristics can perform comparably to, and in some cases better than, more complex decision mechanisms [19, 20]. As Todd and Gigerenzer remark [31]:

> '[W]e show how simple building blocks that control information search, stop search, and make decisions can be put together to form classes of heuristics, including: ignorance-based and one-reason decision making

for choice, elimination models for categorization, and satisficing heuristics for sequential search. These simple heuristics perform comparably to more complex algorithms, particularly when generalizing to new data'. [31, p. 727]

This finding challenges a widely held assumption in cognitive science and elsewhere that performance in decision-making and problem-solving is in direct proportion to the amount of information that is available to cognitive agents — conversely, that when cognitive agents make errors in these domains, this is invariably a consequence of them having insufficient knowledge, information or data at their disposal. That reduced information can lead to enhanced cognitive performance is an important insight into the nature of our rational procedures that was not lost on Walton. It finds expression in Walton's notion of a parascheme in argumentation theory.

According to Walton [44], most of the informal fallacies are associated with an argumentation scheme and a corresponding parascheme. The argumentation scheme is part of a newer (in evolutionary terms) cognitive system which operates in a controlled, conscious and slow manner. This scheme asks critical questions of arguments, questions which are likely to expose logical weaknesses, if such weaknesses exist. The parascheme is a shorter version of the argumentation scheme. It is part of an older cognitive system which uses fast and frugal heuristics to achieve solutions to problems. Some of these heuristics involve jumping to conclusions, a cognitive strategy that can work well enough on some occasions but results in errors on other occasions. Walton demonstrates this view of the fallacies as heuristics in relation to the argument from expert opinion. The parascheme of this argument omits assumptions, exceptions and one ordinary premise that are integral to the corresponding argumentation scheme. By neglecting these aspects, which confer a slow, deliberative character on reasoning, an arguer can employ a fast heuristic to the effect 'if it's an expert opinion, defer to it' [44, p. 170]. This heuristic is depicted in Figure 2.

I have argued that informal fallacies play a role in both systematic and heuristic reasoning in medicine and health [17]. Public inquiries are a type of systematic reasoning par excellence. These inquiries often take many months or even years to complete. The UK's public inquiry into the BSE crisis took three years to complete [1]. They can gather evidence from several hundred witnesses, some of whom submit written statements, while others are directly questioned by the inquiry team. Large volumes of documents are scrutinized at length to obtain answers to questions. The aim is to arrive at the truth of the matter with no stone unturned in the search for truth. Public inquiries of this type are quite common in medicine and health. They are often used to investigate governments' handling of health issues where considerable harm has occurred to patients and their families (e.g. the blood

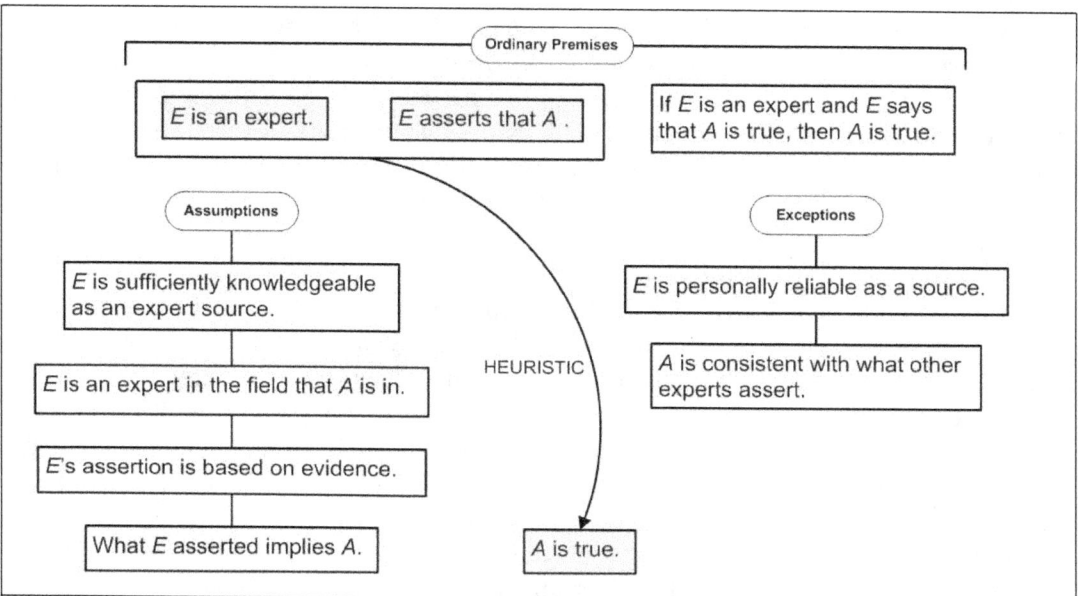

Figure 2: Heuristic of argument from expert opinion, taken from Walton [44, p. 170] (The permission of the editors of Informal Logic to reproduce this diagram is gratefully acknowledged.)

scandal in the UK in which thousands of children and adults received blood products infected with HIV and hepatitis). As I write, there are calls by politicians and health professionals for an inquiry to be conducted into the UK Government's handling of the Covid-19 pandemic [26]. If this inquiry comes about — all indications are that it almost certainly will — its chairperson and members will have powers to call forward and interrogate witnesses and access communications not normally seen by the public. Members of the inquiry team will use critical questioning of the type envisaged by Walton. The outcome will be a set of conclusions that the public can be confident are based on the most detailed examination of the evidence possible.

Public inquiries are not the only context for systematic reasoning in medicine and health (see Cummings for discussion of systematic reviews [17]). But they illustrate very clearly why this type of reasoning is poorly equipped to address many of the most pressing challenges that arise in health. These inquiries are costly in economic and cognitive terms, with significant amounts of money needed to execute them and input from hundreds of government officials and health experts required. Public inquiries also rarely deliver their conclusions in a prompt fashion. In fact, many often exceed by a considerable margin the timeframes to which they were expected to operate. What public inquiries achieve in terms of critical scrutiny and deliberation of an issue is often outweighed, sometimes grievously so, by the costs incurred and time taken to conduct them. These inquiries are better suited to an investigation of historical events (e.g. responses to a pandemic that has passed) rather than events that are still unfolding. The reasoning that health experts must use to make decisions about how to handle the Covid-19 pandemic, as the disease transmits with alarming speed around the world, is more akin to heuristic reasoning. Extended deliberation conducted through detailed critical questioning is a cognitive luxury that many public health agencies dealing with Covid-19 can ill afford. Investigators need to use mental shortcuts in reasoning that can bypass critical questions. I have argued that these shortcuts or heuristics are none other than the informal fallacies [17, 10, 16].

To illustrate what is involved in this view of the fallacies as heuristics, let us return to the Covid-19 pandemic. The Centers for Disease Control and Prevention (CDC) in the USA presented a series of clinical questions on its website. One question examined those groups who were most likely to experience severe clinical outcomes as a result of Covid-19 infection. The CDC acknowledged that the available data on Covid-19 was 'currently insufficient' to address this question but that certain groups could nevertheless be identified 'based on data from related coronaviruses'. The coronaviruses in question are Severe Acute Respiratory Syndrome (SARS-CoV) and Middle East Respiratory Syndrome (MERS-CoV). The CDC used an argument from analogy to draw conclusions about Covid-19 from what was already known about SARS-CoV and MERS-CoV:

Textbox (D)

| Source | Centers for Disease Control and Prevention, USA, 16 April 2020 |
|---|---|//
| **Who is at risk for severe disease from COVID-19?** | |

"The available data are currently insufficient to clearly identify risk factors for severe clinical outcomes. Based on limited data that are available for COVID-19 patients, and data from related coronaviruses such as severe acute respiratory syndrome coronavirus (SARS-CoV) and MERS-CoV, people who may be at risk for more severe outcomes include older adults and persons who have certain underlying chronic medical conditions. Those underlying chronic conditions include chronic lung disease, moderate to severe asthma, cardiac disease with complications, diabetes, or immunocompromising conditions."

Argument from analogy
SARS-CoV and MERS-CoV are human coronaviruses that pose serious risks for older adults and those with underlying chronic medical conditions.
Covid-19 is a human coronavirus.
Therefore, Covid-19 will pose serious risks for older adults and those with underlying chronic medical conditions.

If this argument were part of a process of systematic reasoning based on critical questions, we would interrogate, to the fullest extent possible, the presumed similarity between Covid-19 and the two better known coronaviruses, SARS-CoV and MERS-CoV. We might ask about the genetic composition of these viruses, how they replicate in the cells of a host, and how transmissible they are. We might also ask about the mortality rates associated with each virus, if people infected with these viruses may be asymptomatic, and if people who are asymptomatic can still transmit these viruses to others. We might also consider if it is possible to establish immunity to Covid-19 infection and the other human coronaviruses. Answers to some of these questions may strengthen the presumed similarity between Covid-19 and the SARS and MERS coronaviruses. For example, on 12 January 2020, China publicly shared the genetic sequence of Covid-19, so scientists could be certain of the genetic similarities of this new coronavirus to other human coronaviruses. Answers to other questions may suggest significant differences between these coronaviruses. For example, SARS and MERS have mortality rates of more than 10% and 35%, respectively [30]. Although the exact mortality rate of Covid-19 is still to be determined, it looks likely that it will be lower than that of either SARS or MERS.[2] The reason Covid-19 is taking such a large toll in human life, one much higher than either SARS or MERS, is that it is more transmissible than either of these other human coronaviruses.[3] These differences in mortality and transmissibility may weaken any presumed similarity between Covid-19 and the SARS and MERS coronaviruses.

During systematic reasoning, every possible similarity and difference between Covid-19 and the SARS and MERS coronaviruses can be extensively investigated. Some of these investigations may deliver findings quickly. For example, we already know the respective genetic sequences of these viruses. Other investigations may take much longer to produce findings. We still do not know, for example, if people who develop Covid-19 infection can develop immunity to the disease that might protect

---

[2] A case fatality rate of 2.3% is reported for Covid-19 based on data obtained from the outbreak in Hubei province in China at the start of 2020 [23].

[3] The transmissibility of an infectious disease is indicated by its reproductive number. A reproductive number of 2 indicates that each infected person infects two more people. The reproductive number for Covid-19 is between 2 and 2.5. For SARS, it is between 1.7 and 1.9, while for MERS it is <1 [25].

them from reinfection at a future point in time. But the important point is that the CDC could not await the outcome of these different investigations before issuing public health advice about the groups who are most at risk of severe outcomes of Covid-19 infection. This is because this advice must be disseminated early in the course of the pandemic in order to shield certain groups against infection. Against this urgent backdrop, the CDC carried out its public health role by drawing a tentative analogy between Covid-19 and two better known human coronaviruses. This analogy was not true beyond all doubt. But it was also not without rational warrant. It was already supported, for example, by what was known about the genetics of Covid-19 and some preliminary evidence in other areas, such as that asymptomatic people appeared to transmit the virus [22, 24]]. But with so much still unknown about this novel coronavirus, the CDC could not exclude the possibility that it may need to retract its tentative analogy at a later point in time. However, in the meantime, it provided a rational basis upon which to licence important public health advice, any delay of which could have had serious consequences for human health.

The CDC's analogy functioned as a mental shortcut or heuristic in its reasoning about Covid-19. It allowed scientists to bypass extensive critical questions about the virus that would only serve to delay urgent health advice to the public. That Covid-19 is a coronavirus was enough for the CDC to establish the analogy with SARS and MERS and go on to advise that older adults and those with underlying chronic medical conditions are most at risk of adverse outcomes from this novel virus. This quick judgement based on incomplete evidence has all the hallmarks of a 'fast and frugal' heuristic. Reasoning is not slowed down by extended consideration of evidence but can respond with speed and agility to a serious, emerging health crisis. In the final analysis, Covid-19 may be found to be a coronavirus with properties that are significantly dissimilar from other human coronaviruses. These dissimilarities may substantially shift the extent to which we can base conclusions about the likely behaviour of Covid-19 on other human coronaviruses. This situation arose in the UK's BSE crisis, for example, when an analogy between scrapie (a brain disease in sheep) and BSE in cattle was found to be flawed in a way that had direct relevance to public health — only BSE was transmissible to humans and yet public health advice was based almost entirely on the non-transmissibility of scrapie [6]. But even if an analogy between Covid-19 and other human coronaviruses must eventually be retracted, it nevertheless provides the CDC with a rationally warranted presumption on which to base its public health advice (see Figure 3).

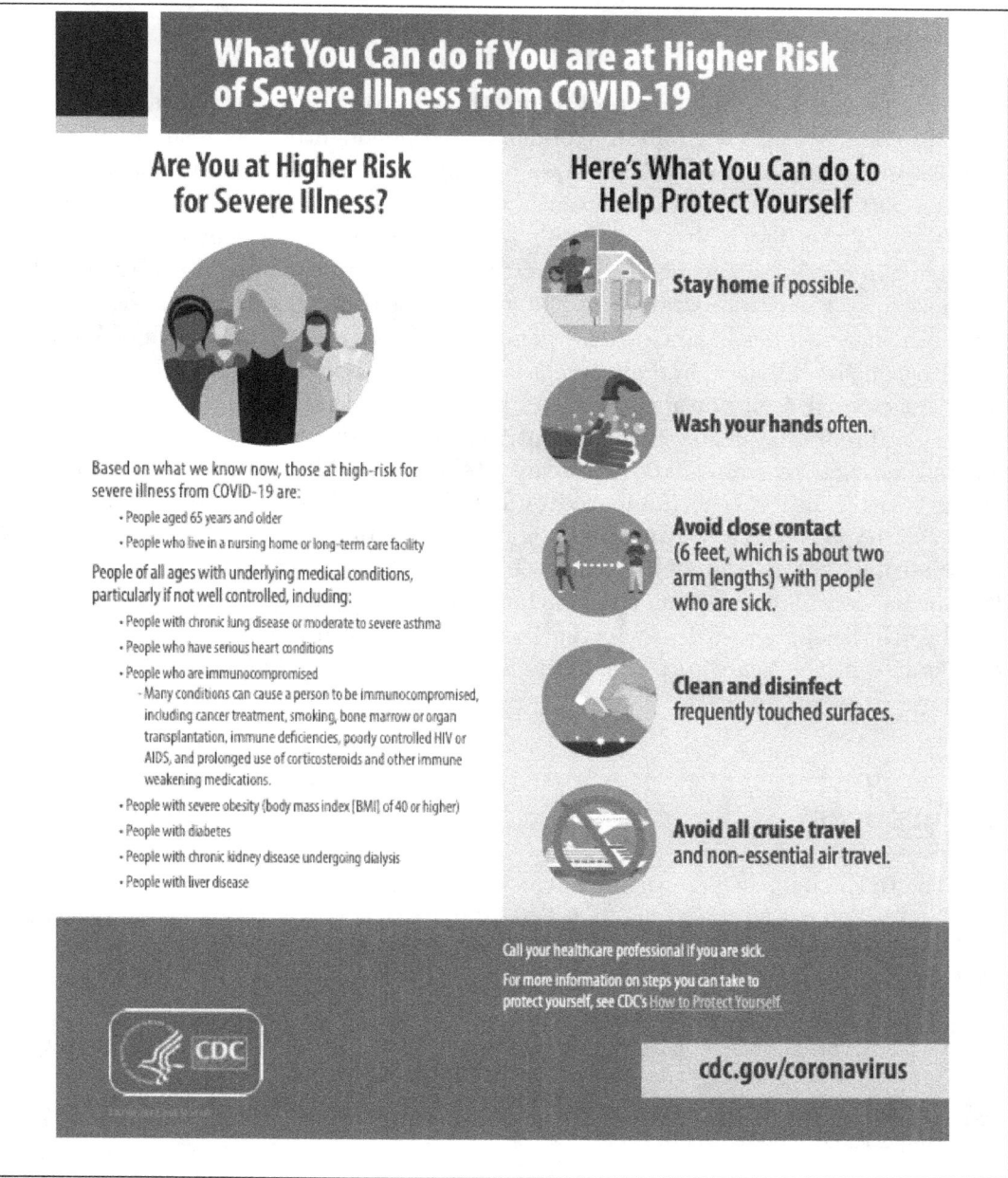

Figure 3: Health advice to people most at risk of severe Covid-19 infection (reproduced courtesy of the Centers for Disease Control and Prevention in the USA)

5 Postscript

It is often the case that ideas can have their most profound impact in places far removed from where they first took root. Douglas Walton's work in fallacy theory, both in collaboration with John Woods and independently, has resulted in one of the richest lines of inquiry into this area of logic that has ever been undertaken. But neither philosopher could have foreseen that they were developing an approach to the study of the informal fallacies that some forty years later would be applied to problems in medicine and health. In fact, not only to medicine and health but also, as this paper has demonstrated, to the greatest health crisis to affect the world in over 100 years, namely, the Covid-19 pandemic. It is a sign of the depth of Walton's thinking on the fallacies, and argumentation theory more generally, that this cross-fertilization with issues in medicine and health has been possible. Douglas Walton has undoubtedly made a significant contribution to logic and argumentation theory as well as legal reasoning and artificial intelligence. All those who have directly worked with him, or been influenced by his ideas, can testify to that contribution. But, as I hope to have conveyed in this discussion, one of Walton's most enduring contributions in the final analysis may be to public health and epidemiology. I have taken considerable inspiration from his ideas when addressing issues in these medical and health disciplines. It is my fervent hope that other argumentation theorists will do likewise in the years to come. This would be a fitting tribute to a truly great scholar.

References

[1] The BSE Inquiry. The report. London: The Stationery Office, 2000.

[2] Louise Cummings. Reasoning under uncertainty: The role of two informal fallacies in an emerging scientific inquiry. *Informal Logic*, 22(2): 113-136, 2002.

[3] Louise Cummings. Analogical reasoning as a tool of epidemiological investigation. *Argumentation*, 18(4): 427-444, 2004.

[4] Louise Cummings. Giving science a bad name: Politically and commercially motivated fallacies in BSE inquiry. *Argumentation*, 19(2): 123-143, 2005.

[5] Louise Cummings. Emerging infectious diseases: Coping with uncertainty. *Argumentation*, 23(2): 171-188, 2009.

[6] Louise Cummings. *Rethinking the BSE crisis: A study of scientific reasoning under uncertainty*. Dordrecht: Springer, 2010.

[7] Louise Cummings. Considering risk assessment up close: The case of bovine spongiform encephalopathy. *Health, Risk & Society*, 13(3): 255-275, 2011.

[8] Louise Cummings. Scaring the public: Fear appeal arguments in public health reasoning. *Informal Logic*, 32(1): 25-50, 2012.

[9] Louise Cummings. Circular reasoning in public health. *Cogency*, 5(2): 35-76, 2013.

[10] Louise Cummings. Informal fallacies as cognitive heuristics in public health reasoning. *Informal Logic*, 34(1): 1-37, 2014.

[11] Louise Cummings. Analogical reasoning in public health. *Journal of Argumentation in Context*, 3(2): 169-197, 2014.

[12] Louise Cummings. Circles and analogies in public health reasoning. *Inquiry*, 29(2): 35-59, 2014.

[13] Louise Cummings. Public health reasoning: A logical view of trust. *Cogency*, 6(1): 33-62, 2014.

[14] Louise Cummings. The 'trust' heuristic: Arguments from authority in public health. *Health Communication*, 29(10): 1043-1056, 2014.

[15] Louise Cummings. The use of 'no evidence' statements in public health. *Informal Logic*, 35(1): 32-65, 2015.

[16] Louise Cummings. *Reasoning and public health: New ways of coping with uncertainty*. Dordrecht: Springer, 2015.

[17] Louise Cummings. *Fallacies in medicine and health: Critical thinking, argumentation and communication*. Houndmills, Basingstoke: Palgrave Macmillan, 2020.

[18] Louise Cummings. Good and bad reasoning about Covid-19. *Informal Logic*, 40(4): 521–544, 2020.

[19] Gerd Gigerenzer. Why heuristics work. *Perspectives on Psychological Science*, 3(1): 20-29, 2008.

[20] Gerd Gigerenzer and Henry Brighton. Homo heuristicus: Why biased minds make better inferences. *Topics in Cognitive Science*, 1(1): 107-143, 2009.

[21] Charles L. Hamblin. *Fallacies*. London: Methuen, 1970.

[22] Anne Kimball, Kelly M. Hatfield, Melissa Arons, Allison James, Joanne Taylor, et al. Asymptomatic and presymptomatic SARS-CoV-2 infections in residents of a long-term care skilled nursing facility — King County, Washington, March 2020. *Morbidity and Mortality Weekly Report*, 69(13): 377-381, 2020.

[23] The Novel Coronavirus Pneumonia Emergency Response Epidemiology Team. Vital surveillances: The epidemiological characteristics of an outbreak of 2019 novel coronavirus diseases (COVID-19) — China, 2020. *China CDC Weekly*, 2(8): 113-122, 2020.

[24] Xingfei Pan, Dexiong Chen, Yong Xia, Xinwei Wu, Tangsheng Li, Xueting Ou, et al. Asymptomatic cases in a family cluster with SARS-CoV-2 infections. *The Lancet Infectious Diseases*, 20(4): 410-411, 2020.

[25] N. Petrosillo, G. Viceconte, O. Ergonul, G. Ippolito and E. Petersen. COVID-19, SARS and MERS: Are they closely related?. *Clinical Microbiology and Infection*. https://doi.org/10.1016/j.cmi.2020.03.026, 2020.

[26] Kate Proctor. Public inquiry into No 10's Covid-19 response inevitable, peer says. *The Guardian*, 3 April 2020.

[27] Raymond Reiter. Nonmonotonic reasoning. *Annual Review of Computer Science*, 2: 147-186, 1987.

[28] Richard Robinson. Arguing from ignorance. *The Philosophical Quarterly*, 21(83): 97-108, 1971.

[29] Jennifer Smith. Coronavirus antibodies do NOT guarantee immunity, doctors warn, as uncertainty over the testing that was supposed to get the world back to work grows. *Mail Online*, 14 April 2020.

[30] Zhiqi Song, Yanfeng Xu, Linlin Bao, Ling Zhang, Pin Yu, Yajin Qu, Hua Zhu, Wenjie Zhao, Yunlin Han and Chuan Qin. From SARS to MERS, thrusting coronaviruses into the spotlight. *Viruses*, 11(1). doi: 10.3390/v11010059, 2019.

[31] Peter M. Todd and Gerd Gigerenzer. Simple heuristics that make us smart. *Behavioral and Brain Sciences*, 23(5): 727-741, 2000.

[32] Amos Tversky and Daniel Kahneman. Judgement under uncertainty: Heuristics and biases. *Science*, 185(4157): 1124-1131, 1974.

[33] Amos Tversky and Daniel Kahneman. Belief in the law of small numbers. In E. Shafir, editor, *Preference, belief and similarity: Selected writings by Amos Tversky*, pages 193-202. Cambridge, MA: MIT Press, 2004.

[34] Douglas N. Walton. Are circular arguments necessarily vicious? *American Philosophical Quarterly*, 22(4): 263-274, 1985.

[35] Douglas N. Walton. *Arguer's position*. Westport, CT: Greenwood Press, 1985.

[36] Douglas N. Walton. The ad hominem argument as an informal fallacy. *Argumentation*, 1(3): 317-331, 1987.

[37] Douglas N. Walton. *Begging the question: Circular reasoning as a tactic of argumentation*. New York: Greenwood Press, 1991.

[38] Douglas N. Walton. Nonfallacious arguments from ignorance. *American Philosophical Quarterly*, 29(4): 381-387, 1992.

[39] Douglas N. Walton. *Plausible argument in everyday conversation*. Albany, NY: SUNY Press, 1992.

[40] Douglas N. Walton. *Slippery slope arguments*. Oxford: Clarendon Press, 1992.

[41] Douglas N. Walton. *Appeal to expert opinion*. University Park, PA: The Pennsylvania State University Press, 1997.

[42] Douglas N. Walton. *Appeal to popular opinion*. University Park, PA: The Pennsylvania State University Press, 1999.

[43] Douglas N. Walton. Abductive, presumptive and plausible arguments. *Informal Logic*, 21(2): 141-169, 2001.

[44] Douglas N. Walton. Why fallacies appear to be better arguments than they are. *Informal Logic*, 30(2): 159-184, 2010.

[45] Douglas N. Walton. The basic slippery slope argument. *Informal Logic*, 35(3): 273-311, 2015.

[46] John Woods. Appeal to force. In Hans V. Hansen and Robert C. Pinto, editors, *Falla-*

cies: Classical and contemporary readings, pages 240-250. Pennsylvania: The Pennsylvania State University Press, 1995.

[47] John Woods. *The death of argument: Fallacies in agent-based reasoning.* Dordrecht: Kluwer Academic Publishers, 2004.

[48] John Woods. Lightening up on the ad hominem. *Informal Logic*, 27(1): 109-134, 2007.

[49] John Woods. Begging the question is not a fallacy. In Cédric Dégremont, Laurent Keiff and Helge Rükert, editors, *Dialogues, logics and other strange things: Essays in honour of Shahid Rahman*, pages 523-544. London: College Publications, 2008.

[50] John Woods and Douglas Walton. Ad baculum. *Grazer Philosophische Studien*, 2(1): 133-140, 1976.

[51] John Woods and Douglas Walton. The fallacy of 'ad ignorantiam'. *Dialectica*, 32(2): 87-99, 1978.

Argumentation Schemes for Composition and Division Arguments: A Critique of Walton's Account

Maurice A. Finocchiaro
Department of Philosophy, University of Nevada - Las Vegas, 4505 Maryland Parkway, Las Vegas, NV 89154-5028, USA.
maurice.finocchiaro@unlv.edu

Abstract

This essay begins with a description of my acquaintance with Douglas Walton's scholarly work: this acquaintance goes back some five decades, but it is relatively meager given his enormous output, and yet I recently renewed and deepened it for the purpose of the present contribution. For this purpose, I decided to focus on a topic at the borderland of two things: the notion of argumentation schemes which seems to have earned Walton the greatest notoriety lately, and the fallacy of composition on which I have focused in the last several years. Thus, next I summarize Walton's account of argumentation schemes for the fallacy (and argument) of composition (and of division); unfortunately, it seems to be highly unsatisfactory. I also examine Chaim Perelman's account of the same topic since Walton refers to his work; Perelman's account is terminologically anomalous but seems to make some conceptual sense, and yet it magnifies further the inadequacy of Walton's account. Finally, I undertake a more constructive effort and sketch what I feel is a promising and more adequate account, elaborating several argumentation schemes and several evaluative principles, based on realistic examples.

1 My Acquaintance with Walton's Work

My acquaintance and involvement with Douglas Walton's scholarly work goes back a long time. In the 1970's, one of my main lines of research was the nature of fallacies, and so I read several of the articles on the topic that were being co-authored by him and John Woods; thus, one of my own major articles stemming from that period included a critical appreciation of their essay on the *post hoc ergo propter hoc* ([56]; cf. [9, pp. 18–20]). In 1985, Walton published an important book on *ad hominem* arguments entitled *Arguer's Position: A Pragmatic Study of Ad Hominem Attack, Criticism, Refutation, and Fallacy*; and upon the invitation of Henry Johnstone Jr, editor of the journal *Philosophy and Rhetoric*, I wrote a review of it that was extremely favorable, although not uncritical ([37]; cf. [10]). Next, in the 1990's, one strand of my work was a critique of the dialogical and dialectical approaches to the study of argumentation, and at one point I criticized Walton's own approach, as he had elaborated it in the first two chapters of his then-recent *Argument Structure: A Pragmatic Theory* ([42]; cf. [11, pp. 270–71]). More recently, in the context of my work on the fallacy of composition, I found myself studying a theoretical article on the topic co-authored by Woods and Walton, and a practical application of the concept to economic reasoning, as elaborated in their textbook *Argument: Critical Thinking, Logic and the Fallacies*; I found their work on the fallacy of composition in economics inspiring, fruitful, and important, but criticized them for not having continued to pursue such a project, and committed myself to doing so ([55]; [54, pp. 250–67]; cf. [14, 15, 16]).

Besides the occasions just mentioned, which left a published record, I had other encounters with Walton's work which did not, but were also important. I certainly read his *Informal Logic: A Handbook for Critical Argumentation* [39], and I could see that the book had some value for students and outsiders to the field. However, from the point of view of a specialist who had contributed to the development of the field, my overall impression was that (to use a cliché) what was new was not true and what was true was not new.

Analogous but appropriately different was my reaction to *A Pragmatic Theory of Fallacy* [40]; again I avidly read it when it first appeared, but I was disappointed to have to conclude that this book was (to paraphrase Voltaire) neither pragmatic, nor a theory, nor a work on fallacies. Let me elaborate. First, the work did not strike me as being pragmatic because the material and examples used to illustrate concepts were not sufficiently realistic and down-to-earth; to be sure, they were more realistic than the typical examples used in most textbooks, but not enough. Second, I did not find the book to be advancing a theory because it seemed to lack any ideas or principles that had the requisite simplicity, systematicity, unity, or explanatory power; that is, the book was too unfocused to be a theory, and too many topics came under discussion which had little connection with one another. Third, the reason why the subject matter of the work did not seem to be fallacy is the following: the book focused on various types of argument (e.g., *ad hominem*, analogy, and affirming the consequent) which, although deductively invalid, are not always logically incorrect; in fact, Walton himself tried to find the conditions under which such arguments are correct, and so such arguments are not always fallacies; thus, what we really had here was a study of certain argumentation techniques that are of special interest in informal logic, and a book that was not essentially different from the author's previous *Informal Logic*.

The next book by Walton that attracted my attention was *Abductive Reasoning* [45]. This topic was obviously much more focused than that of Walton's two previously mentioned works. Moreover, the subject matter happened to be much closer to my own interests, in light of my own long-standing work in the theory of explanation and the history and philosophy of science (cf., e.g., [7]). Unfortunately, this ensured that my disappointment would be greater as a result of reading the book. In fact, the examples used continued to be insufficiently realistic (as previously encountered), but also insufficiently scientific; and by the latter I mean that for the most part they were not taken from science, but from the law and from everyday reasoning. Moreover, at the conceptual level the book advanced a muddled account of the meaning of the notions of abductive, deductive, inductive, probable, plausible, and presumptive arguments; and as far as I could tell, a root cause of this confusion was the failure to be clear and critical about the distinction between interpretation aimed to understand arguments and evaluation aimed to determine their strengths or weaknesses.

Finally, there was a fourth book with which I became acquainted, *Argumentation Schemes* ([50]; cf. [49]). Its contents included discussions of almost all types of arguments, and so it seemed relevant to several of my own research projects. Moreover, because of the amount of scholarly attention and citations it received, this work was hard to miss. I concluded that this book would be worth consulting as needed in the future.

Now, despite the fact (just recounted) that my acquaintance with Walton's scholarship is long-standing and two-fold, I could readily admit that it was also meager and incomplete, for I was also aware of the obvious fact that his scholarly output was massive. Thus, for the purpose of the present contribution, I decided to become better acquainted with it. However, again, because of the magnitude of Walton's work, my plan was not to study or read all of it, which would have necessitated the unrealistic abandonment of all my other scholarly interests and involvements; rather, my plan was to learn more of, and about, his work in the hope of finding a manageable topic amenable to discussion in a brief essay appropriate to the present context.

Accordingly, I read various reviews of Walton's book. To begin with, regarding Walton's *Pragmatic Theory of Fallacy* [40], Ralph Johnson's [20] review was mostly critical, and thus reinforced my own unpublished opinion of this book. On the other hand, I also read Woods's [51] review of Walton's [44] *Ad Hominem Arguments*; I was pleased to discover that the review was highly positive, thus confirming my own judgment about Walton's earlier book on this particular topic. I also read

Tony Blair's [2, 3] review of *Argumentation Schemes for Presumptive Reasoning* [41], which is mostly critical despite some deferential lip service. For the *Fundamentals of Critical Argumentation* [46], I found Marcin Lewinski's [23] review very valuable. Regarding *Argumentation Schemes* [50], I found the critical analysis by Christoph Lumer [26, 27] very informative, insightful, and incisive. For *Methods of Argumentation* [48], the review by Corina Andone [1] was very useful.

As a result of this increased acquaintance with Walton's work, and of the evolution of my own scholarly involvements, it became increasingly clear to me that I should try to write something on the connection between the notion of argumentation scheme and the fallacy of composition. To this end, one more piece of research I undertook was to see whether this topic had been discussed by Walton in places other than those I was already acquainted with, such as the 1977 article co-authored with Woods, the co-authored textbook *Argument*, the book *Informal Logic*, and the book *Argumentation Schemes*. Although I found no additional sustained discussions, a few other minor ones emerged. For example, there is a discussion of "composition and division" in a very brief section of Chapter 8 of *Informal Fallacies: Towards a Theory of Argument Criticisms* [38, pp. 214-15]; there is a one-paragraph mention in *Fallacies Arising from Ambiguity* [43, pp. 99-100; 274-75]; and there is a repetition of the four-page discussion from the first edition of *Informal Logic* in the second edition of this book ([39, pp. 128–31]; [47, pp. 156–58]).

2 Walton on Argumentation Schemes for Arguments from Composition and Division

Let us now try to reconstruct Walton's account of argumentation schemes as they apply to the fallacy of composition, including the related concepts of argument from composition, argument from division, and fallacy of division.

To begin with, Walton is clear that we must make a distinction between the traditional concept of the fallacy of composition stemming from Aristotle and the current conception. The Aristotelian notion is based on the distinction between the distributive and the collective meaning of words. To say that a term is used distributively means that it refers to each entity described by the term; whereas to say that a term is used collectively means that it refers to the whole set of entities it describes. Thus, one possible error (a "fallacy") is to argue from premises that use a term distributively to a conclusion that uses it collectively. One common example of this, repeated by Walton [39, p. 129] is the following argument: "A bus uses more gas than a car. Therefore, all buses use more gas than all cars." Here, the premise uses the words 'bus' and 'car' distributively: it is talking about each and every bus and car, and saying that if you take any one bus and any one car, and compare their gasoline consumption for the same distance, the bus will be using more gasoline than the car; in this sense the premise is true. However, the conclusion is using the same two words in a collective sense, since it is claiming that the entire class of buses consumes more gasoline than the entire class of cars; this is not true for the simple reason that there happen to be many fewer buses than cars.

Instead, the concept Walton has in mind (and discusses in more detail) is that of a fallacy of composition as an argument which erroneously reasons from parts to the whole, in the sense that the premises assert that something is true of the parts and the conclusion infers that the same thing is true of the whole. A common example, also mentioned by Walton [39, p. 129], is this: "All the parts of this machine are light. Therefore, this machine is light." Obviously, there are many machines which are heavy (not light) even though all its many parts are light; being light-weight is not a property that can be transferred from parts to whole; weight adds up, so to speak.

A second important point which Walton hastens to add is that not all arguments having this form from parts to whole are fallacious; some are deductively valid. He gives the following example: "All the parts of this machine are iron. Therefore, this machine is made of iron" [39, p. 130]; indeed, the property of chemical composition does transfer from parts to whole. In this connection, Walton wisely introduces the term "argument from composition" [50, p. 113] to refer to an argument having this form from what is true of the parts to the same thing being true of the whole; thus, arguments from composition are sometimes correct and sometimes incorrect. It is only when such an argument is incorrect that one may speak of a fallacy of composition. Thus, the present point may be formulated by saying that Walton makes a second important distinction — between fallacy of composition and argument from composition.

Next, a third distinction is also worth emphasizing, namely that between composition and division. Walton is clear that there is a type of reasoning which is the reverse of composition: "an argument from division" is one that reasons from what is true of the whole to the same thing being true of the parts [50, p. 113]. It too is sometimes correct and sometimes incorrect. In the latter case it may be called a fallacy of division, for example: "This machine is heavy. Therefore, all the parts of this machine are heavy" [39, p. 130].

Using the notion of an argumentation scheme, the three points elaborated above may be formulated as follows. An argument from composition is one whose form fits the following scheme:

Premise: All the parts of X have property Y.

> *Conclusion*: Therefore, X has property Y.
> [39, p. 130]; [50, p. 113, 316]

An argument from division is one whose form fits the following scheme:

> *Premise*: X has property Y.
> *Conclusion*: Therefore, all the parts of X have property Y.
> [50, p. 114, 317]

In my opinion, these definitions in terms of these schemes are essentially correct. However, they are over-simplifications,[1] as we shall begin to see below when we discuss other versions of these arguments which Walton mentions, and also when I undertake a constructive and empirically based account. For the time being, I want to focus on a number of other issues.

The most immediately relevant issue pertains to the conditions under which such arguments are fallacious, or at least incorrect. Whether or not such arguments are incorrect, depends, according to Walton, on the answer to some corresponding "critical questions." For arguments from composition, the critical question is:

> Is property Y compositionally hereditary with regard to aggregate X? That is, when every part that composes X has property Y, does X (the whole) have property Y?
> [50, p. 113]

For arguments from division, the critical question is:

> Is property Y divisionally hereditary with regard to aggregate X? That is, when X (the whole) has property Y, does every part that composes X have property Y?
> [50, p. 114]

By way of criticism, I would like to point that these critical questions are unsatisfactory. The most striking flaw, which applies equally to both, is that they are completely useless and unhelpful. Each is merely a restatement of what it means to advance the corresponding argument; to infer a conclusion C from a premise P is to claim that when P is true, so is C.

Moreover, I would point out that both critical questions contain implicit definitions of technical terms, respectively, 'compositionally hereditary' and 'divisionally

---

[1] Without intending to make invidious comparisons or to sow discord among friends, I should mention that such criticism seems to me to corresponds to John Woods's charge of "over-abstraction" against such a scheme ([53]; cf. [52, 15, 16]).

hereditary'. This is worse than unnecessary; it is distracting. They should be restated more simply. The first should read: "when every part that composes X has property Y, does X (the whole) have property Y? The second should read: "when X (the whole) has property Y, does every part that composes X have property Y?".

A third criticism I would make involves the formulation of these critical questions in the so-called "User's Compendium of Schemes" of the book *Argumentation Schemes*. There, the critical question for the argument from composition reads: "Is property Y compositionally hereditary with regard to aggregate X (when X [the whole] has property Y, then every part that composes X has property Y)?" [50, p. 316]. And the critical question for the argument from division reads: "Is property Y divisionally hereditary with regard to aggregate X (when every part that composes X has property Y, then X [the whole] has property Y)?" [50, p. 317]. Obviously, these formulations are incorrectly reversing the definitions of 'compositionally hereditary' and 'divisionally hereditary'. This reversal could be merely a trivial slip of the pen or typographical error, rather than a conceptual confusion. However, even so, I believe the reversal is significant evidence that Walton himself is not taking seriously these critical questions — that he too really regards them as useless.

Later, I shall try to be more constructive with regard to such critical questions, just as I shall also be regarding the form of the schemes. Before that, however, some more criticism is in order, which involves some of what Walton allegedly derives from, and attributes to, Chaim Perelman's *New Rhetoric* [31].

3 Walton on Perelman on Composition and Division

The above-mentioned "User's Compendium of Schemes" in the book *Argumentation Schemes* contains 60 sections each of which summarizes the definition and critical questions of a major type of argument. However, most such sections also include subsections that summarize subtypes of major arguments, to yield a grand total of more than 100 argument schemes. Thus, the section dealing with the argument from composition [50, p. 316] includes partly the scheme discussed above, which is labeled "generic," and for which the only reference given is to Walton's [39, p. 130] *Informal Logic*; but that same section also discusses another scheme labeled "inclusion of the part in the whole," for which the only reference is a 10-page section of Perelman and Olbrechts-Tyteca's *New Rhetoric* [31, pp. 231–41]. Similarly, the section dealing with the argument from division [50, p. 317] includes partly the scheme discussed above, which is also labeled "generic," and for which the only reference given is to Woods and Walton's [57, pp. 206–208] textbook *Argument: The Logic of Fallacies*; but this same section also discusses another scheme labeled "division of the whole into its parts," for which the only reference is once again the same 10-page section of Perelman and Olbrechts-Tyteca's *New Rhetoric* [31, pp. 231–41]. In short, under the general labels of composition and division, besides the two types of argument on which I commented earlier, Walton discusses two other (sub)types, labeled respectively "inclusion of the part in the whole" and "division of the whole into its parts," and attributed basically to Perelman's *New Rhetoric*. Obviously, these two other types also need some discussion here.

Walton defines these two schemes as follows:

> Inclusion of the Part in the Whole
> *Premise 1*: y is a species (part) of X.
> *Premise 2*: X is A.
> *Conclusion*: y is A (is less A than X, because it is part of it; it is less A than X because it is a smaller part of it).
> [50, p. 316]

> Division of the Whole into its Parts
> *Premise 1*: X is the whole of $x_1, x_2, \ldots, x_n$ ($x_1, x_2, \ldots, x_n$ are the parts of the whole X).
> *Premise 2*: Only if x_1, or x_2, or $\ldots x_n$ is A, X is A.
> *Premise 3*: x_1 is A (no x is A).
> *Conclusion*: X is A (X is not A).
> [50, p. 317]

No critical questions are listed for either one of these schemes. For this reason, and also because of various difficulties with these definitions, it is only natural to want to consult Perelman's *New Rhetoric*, to which Walton refers. But before we do that, let me add some comments.

First of all, it is undeniable that these schemes are a step in the right direction of correcting the over-simplification of Walton's "generic" versions of the arguments. The main improvement is the addition (in both schemes) of premise 1, which specifies which entities are parts of which whole.

On the other hand, from the point of view of simplicity vs. complexity, these schemes are unnecessarily complicated by the inclusion of additional possibilities added in parenthesis; this happens in the conclusion of the "inclusion of the part in the whole" scheme and in premise 3 and conclusion of the "division of the whole into its parts" scheme. Moreover, in Walton's definition of "division of the whole into its parts," it's unclear whether the 'only if' of the second premise is meant literally, or whether it is to be understood as 'if and only if'; if meant literally, then the third premise corresponds to the consequent of the conditional second premise, and so this scheme becomes a version of affirming the consequent, and the definition becomes whimsical and arbitrary.

Finally, there is what is perhaps a more serious difficulty. That is, Walton regards the "inclusion of the part in the whole" as a special case of the argument from composition. This is certainly a misconception because, according to his own definition, such "inclusion" is an argument from what is true of the whole to the same thing being true of the parts, namely an argument from division. Similarly, but in reverse, Walton's "division of the whole into its parts" is reasoning from what is true of parts to the same thing being true of the whole; thus, it is a version of the argument from composition, and not of the argument from division. It is now time to look at Perelman's account to see whether it is to blame for Walton's difficulties.

Let us begin with what Perelman labels "division of the whole into its parts" [31, p. 234–41]:

> The concept of the whole as the sum of its parts provides the basis for a series of arguments that can be called arguments of *division* ... [p. 234]. We shall consider that in the argument by *division* the parts must be exhaustively enumerable, but that they can be chosen at will in a variety of ways on condition that by adding them up the given whole may be reconstituted. [p. 235] ... the argument by division presupposes that the sum of the parts equals the whole and that the situations being considered exhaust the possibilities ... [p. 238]. All the arguments by division obviously imply relations between the parts such that their sum can reconstitute the whole. [p. 239]

In other words, by "division of the whole into its parts," or more specifically by "argument of division," Perelman means what is commonly called argument from composition!!!

This point is reinforced by the clearest example he gives of such an argument: "one might prove to someone who doubts it that a city has been completely destroyed by enumerating exhaustively the districts that have been destroyed" [31, p. 236]. Here, I would add that this is a good example of a non-fallacious argument from composition.

Obviously, Perelman's terminology is linguistically deviant[2] and conceptually confusing (as it seems to have confused Walton and/or his co-authors). However, Perelman is at least consistent and does not seem to be himself confused, since the other type of argument which he labels "inclusion of the part in the whole"

---

[2] I am aware, of course, that Perelman and Olbrechts-Tyteca's *New Rhetoric* [31] is a translation from the French [30], and that perhaps the original French is not beset by this oddity. It would be interesting to check, but that is beyond the scope of the present investigation.

corresponds to what is normally called argument from division. In fact, his basic definition of "inclusion of the part in the whole" makes it clear that he is talking about "arguments ... which are based on the principle 'what is true of the whole is true of the part'" [31, p. 231]. And as an illustration, he quotes "this assertion of Locke: For whatsoever is not lawful to the whole Church cannot by any ecclesiastical right become lawful to any of its members" [31, p. 231]. As it stands, this assertion is certainly cryptic,[3] but fortunately it can be clearly deciphered by consulting the original passage in Locke.

Perelman tells us in a note that this quotation comes from John Locke's "Letter Concerning Toleration," published in 1689. The context of this assertion is the following argument [25, pp. 4–7]. Locke first argues that a Church is a free and voluntary society whose aim is to worship God and to acquire eternal life; it follows that a Church cannot use force to deprive its members of civil rights like liberty and private property; rather the only thing which a Church can do against persons who do not follow its rules is to expel them from membership in the Church; from this Locke thinks it also follows that clergymen, "whether they be bishops, priests, presbyters, ministers, or however else dignified or distinguished" [25, p. 7] also cannot deprive Church members of liberty or property; "for whatsoever is now lawful for the whole Church cannot by any ecclesiastical right become lawful to any of its members" [25, p. 7]. In this sequence, the second, fourth, and fifth assertions make up a subargument with the form commonly termed "argument from division" and here labeled by Perelman "inclusion of the part in the whole." And, I would add, this is an interesting, plausible, and nonfallacious argument, although also not deductively valid.

4 Some Constructive Suggestions

Let us now try to move in a more constructive direction. Let us begin with the oversimplified and overly abstract (though essentially correct) scheme for the argument from composition: (P1) All parts of W have property Y; (C) Therefore, W has property Y. The first improvement to make here might be to split the premise into two parts: (P11) all a's have property Y; and (P12) all parts of W are a's. One reason for this is that in such argumentation one seldom asserts explicitly a claim such as (P1). Instead, one is more likely to explicitly assert (P11) and leave (P12)

---

[3]Perelman's *New Rhetoric* has the merit of frequently giving illustrations consisting of texts quoted from classical sources, but also the demerit that such quotations are usually so cryptic as to require further analysis for an adequate understanding. Another example of such a double-edged presentation by Perelman involves the concept of begging the question and a quotation from Antiphon's speech on the murder of Herodes; this is criticized by Finocchiaro [8, pp. 273–77].

implicit. Similarly, regarding the connection between the parts and the whole, one needs to assume, and perhaps leave implicit, a third claim: (P13) if all parts of W are Y, then W is Y. Thus, we get the following scheme:

Scheme 4.1:
(P11) All a's are Y.
(P12) All parts of W are a's.
(P13) If all parts of W are Y, then W is Y.
Therefore, (C) W is Y.

This may also regarded as a cleaned-up or simplified version of what Walton, allegedly following Perelman, labels "division of the whole into its parts" [50, p. 317]. The (P11) here corresponds to "Premise 3" there; (P12) here to "Premise 1" there; and (P13) here to "Premise 2" there.

Actually, various nuances may be added to this scheme. One is that sometimes there are two main subsets of W that have the property Y; besides the already mentioned a's, we have what we shall call b's. The above scheme would then become:

Scheme 4.2:
(P11$'$) All a's and b's are Y.
(P12$'$) All parts of W are a's or b's.
(P13) If all parts of W are Y, then W is Y.
Therefore, (C) W is Y.

More importantly, with regard to the first mentioned claim (P12), one is more likely to think of it as, or to formulate it as: W is a whole whose *relevant* parts are the a's. A similar qualification should be made for (P11), so that it does not sound like a universal generalization, but rather like a generic or normic generalization, namely: a's, normally, or typically, have property Y. (P11) might also have to be replaced by a statistical generalization, to the effect that: most a's are Y. And (P13) might have to be formulated as a probabilistic claim. Then we would get a scheme such as the following:

Scheme 4.3:
(P11$''$) a's are, normally, Y.
(P12$''$) The relevant parts of W are a's.
(P13$''$) If all relevant parts of W are Y, then W is probably Y.
Therefore, probably, (C) W is Y.

Let me give some examples, which are not merely illustrations of the schematic concepts just presented, but rather actual argumentative situations from which I have derived these concepts.

Consider the problem of public vs. private deficits and debts in economics (cf. [34, pp. 426–31]; [36, pp. 357–64]; [21]; [22]; [11, pp 138–44]). One of the most popular arguments on this topic is the following. It would obviously be wrong (irresponsible, unsustainable, and unacceptable) for a family to constantly live beyond its means by always spending more that it earns, borrowing money, and accumulating a growing debt. Therefore, it is wrong (irresponsible, unsustainable, and unacceptable) for the national government to constantly have unbalanced budgets, run deficits, and maintain a growing national debt. Without worrying for the moment about evaluating this argument, the focus now is on understanding that this is an argument from composition: the property Y is the irresponsibility of the practice of operating constantly with a deficit and accumulating a growing debt; the W is the national government; and the parts are families. Furthermore, one does not bother saying explicitly that families are parts of a national economy. But note also that one is assuming that families are the crucial or relevant parts of a national economy.

However, in this case, families ("the a's") are not the only relevant parts of a national economy. Business firms ("b's") are equally important and relevant. Now, it so happens that the same thing ("Y") is true of them as it is of families: it is wrong (and unsustainable, etc.) for a business firm to operate with constant deficits and a growing debt. Thus, the attribution of the same requirement to a national government can also be based on these parts, and the conclusion of the compositional argument is thereby strengthened.

Another example involves a topic widely discussed in political science and political sociology, the so-called "iron law of oligarchy" (cf. [28]; [29]; [14, pp. 34–36]). This is the claim that a democratic society has an irresistible tendency to become oligarchical or anti-democratic. The evidence for this claim is the fact that, if one studies some crucial institutions advocating democratic values (e.g., political parties and labor unions), one finds that they inevitably become oligarchical in their own internal operations. Now, in such a context, political parties and labor unions may be regarded as the relevant parts of the whole society. Then assuming that whatever holds for the parts also holds for the whole, the argument infers the conclusion attributing the same property to the whole society.

Next, it may be of some interest to sketch the sub-argument supporting the above mentioned crucial factual premise about the oligarchic tendency of political parties and labor unions. That is: it is technically impossible for the majority of members to directly administer such institutions; they have to elect leaders; leaders get constantly re-elected, partly because at first they have an advantage over newcomers; moreover, they control party machinery, such as the press; and they change psychologically in their attitude due to the salary they receive, the power they exercise, their interaction with the ruling class, their age, and their attachment to their own accomplishments.

In this overall argument, the *a*'s and *b*'s of the scheme are political parties and labor unions. The W is the whole society. There is no pretension that they make up all parts of the whole society, but only the relevant parts. The property Y is the development of unavoidable anti-democratic tendencies.

Let us now try to formulate some principles for the evaluation of such arguments from composition. This corresponds to what Walton and his followers call "critical questions," but I shall call them "evaluative principles" or "principles of evaluation." Recall that the only such principle formulated by Walton was completely useless, being merely a definition of the term 'compositionally hereditary'.

The first principle I would formulate is one that makes clear that we are concerned primarily, not with the question of the deductive validity of the argument, but with the question of whether the inference is reasonable, plausible, probable, cogent, or strong; and if it is, how much. These terms are not given a precise or explicit definition, but are taken in their ordinary meaning. The point is primarily to provide an alternative to deductive evaluation. This is my formulation of a principle that has also been suggested by Juho Ritola [33], in the context of a commentary to a paper by James Gough and Mano Daniel [18]. Thus, we have:

Evaluative Principle 1: Independently of the deductive validity of an argument from composition, the primary aim is to determine whether the inference from the premises to the conclusion is reasonable, plausible, probable, cogent, or strong, and, if so, how much.

A second principle may be gathered from some suggestions by Frans van Eemeren and Rob Grootendorst ([5, p. 177]; [6]). It is based on two distinctions. One is between absolute and relative properties, for example, square vs. heavy. The second distinction is between structured or heterogeneous and unstructured or homogeneous wholes, for example a basketball team and a pile of sand. The principle asserts that properties can be transferred from parts to the whole if only if the properties are absolute and the whole is unstructured. This means that a property cannot be transferred from parts to whole when the properties are relative or the whole is structured; in all such cases, the corresponding argument from composition would be incorrect. Thus, for example, it would be correct to argue that this pile of sand is white because all its grains of sand are white; for in this case the property of being white is absolute and the whole pile is unstructured. On the other hand, in the other three possible cases the arguments would be incorrect: this figure, a rectangle consisting of two squares side by side, is square because all its parts are square (absolute property and structured whole); this pile of sand, from several truck loads, is light because all its grains of sand are light (relative property and unstructured whole); and this football team is good because its players are good (case of relative property and structured whole).

This principle is of some use, at least as long as one does not interpret it in too precise a manner. However, as Eemeren and Grootendorst themselves recognize, aside from simple cases, it is difficult to determine whether a given property is absolute or relative, and whether a given whole is homogeneous or heterogeneous; frequently, this cannot be determined prior to, or independently of, knowing the correctness of the corresponding compositional arguments. In any case, we have:

Evaluative Principle 2: In an argument from composition, determine whether the property ("Y") to be transferred is absolute or relative, and whether the whole ("W") is structured (heterogeneous) or unstructured (homogeneous). The argument is basically correct if and only if the property is absolute and the whole is unstructured (homogeneous).

There is a third evaluative principle, which I derive from the evaluative practice of the social scientists who have criticized the two arguments from composition presented above, dealing with private and public deficits and debts and with the iron law of oligarchy. Let us begin with the latter.

Recall that the argument for the iron law of oligarchy derives a claim about the unavoidable anti-democratic tendencies of a democratic society from the inevitable anti-democratic tendencies of political parties and labor unions, despite the democratic aims of the latter. Critics have objected that this inference from these parts to the societal whole is illegitimate because there are some crucial differences or dissimilarities between parties and unions on the one hand and the society as a whole. Political theorist Robert Dahl [4, p. 276] (cf. [14, pp. 34-36]) has focused on the phenomenon of competition: he has argued that a democratic political system usually allows competition among different political parties and labor unions, and such competition enables it to counteract the oligarchical tendencies at the societal level; however, political parties and labor unions are usually founded and run in a one-sided or partisan manner, which seeks to defend and foster the particular interests and aims of the members. And political sociologist Seymour Martin Lipset [24] (cf. [14, pp. 34–35]) has objected that there exists a crucial condition in democratic societies which is absent in undemocratic societies and in particular institutions of democratic ones: a constitutional stipulation or a traditional practice banning any one group from exercising tyrannical power over opposing groups.

Let us now look at the criticism of the argument from private to public debt. Economists (e.g., Samuelson and Temin [35, pp. 365–76]; cf. Finocchiaro [17, pp.

138–44]) usually point out that there are significant dissimilarities between private and public debts, and that is the main reason why one cannot argue from what is true of the former to the same thing between true of the latter. To begin with, debts by families and firms are usually "external," whereas public debts are mostly "internal." That is, private debts usually involve money owed by families or firms to other entities, such as banks, or other families, or other firms. On the other hand, national debts mostly involve money which the citizens of a country owe to themselves. (Note the qualifications denoted by "mostly" here; for to some extent, many nations also borrow money from other countries; and insofar as such external debt grows, so does the analogy between public and private.) The next point to understand is that a public (internal) debt has many benefits which a private one does not: one of these benefits is that a public debt generates government bonds, and the private wealth and the consumption of the citizens who own such bonds increases; another is that the management of the public debt enables a government to manage such things as interest rates and the printing of money, and thus improve the economy.

Now, in the present context, the upshot of such criticism seems to be that an argument from composition is weakened insofar as there are important dissimilarities between the parts and the whole. However, by evaluating the argument in this manner, we seem to be interpreting it as an argument from analogy, rather than as an argument from composition. Now, even if this were true, perhaps the point to make would be to say that sometimes arguments that seem to be compositional are really analogical. And indeed, this point has already been suggested by Trudy Govier and partly endorsed by others [19, 13, 52]. However, in the present context, I would like to explore whether such an argument can retain some aspect of compositionality while also having an aspect of analogy. The following scheme might do the trick:

Scheme 4.4:
(1) a's and b's are parts of W (perhaps the only parts, or the only relevant parts).
(2) a's and b's are Y.
(3) a's and b's are analogous to W (they have many properties in common).
(4) If two entities share many known similarities, they are likely to share additional ones.
(5) Therefore, probably, W is Y.

In such a scheme, one is still reasoning from what is true of parts to the same thing being true of the whole, which would amount to an argument from composition. But one is also claiming that there is an analogy between two entities, and that this analogy justifies attributing to one of them a property known to belong to the other. This scheme also seems to embody a connection between the two aspects; in fact, the analogy (claim no. 3) is being grounded mostly, and perhaps exclusively, on the part-whole relationship claim (no. 1). Moreover, just as in the usual argument from composition the part-whole claim is often not explicitly asserted, but implicitly assumed, the same thing happens here with the analogy claim. With these provisos and qualifications, we can assert that both the argument from private to public debts and the argument for the iron law of oligarchy fit this scheme.

This scheme also enables us to formulate the evaluative principle which the critics of these arguments were implicitly using. That is, we can interpret their criticism as an attempt to undermine claim no. 4 of this scheme. As stated this claim is not true: its truth depends on the absence of significant dissimilarities between the two alleged analogues. Thus, the claim should be stated as follows: if and only if, two entities share many known similarities, and they do not embody significant dissimilarities, are they likely to share additional similarities; in other words, two entities are likely to share additional similarities if and only if they share many known similarities, and they do not embody significant dissimilarities. Moreover, it should be noted that the other schemes (4.1, 4.2, and 4.3) discussed above also contain or assume a premise (the third one) that can be undermined by such dissimilarities. Thus, the evaluative principle we are searching for can also be applied to compositional arguments that do not explicitly have an analogical component. The principle might be stated as follows:

Evaluative Principle 3: When evaluating an argument from composition, it is always relevant and important to check whether or not there exist significant dissimilarities between the parts and the whole. If so, the argument is thereby weakened; if not, the argument is strengthened to some extent.

5 Epilogue

In this essay, I began by giving a general descriptive account of Douglas Walton's work in logic and argumentation theory. It is obvious that his work is impressive for its quantity and variety. Indeed, this point was further strengthened by the fact that, on the one hand, I have followed his work for about five decades, publishing several discussions of some parts of it, and privately studying other parts; on the other hand, I readily admitted that my acquaintance with it has been relatively meager. Thus, in the present context I undertook some further study of Walton's work. As a result, I decided to focus on the issue of argumentation schemes for the argument from composition and the argument from division. My reason for this choice was not that Walton had written a whole book on this type of argument; in fact, although (by one count) he wrote at least sixteen books on various special types of arguments (besides even more books on argumentation in general), and although he authored or co-authored several articles and chapters on arguments from composition and division, it so happens that his work did not include a whole book on such particular arguments. Rather my reason was partly that Walton wrote two books and many chapters and articles on argumentation schemes, and tried to apply the concept to every type of argument; partly that this aspect of his work has been very widely discussed in the scholarly literature; partly that in the last several years I have myself worked on the fallacy of composition; and partly that this fallacy continues to be widely regarded as extremely common and extremely important [14, pp. 25–26].

Thus, my next self-appointed task was to understand and reconstruct Walton's account of argumentation schemes as they applied to arguments from composition and from division. This account is found primarily, although not exclusively, in his book *Informal Logic* and his co-authored work *Argumentation Schemes*. The account is relatively brief and involves one scheme and one critical question for each the argument from composition and the argument from division. Unfortunately, Walton's account is highly unsatisfactory. The main difficulties are over-simplification, uselessness, and muddled confusion.

Walton's account also includes some references to Chaim Perelman's account of "inclusion of the part in the whole" and "division of the whole into its parts," as he labels them in *The New Rhetoric*. These references motivated me to examine Perelman's account. This examination revealed that what Perelman calls "division of the whole into its parts" corresponds to what is usually called argument from composition, and what he calls "inclusion of the part in the whole" corresponds to what is usually called argument from division; and Perelman also gives at least two interesting, clear, and plausible illustrations. Unfortunately, Walton's account misinterprets these correspondences as being the reverse of what they really are.

Finally, I attempted to sketch a constructive account of how the notion of argumentation scheme might be applied to arguments from composition and from division. My account elaborates four distinct (but interrelated) schemes for the argument from composition; such schemes are meant primarily to understand or interpret such arguments. Moreover, I also elaborate three principles of evaluation for such arguments, which are meant to correspond to Walton's "critical questions," whose terminology I wish to avoid; as my own terminology tries to make clear, such principles are meant primarily to evaluate or assess such arguments. Thirdly, my account is based on a presentation of some realistic examples, specifically the argument from private to public debts in economics, and the argument for the iron law of oligarchy in political science and political sociology. Fourthly, an interesting point that emerges in my account is that some arguments from composition are also simultaneously arguments from analogy.

Much more remains to be done, not only from the point of view of a general study of the fallacy of composition, as I have had the occasion of pointing out before [14, pp. 36–41]. However, even from the point of view of the present focus (argumentation schemes for compositional arguments), further studies are needed. For example, perhaps more nuances need to be elaborated for the four schemes in my constructive account. Perhaps additional, although interrelated, schemes may have to be defined. Perhaps the same two points apply also to the three evaluative principles in my account; that is, more nuances for the principles already mentioned and additional principles to be formulated. And since such schemes and principles should be grounded on realistic examples, the search for the latter must continue.

References

[1] Andone, Corina. 2014. Review of Douglas Walton's *Methods of Argumentation. History and Philosophy of Logic* 35: 304-306; DOI: 10.1080/01445340.2014.894711.

[2] Blair, J. Anthony. 1999a. Review of D.N. Walton's *Argumentation Schemes for Presumptive Reasoning. Argumentation* 13: 338–43.

[3] Blair, J. Anthony. 1999b. "Walton's Argumentation Schemes for Presumptive Reasoning." In *Proceedings of the Fourth International Conference of the International Society for the Study of Argumentation*, ed. Frans H. van Eemeren, Rob Grootendorst, Anthony Blair, and Charles A. Willard, pp. 56-61. Amsterdam: SicSat.

[4] Dahl, Robert A. 1989. *Democracy and Its Critics*. New Haven: Yale University Press.

[5] Eemeren, Frans H. van, and R. Grootendorst. 1992. *Argumentation, Communication, and Fallacies*. Hillsdale: Lawrence Erlbaum Associates.

[6] Eemeren, Frans H. van, and R. Grootendorst. 1999. "The Fallacies of Composition and Division." In *JFAK: Essays Dedicated to Johan van Benthem on the Occasion of His 50th Birthday*, ed. J. Gerbrandy, M. Marx, M. de Rijke, and Y. Venema. Amsterdam: University of Amsterdam, Institute for Logic, Language, and Computation. At www.illc.uva.nl/j50/, consulted on 18 June 2013.

[7] Finocchiaro, Maurice A. 1973. *History of Science as Explanation*. Detroit: Wayne State University Press.

[8] Finocchiaro, Maurice A. 1980. *Galileo and the Art of Reasoning: Rhetorical Foundations of Logic and Scientific Method*. (Boston Studies in the Philosophy of Science, vol. 61.) Boston: Reidel.

[9] Finocchiaro, Maurice A. 1981. "Fallacies and the Evaluation of Reasoning." *American Philosophical Quarterly* 18: 13-22. Reprinted in [12, Chapter 6, pp. 109–27].

[10] Finocchiaro, Maurice A. 1987. Review of Douglas Walton's *Arguer's Position: A Pragmatic Study of Ad Hominem Attack, Criticism, Refutation, and Fallacy*. *Philosophy and Rhetoric* 20: 63-65.

[11] Finocchiaro, Maurice A. 1999. "A Critique of the Dialectical Approach, Part II." In *Proceedings of the Fourth International Conference of the International Society for the Study of Argumentation*, ed. Frans H. van Eemeren, Rob Grootendorst, Anthony Blair, and Charles A. Willard, pp. 195-99. Amsterdam: SicSat. Reprinted in [12, Chapter 15, pp. 265-76].

[12] Finocchiaro, Maurice A. 2005. *Arguments about Arguments: Systematic, Critical, and Historical Essays in Logical Theory*. New York: Cambridge University Press.

[13] Finocchiaro, Maurice A. 2013. "Debts, Oligarchies, and Holisms: Deconstructing the Fallacy of Composition." *Informal Logic* 33: 143-74.

[14] Finocchiaro, Maurice A. 2015. "The Fallacy of Composition: Guiding Concepts, Historical Cases, and Research Problems." *Journal of Applied Logic*, vol. 13, issue 2, part B, June 2015, pp. 24–43.

[15] Finocchiaro, Maurice A. 2016a. "Economic Reasoning and Fallacy of Composition, Part I: The Problem." *Eris: Rivista internazionale di argomentazione e dibattito*, vol. 1, no. 2, pp. 17-38. ISSN 2421-6747; at http://eris.fisppa.unipd.it/Eris.

[16] Finocchiaro, Maurice A. 2016b. "Economic Reasoning and Fallacy of Composition, Part III: Response to John Woods's Comments." *Eris: Rivista internazionale di argomentazione e dibattito*, vol. 1, no. 2, pp. 46-56. ISSN 2421-6747; at http://eris.fisppa.unipd.it/Eris.

[17] Finocchiaro, Maurice A. 2019. "Samuelson on the Fallacy of Composition in Economics: A Woodsian Critique." In *Natural Arguments: A Tribute to John Woods*, ed. Dov Gabbay, Lorenzo Magnani, Woosuk Park, and Ahti Veikko Pietarinen, pp. 125-72. London: College Publications.

[18] Gough, James E., and Mano Daniel. 2009. "The Fallacy of Composition." In Ritola 2009a.

[19] Govier, Trudy. 2009. "Duets, Cartoons, and Tragedies: Struggles with the Fallacy of Composition." In *Pondering on Problems of Argumentation*, ed. Frans H. van Eemeren and B. Garssen, pp. 91-104. Dordrecht: Springer.

[20] Johnson, Ralph H. 1998. Review of Douglas Walton's *A Pragmatic Theory of Fallacy*. *Argumentation* 12: 115–23.

[21] Krugman, Paul. 2013a. "Austerity Wrought Pain, No Gain." *Las Vegas Sun*, January 8, p. 3; available at: http://lasvegassun.com/news/2013/jan/08/austerity-wrought-pain-no-gain/#.VHPVyW1c75E.gmail.

[22] Krugman, Paul. 2013b. "The Punishment Cure." *The New York Times*, December 8; available at: http://www.nytimes.com/2013/12/09/opinion/krugman-the-punishment-cure.html?_r=0.

[23] Lewinski, Marcin. 2009. Review of Walton's *Fundamentals of Critical Argumentation*. *Argumentation* 23: 123–26; DOI 10.1007/s10503-008-9111-1.

[24] Lipset, Seymour Martin. 1962. Introduction to Michels 1962, 15-39.

[25] Locke, John. 1952. "A Letter Concerning Toleration." In *Great Books of the Western World*, vol. 35, Locke, Berkeley, Hume. Chicago: University of Chicago Press.

[26] Lumer, Christoph. 2011. "Argument Schemes — An Epistemological Approach." In *Argumentation, Cognition and Community: Proceedings of the 9th International Conference of the Ontario Society for the Study of Argumentation (OSSA)*, ed. Frank Zenker. Windsor, University of Windsor. http://scholar.uwindsor.ca/cgi/viewcontent.cgi?article=1016&context=ossaarchive.

[27] Lumer, Christoph. 2016. "Walton's Argumentation Schemes." In *Argumentation, Objectivity, and Bias. Proceedings of the 11th International Conference of the Ontario Society for the Study of Argumentation (OSSA)*, ed. P. Bondy and L. Benaquista. Windsor, University of Windsor. Online: https://scholar.uwindsor.ca/ossaarchive/OSSA11/papersandcommentaries/110.

[28] Michels, Robert. 1915. *Political Parties: A Sociological Study of the Oligarchical Tendencies of Modern Democracy*. Trans. E. Paul and C. Paul. Glencoe: The Free Press.

[29] Michels, Robert. 1962. *Political Parties: A Sociological Study of the Oligarchical Tendencies of Modern Democracy*. Trans. E. Paul and C. Paul. New York: Collier.

[30] Perelman, Chaim, and L. Olbrechts-Tyteca. 1958. *La nouvelle rhétorique: Traité de l'argumentation*. Paris: Presses Universitaires de France.

[31] Perelman, Chaim, and L. Olbrechts-Tyteca. 1969. *The New Rhetoric: A Treatise on Argumentation*. Trans. J. Wilkinson and P. Weaver. Notre Dame: University of Notre Dame Press.

[32] Ritola, Juho, ed. 2009a. *Argument Cultures: Proceedings of the 8th Biennial Conference of the Ontario Society for the Study of Argumentation (OSSA, 2009)*. Windsor, ON: Ontario Society for the Study of Argumentation.

[33] Ritola, Juho. 2009b. "Commentary on James E. Gough and Mano Daniel's 'The Fallacy of Composition'." In Ritola 2009a.

[34] Samuelson, Paul A. 1948. *Economics*. 1st edn. New York: McGraw-Hill.

[35] Samuelson, Paul A, and Peter Temin. 1976. *Economics*. 10th edn. New York: McGraw-Hill.

[36] Samuelson, Paul A., and William D. Nordhaus. 1985. *Economics*. 12th edn. New York: McGraw-Hill.

[37] Walton, Douglas N. 1985. *Arguer's Position: A Pragmatic Study of Ad Hominem Attack, Criticism, Refutation, and Fallacy*. Westport, CT: Greenwood Press.

[38] Walton, Douglas N. 1987. *Informal Fallacies: Towards a Theory of Argument Criticisms*. Amsterdam and Philadelphia: John Benjamins.

[39] Walton, Douglas N. 1989. *Informal Logic: A Handbook for Critical Argumentation*. New York: Cambridge University Press.

[40] Walton, Douglas N. 1995. *A Pragmatic Theory of Fallacy*. Tuscaloosa, AL: University of Alabama Press.

[41] Walton, Douglas N. 1996a. *Argumentation Schemes for Presumptive Reasoning*. Mah-

wah, NJ: Lawrence Erlbaum Associates.

[42] Walton, Douglas N. 1996b. *Argument Structure: A Pragmatic Theory*. Toronto: University of Toronto Press.

[43] Walton, Douglas N. 1996c. *Fallacies Arising from Ambiguity*. Dordrecht: Kluwer.

[44] Walton, Douglas N. 1998. *Ad Hominem Arguments*. Tuscaloosa, AL: University of Alabama Press.

[45] Walton, Douglas N. 2004. *Abductive Reasoning*. Tuscaloosa, AL: University of Alabama Press.

[46] Walton, Douglas N. 2006. *Fundamentals of Critical Argumentation*. New York: Cambridge University Press.

[47] Walton, Douglas N. 2008. *Informal Logic: A Pragmatic Approach*, 2nd ed. New York: Cambridge University Press.

[48] Walton, Douglas N. 2013. *Methods of Argumentation*. New York: Cambridge University Press.

[49] Walton, Douglas N., and F. Macagno. 2016. "A Classification System for Argumentation Schemes." *Argument and Computation* 6: 219-45.

[50] Walton, Douglas N., Chris Reed, and Fabrizio Macagno. 2008. *Argumentation Schemes*. New York: Cambridge University Press.

[51] Woods, John. 2001. Review of Douglas Walton's *Ad Hominem Arguments*. *Argumentation* 15: 503–507.

[52] Woods, John. 2016a. "Economic Reasoning and Fallacy of Composition, Part II: Comments on Maurice Finocchiaro's Paper." *Eris: Rivista internazionale di argomentazione e dibattito*, vol. 1, no. 2, pp. 39-45. ISSN 2421-6747; at http://eris.fisppa.unipd.it/Eris.

[53] Woods, John. 2016b. "Economic Reasoning and Fallacy of Composition, Part IV: Some Parting Words." *Eris: Rivista internazionale di argomentazione e dibattito*, vol. 1, no. 2, pp. 57-61. ISSN 2421-6747; at http://eris.fisppa.unipd.it/Eris.

[54] Woods, John, A. Irvine, and D. Walton. 2004. *Argument: Critical Thinking, Logic and the Fallacies*, 2nd edn. Toronto: Prentice-Hall.

[55] Woods, John, and Douglas Walton. 1977a. "Composition and Division." *Studia Logica* 36: 381-406. Reprinted in [58, pp. 93-119, 279-81].

[56] Woods, John, and Douglas Walton. 1977b. "*Post Hoc, Ergo Propter Hoc*." *The Review of Metaphysics* 30: 569-93.

[57] Woods, John, and Douglas Walton. 1982. *Argument: The Logic of Fallacies*. Toronto: McGraw-Hill Ryerson Limited.

[58] Woods, John, and Douglas Walton. 1989. *Fallacies: Selected Papers 1972-1982*. Dordrecht: Foris Publications.

Appealing to Ignorance? De-extinction and Accounts of a Fallacy

Trudy Govier
University of Lethbridge, Canada
`Trudy.govier@uleth.ca`

In this essay I will consider two topics. The first is that of the fallacy of appealing to ignorance. The second is that of de-extinction, the project of bringing back into existence animals of species that have become extinct. Species considered in this context include the woolly mammoth, the passenger pigeon, the Tasmanian (thylacine) tiger, the auroch, the pyrenian ibex, and the dodo bird. I will explore each issue separately before spelling out my interest in their intersection. But first, some words about Douglas Walton, in whose memory this journal issue has been commissioned.

Although I did not know Doug Walton well, I had known him for many years — ever since the 1970s, in fact. We were both graduates of the University of Waterloo, though he had left there several years before I arrived. As many will attest, Doug was a humble person despite his extensive professional accomplishments. His energy and its products were prolific and for that he was renowned. Doug was very generous with his help to more junior scholars, with many of whom he collaborated on projects and articles. He was greatly appreciated for his creative ideas, productivity and helpfulness. Doug's common sense and sensitivity to both ordinary and specialist contexts made his talents especially suitable for the developing field of informal logic. His death was untimely and unexpected, and he will be sadly missed by students, colleagues, and friends.

On the fallacies, Doug Walton's accounts developed over several decades. But there was a common theme: he maintained that some lines of reasoning standardly deemed fallacious were not always so; rather, they were reasonable in some contexts. As we will see, this stance characterized Doug Walton's approach to the fallacy of appealing to ignorance.

Appeals to Ignorance

The classic form of such an appeal goes like this.

Model I:

1. X is not known to be true.
   Therefore

2. Not-X.

Clearly any argument that can be appropriately represented in this way is fallacious because the fact that we do not know something does not show that it is false. An argument on these lines is neither deductively valid nor inductively strong. Our lack of knowledge is simply that, and does not establish a case one way or the other.

Jonathan Adler offered a slightly different account.[1] Adler's account of appeals to ignorance considered a line of argument advanced by the journalist C.D.B. Bryan,[2] who had maintained that since there was no evidence that aliens did not exist, it was possible they did exist; thus one should keep an open mind on the matter. On the contrary, Adler maintained, an open mind on this matter would be appropriate only if the existence of alien beings was a serious possibility. He argued that its mere possibility, inferred from lack of disproof of the opposite, was not sufficient to establish its status as a serious candidate for belief; accordingly there was a crucial equivocation in Bryan's argument between a mere possibility and a possibility to be taken seriously.

If a claim X has not been disproven and is not contradictory, one may infer from the lack of disproof that there is a sense in which it is possible that X is true. That is to say, the truth of X is a *mere or bare possibility*. One may deem an argument from ignorance to bare possibility to be non-fallacious. But suppose — as would normally be the case — that one's concern is with a different sense of possibility, one in which the claim that X is *possible* has a real bearing on one's projects and decisions, with the implication that X is the sort of thing about which one should keep an open mind. That conclusion is not supported by a claim that one does not know the opposite of X to be true. An argument from lack of knowledge to a serious possibility amounts to a fallacious appeal to ignorance. Adler argues that if such an appeal to ignorance seems reasonable, it is due to a failure to clarify what sort of possibility is intended.

---

[1] Jonathan Adler, "Open Minds and the Argument from Ignorance," *Skeptical Inquirer* (Jan/Feb 1998) 22(1).

[2] C.D.B. Bryan, *Close Encounters of the Fourth Kind: Alien Abduction, UFOs and the Conference at MIT* (New York: Knopf 1995)

Adler attributed to Bryan an argument along these lines:

1. It has not been proven false that X.
   so
2. It is possible that X.
   so,
3. One should keep an open mind as regards X.

Ignorance as in (1) here is never sufficient for belief, Adler maintains. There are problems in the inferential move from (1) to (2) if the term "possible" refers to a *serious* possibility. But only if "possible" is understood in this sense could we reason from (2) to (3). There is an equivocation in (2) as between mere possibility and a serious possibility.

Adler maintains that only (some) supporting evidence will do if we seek to establish that it is reasonable to believe a claim. With (some) supporting evidence, we have left the realm of the mere possibility that a conclusion claim is true; we have gone beyond that to serious possibility. That is a mistake. If we have enough supporting evidence to support a serious possibility in (2) and we reason from (2) to (3), we have a different argument, potentially a cogent argument. And it is not an argument from ignorance.

A Textbook Account

My own account in *A Practical Study of Argument* offers a broader interpretation of the fallacy, as follows:

Model III:

1. X is not known to be true.
   Therefore
2. X is not true.
   Therefore
3. A.

This is clearly different from the classical model, which is also acknowledged in the text. Model III fits some cases commonly discussed as appeals to ignorance — most notably reasoning arguing for God's existence and contested as reasoning only to a 'God of the gaps.' (It also fits cases in which medical people, not knowing the cause of symptoms, argue that they are due to stress.) In the context of theology, one may assert that *either* a proposed naturalistic explanation *or* a supernatural explanation (God or angels) of phenomena is true. Then one may insist that one will never be in a position to know that or even to have good reasons for the naturalistic explanation and hence a supernaturalistic one must be correct. In arguments on this type, failure to find a naturalistic explanation of some phenomenon or other is cited as a strong reason to believe that the phenomenon has a supernatural explanation, being caused by God.[3] Such reasoning is often used to support belief in miracles or, for that matter, divine creation of the earth. Thus:

Model III*:

1. X is not known to be true.
   Therefore

2. X is not true.

3. Either X is true or A is true.
   Therefore

4. A is true.

The classic fallacy of ignorance appears in the inference from (1) to (2). The argument from (2) and (3) to (4) is rendered deductively valid, and non-fallacious, due to premise (3), often not stated. The problem is with that premise. Introduced to create formal validity, a premise along these lines will typically be problematic; to say the least, it will need support. Often it will amount to a false dichotomy. That the proferred naturalistic explanations OR a theological account are correct is a false dichotomy. So too is an assumption to the effect that a symptom is caused EITHER by physically recognizable causes OR by stress.

Walton on Appeals to Ignorance

Over several decades Douglas Walton published a number of articles about appeals to ignorance. He also wrote a book about the topic.[4] Walton noted that reasoning

---

[3]Trudy Govier, *A Practical Study of Argument.* Seventh Edition enhanced. (Wadsworth/Cengage: San Francisco 2010). The expansion so that one is arguing for a claim distinct from the claim X said to be unknown is perhaps unorthodox, but so far as I know it has not been subject to criticism.

[4]Douglas Walton wrote about this fallacy many times over the years. In addition to the paper discussed here, some other writings are "Nonfallacious Arguments from Ignorance," *Ameri-*

based on our lack of knowledge is not always referred to as appealing to ignorance but may be given other names such as 'negative evidence' or 'ex silentio'. His accounts varied in some details, incorporating shifts in his underlying views about the nature of argument and argumentative dialogues. Yet through these shifts, several broad themes persist, the central one being that there are appeals to ignorance that are not fallacious.

Here I discuss Walton's 2006 account.[5] Walton begins this account by noting that in some artificial intelligence systems, if one searches for a claim and does not find it listed as true, one is advised to treat that claim as false. Clearly those constructing such artificial intelligence systems did not regard appeals to ignorance as fallacious. Now one might maintain that those persons were simply mistaken. Characteristically, Walton was unwilling to take that stance. On his account, although some appeals to ignorance are fallacious, others are not.

Now if some appeals to ignorance are fallacious and some are not, how do we tell the difference? Walton's approach makes use of a supplementary premise to the effect that if the claim considered were true, we would know it to be true. If we posit a conditional premise to be inserted in a putatively fallacious appeal to ignorance, and give reasons to support that conditional premise, then we can construct a reasonable argument to support a conjecture based on lack of evidence. At this point I should note my own general scepticism about the strategy of improving arguments by reconstructions involving supplementary premises. (After all, any flawed argument can be turned into a good argument by adding suitable premises to it. Whether that approach to reconstruction shows the original argument to have been cogent (in any sense) is disputable.) I have argued the point elsewhere but cannot pursue it here.[6] I presume that Walton regarded the premises he suggested as claims that the arguers appealing to ignorance would have accepted, and just failed to make explicit. In context, he would have claimed that it is interpretively plausible to add them to an argument that would, on a less generous interpretation, have amounted to a fallacious appeal to ignorance.

Model IV (Walton 2006):

1. We do not know A to be true.

---

can Philosophical Quarterly 29(4) 1992; *Arguments from Ignorance* (University Park Pennsylvania: Pennsylvania State University Press 1996; "Profiles of Dialogue for Evaluating Arguments from Ignorance", *Argumentation* 1998, 13(3), 53-71; and "The Appeal of Ignorance; or *Argumentum Ad Ignorantium*," *Argumentation* 1999, 13(4) 367-377.

[5]Douglas Walton, "Rules for Reasoning from Knowledge and Lack of Knowledge." 2006. (*Philosophia* 34, 355 – 376).

[6]Trudy Govier, *Problems in Argument Analysis and Evaluation.* Updated Edition. 2017, Chapter Five. Windsor, Ontario: Windsor Studies in Argumentation.

2. If A were true, we would know A to be true.
   Therefore

3. A is not true.

This argument is deductively valid. There is a statement of ignorance in the first premise. The second premise is an add-on to that appeal. The argument is a form of *modus tollens*. However, such an argument can support its conclusion only if both premises are reasonably acceptable. Without (2), an argument from (1) to (3) will be a classically fallacious appeal to ignorance. When one comes to assess such an argument, questions will arise concerning the added premise, (2). Why is it that we would know the truth of A, if it were true? (2) needs support. And indeed such support will sometimes be available.

Consider, for instance, a case in which the putative evidence for A would be conspicuous and easy to find. Suppose, for instance, that A were the claim that a bulldozer is on Kitsilano Beach (Vancouver). Now a bulldozer is the sort of thing that one would easily spot on a beach. If people have looked all over the beach for a bulldozer and not found one there, they have good reason to support their claim that if a bulldozer were to be on the beach, they would know it. In such a case, a conditional premise along the lines proposed by Walton could be supported by a cogent sub-argument. Indeed, *if* the bulldozer were there, people would know it. We can say why because of their search and because of the conspicuousness of bulldozers. Consider premise (1) together with (2) here. From (1) and (2), we may validly infer (3) that there is not a bulldozer on the beach. The argument is deductively valid (*modus tollens*), its premises are true, and the conclusion is established. No fallacy is involved, even though although in the first premise there is an appeal to ignorance.

The key issue here is that the added premise (2) needs support, which (in the bulldozer case) it can receive in an appropriate sub-argument. One can thus construct an argument that incorporates an appeal to ignorance but is not fallacious. Model IV* represents this need for a sub-argument defending (2).

Model IV:

a. If evidence for A existed we would have found that evidence.

b. We have not found evidence for A.
   therefore

c. There is no evidence for A.

1 We do not know A to be true.
   Therefore,

2 A is false.

What I draw from this account, building on Walton's 2006 paper, is that we can distinguish fallacious from non-fallacious appeals to ignorance by constructing a conditional premise and sub-arguments to be added, as exemplified in Model IV*. Walton calls such reasoning 'presumptive'. But the argument from the conjunction of (a) and (b) to (c) is deductively valid. In view of its deductive validity, the appropriacy of the term 'presumptive' might be questioned. In the inference from (1) to (2), we see a classic appeal to ignorance. The term "presumptive" does accord with the fact, noted by Walton, and applying obviously to (a) and (b), that the search for evidence will likely be ongoing. Our failure to have found evidence for claim A at a given time only establishes a *presumption* that we will never find evidence to support claim A. Evidence was sought and was not found and it was the sort of evidence that would have been found had it been present. Support for (1) depends on a presumptive claim (b) and support for the conclusion that A is false depends on support for (1) offered by (a) and (b). The real issue regarding Model IV* concerns the cogency of the sub-argument and its proffered support for the supplementary conditional premise (a).

Moving On: Issues of De-extinction

When I began to read about de-extinction projects, my concerns lay in the area of risks. I was worried that one would not know, and could not know in advance of any 'success' of such projects. What would result if extinct species were resurrected and new-to-our-times animals began to live in contemporary environments? I then wondered whether the line of argument that I found so tempting amounted to a fallacious appeal to ignorance. Coming as they did at the time of Doug Walton's untimely death, these thoughts led me to consider again his reflections on appeals to ignorance.

De-extinction would be the bringing back into existence species of animals and plants that have become extinct. As for plants, the chestnut tree is often mentioned. The essays considered here focus on animal species. Suggested candidates include the woolly mammoth, the passenger pigeon, the quagga (a species of zebra), the Tylacine (or Tasmanian) tiger, the gastric-brooding frog, and the moa, and the dodo. And there are many others. De-extinction has not quite happened yet, though efforts are in progress, some of them serious, advanced, and defended by established scientific researchers. Prominent are the efforts of George Church, who is working at no less a center than Harvard University on efforts to revive the woolly mammoth. His efforts are intended to support a Pleistocene Park project pursued by Sergei and

Nikita Zimov, who seek to slow the thawing of Arctic permafrost, by establishing large mammoths who will trample trees so as to recreate grassland. Slowing this Arctic thawing would make a substantial contribution to the slowing of climate change.[7] The woolly mammoth (to be rescued from extinction) would be created by extracting mammoth DNA from frozen dead animals and combining it with DNA from Asian elephants.

Reviewing a number of essays by scientists and commentators, I found an array of arguments for and against projects of de-extinction. Favoring the projects were considerations of novelty and wonder; human interest due to educational and cultural value; challenges for scientific research; notions of atonement or justice to species lost due to human intervention; and potential environmental benefits. Against were prospects of unwise expenditures; concern that scientists would be 'playing God'; distraction from needed conservation efforts involving extant but threatened species; suffering likely imposed on individual animals such as surrogate breeders or candidate members of the species to be recreated; potential environmental damage; and the possibility of serious pathogens emerging from material taken from long-dead animals. Given that I was writing in the midst of a global pandemic of COVID-19, this last matter quite naturally captured my attention.

Projects of de-extinction should certainly attract the attention of philosophers, particularly in areas of ethics and environmental ethics. But as yet philosophical accounts have been relatively few. Outstanding among them is that of Shlomo Cohen.[8] Cohen explains that he wrote about de-extinction not to take a stance for or against various de-extinction projects, but rather to explore lines of argument about them. He does this with great care and thoroughness. Closest to my own concerns were the matters described under Cohen's sub-title 'negative utility,' and here I restrict my attention to them. In this section of his paper, after noting the possibility of unwise expenditures, Cohen goes on to consider environmental and health concerns. Cohen notes that the resurrected species could harbour unrecognized and dangerous retro viruses, introducing health threats to humans and other animals. But he does not give decisive weight to these themes about possible negative consequences, noting in his discussion that utilitarianism can never deliver on its promise, given that the consequences of actions are never fully known.

Environmentally, Cohen acknowledges that a reintroduced species could become a pest, damaging environments and extant species in ways analogous to those resulting in some cases from invasive species. At this point it is fundamentally important to note that for various reasons 'resurrected' species would not be exactly similar to

---

[7] For a description see Angela Chen, with Ben Mesrich, "Will bringing back the woolly mammoth save humanity from itself?", *The Verge*, July 27, 2017.

[8] Shlomo Cohen, "The ethics of de-extinction," *Nano-ethics* 2014 8(10), 51 – 78)

the original ones. For an illustration of this point, consider the role of the Asian elephant in George Church's project. And even if the resurrected animals were exactly similar to the extinct ones, the eco-systems into which they would be introduced have changed, many of them considerably, since animals of that species lived within them.

David Blockstein writes as a scientist concerned with ethical questions.[9] He stresses that resurrected animals *would not be identical* to original members of a species, and given that fact, emphasizes that those who believe an extinct species had been brought back to life would be deceived. The brought-to-life animal is a (perhaps slightly) different species. Blockstein maintains that scientists pursuing de-extinction are in fact conducting synthetic biology when they use DNA extracted from museum specimens, compare it with DNA from the nearest surviving relative species, and insert the former into the latter. When this is done, a *new species* is created. However much it may look like the extinct species, it is not identical to it. Nor would one get the original species from back-breeding, beginning with the most similar extant species. Even cloning would not do the trick, because the embryo resulting from it would have to be implanted in another species and its development would be affected by its pre-birth environment.

In his essay Blockstein concentrates on the case of the passenger pigeon. Passenger pigeons once flew in enormous flocks over eastern North America; their numbers were estimated at three to five billion. The species became extinct as a result of its use by humans for food, commerce, and sport. It cannot be resurrected just as it was, to function in the environment that existed in eastern North America in the nineteenth century, because in addition to the factors already mentioned, the environment of the past no longer exists. Blockstein emphasizes that to think that such a 'resurrected' species could survive and amount to a de-extinction (resurrection) of the passenger pigeon is to be deluded. He argues that the delusion is a dangerous one.

A number of essays about de-extinction hint at arguments from ignorance, while not taking them to be decisive. Lynn Rothschild states that it is impossible to know how a local ecology would change if a newly resurrected animal were introduced into it.[10] She also notes that there would be a new microbiome with unknown qualities and effects. Unintended consequences on ecosystems could emerge: she states that a Pleistocene Park in Siberia for woolly mammoths would be just as unwise as the fictional Jurassic Park. H. Nicholls states that re-introducing an extinct species

---

[9]David Blockstein "We Can't Bring back the Passenger Pigeon: the Ethics of Deception around De-Extinction", *Ethics, Policy and Environment* 2017 20(1), 33 – 37.

[10]Lynn Rothschild, "Seven reasons we shouldn't bring extinct animals back to life," *Quartz*, March 15, 2019.

could pose unknown threats.[11] In "Should we bring extinct species back from the dead?", David Schultz argues that negative effects of de-extinction would certainly be possible.[12] Human beings are fallible and should not be arrogant, states Bruce Jennings in a 2017 paper.[13] Our efforts might backfire. We are part of the natural world and should not presume a capacity to manipulate it wisely. Other authors note that de-extinction would be risky and that projects of de-extinction are by no means urgent. David Sprayer maintains that ecological effects of introducing resurrected species could be costly and difficult, citing experiences with introduced species that turned out to be invasive and difficult to remove.[14]

For Anne and Paul Ehrlich the strongest argument against de-extinction projects is misallocation of effort.[15] They maintain that projects to reintroduce 'resurrected' species into contemporary environments would be complex, expensive, and unlikely to succeed. They urge that efforts to save animal species should be directed to preserving species that survive today but are threatened. Introduced supposedly de-extinct species could become pests in their new environment and might prove ideal reservoirs or vectors for nasty plagues. They could harbour nasty retroviruses in their genes. T.J. Kasperbauer argues against such a project, pointing out that the species in question could become invasive.[16] Those seeking 'resurrection' would wish their creations to be resilient, but resiliency argues against removability in the event that problems occur. Kasperbauer argues that there is no human obligation to redress harms done to a species. A species as such does not have interests and cannot be harmed or benefitted.

Bringing Themes Together: A Fallacious Argument from Ignorance?

Building on these materials but not claiming to paraphrase any one text, let me construct an argument to consider in this context. Let 'S' be the statement: Introducing a resurrected species into a contemporary environment is safe. By 'resurrected species' here I mean a species nearly but not in every regard identical to a species

---

[11] H. Nicholls, "Ten extinct beasts that could walk the earth again," *New Scientist* 201 (2009), 24 – 28).

[12] David Schultz, "Should we bring extinct species back from the dead?" *Science*, Sept 26, 2016.

[13] Bruce Jennings 2017 "The Moral Imagination of De-extinction" (*Hastings Centre Report* 7(5).

[14] David Sprayer, "De-extinction, a risky ecological experiment," *The ESA*, February 12, 2016.

[15] Anne and Paul Ehrlich, "The Case against De-extinction: it's a fascinating but dumb idea", *Yale Environment* 360; January 13, 2014.

[16] T.J. Kasperbauer, "Should we bring back the passenger pigeon? The ethics of de-extinction." *Ethics, Policy and Environment* 2017 29(1), 1 – 14.

that became extinct. Now consider the following argument.

1. We do not know that S is true.
   Therefore

2. S is false.

We do not know that the introduction would be safe; therefore it would not be safe to introduce a resurrected species into a contemporary environment. This would be a fallacious appeal to ignorance.

Although premise (1) is true, this argument is not deductively valid or inductively strong.

Consider now a variant along the lines discussed by Adler.

1. We do not know that S is true.
   Therefore

2. S is possibly false.
   Therefore

3. We should keep an open mind toward not-S.

In other words, introducing a resurrected species into a contemporary environment is possibly not safe. (We might gloss this as 'possibly dangerous'.[17]) If we interpret "possibly" in the conclusion here to refer to a significant possibility, one that needs to be taken seriously in the context of practical or theoretical reasoning, this variant is also fallacious, as explained by Adler.

So let us move on to consider a model of the type Model IV, based on the work of Douglas Walton.

1. We do not know S to be true.

2. If S were true, we would know S to be true.
   Therefore

3. S is not true.

---

[17]There are interesting questions to be raised about 'safe' and 'dangerous' as opposites. Is something that is *not safe* thereby *dangerous*? We often speak that way. But both safety and dangerousness are matters of degree. It would be better to speak of a spectrum, of degrees of safety and of danger. I cannot explore these matters here.

As noted, an argument along these lines is deductively valid and does not exemplify a fallacy. However it will not qualify as a cogent argument unless the second premise can be supported. To be warranted in asserting (2) in this argument, it would need to be the case that evidence for S would be available if we had looked for it, and we have looked for it. As explained earlier, conditional claim (2) requires support in a sub-argument. Stating such an argument, consider the following:

a. We have sought evidence for S.

b. If evidence for S existed, we would have found it.
so

2 If S were true, we would know S to be true.

1 We do not know S to be true.
Therefore

3 S is not true.

But this sub-argument strategy will not be viable in contexts of de-extinction. Claim (b) would not be available for such a case. Arguments based on predicted negative consequences are about whether the introductions understood as de-extinctions *should* be carried out. Such arguments are offered and are relevant to decision-making *in advance of such efforts*. What would be needed for evidence about the effects of de-extinction introductions to be discoverable would be facts after the actual introduction of the resurrected species. But in the nature of the case, the introductions have not happened yet. At the time they are made, we cannot search for evidence about the effects of re-introductions; our failure to find such evidence is not significant because that failure is based on the impossibility of gaining it in real time. The failure is just due to the temporal framework and has no implication that such evidence would not exist in different circumstances. For these reasons, in the context of deliberations about the safety or danger of introducing 'de-extinct' species, the supplementary conditional premise (b) cannot be defended with cogent sub-arguments.

I conclude that a strategy along the lines suggested by Walton is not feasible in this sort of case. A cogent *modus tollens* argument cannot be constructed because the conditional it requires cannot be supported by cogent sub-arguments. We are left with an argument to the effect that we do not know S to be true; therefore it is false. The argument we are left with is that because we don't know that resurrection introductions are safe, they are not safe. In other words, we conclude from our lack of knowledge that they are safe that these introductions are dangerous. That is a classic appeal to ignorance and is fallacious. We cannot produce an adapted arguments along the lines of Model IV*. Our failure to know that resurrection introductions are safe does not show that they are not safe.

Another Approach

As explained, in the nature of the case, we are considering the pros and cons of introductions before those introductions have occurred, and thus we cannot obtain direct evidence about their safety or lack of safety. However there are simpler approaches to questions of safety and danger in the context of de-extinction. These approaches involve citing inductive analogical evidence for claims to the effect that species introductions under the guise of de-extinction are risky. Reasonable inductive arguments based on analogous situations can be developed. Such arguments involve invasive species. Indeed, they are mentioned in some accounts of de-extinction projects. They hinge on two basic points. The first point is that a contemporary environment into which a de-extinct species would be introduced would be different, in all likelihood *very different*, from the past environment in which the now-extinct species had survived and thrived. The second point is that it is a mistake to think literally of de-extinction. The thought-to-be resurrected species amounts to a *new species*, because due to genetic, congenital, and environmental circumstances, scientific efforts will not suffice to create an exact duplicate of the supposed ancestor. Introducing a 'resurrected' species into a contemporary environment would, in effect, be introducing a new species into an environment in which neither that species or its supposed ancestor had previously existed.[18]

In many cases in which species have been introduced into environments, they have become invasive, bringing considerable damage to eco-systems and species native to that environment. These facts provide the basis for an inductive analogical argument against introducing new species into an environment. This basis is not to be found in reflections on our ignorance; that is not the point. It is not that we do not know that such an introduction would be safe. Rather, it is we have relevant inductive evidence for the claim that their introduction would be dangerous. Notoriously, there is evidence for such a claim, as for instance with rabbits in Australia; Africanized bees in California; Burmese pythons in Florida; and Canadian beavers in Tierra del Fuego. (In the nineteen forties, 25 pairs of Canadian beavers were introduced in southern Chile and Argentina, the rationale being that these animals would support jobs in a fur trade and also be an attraction to tourists. The beavers

---

[18] Jorge Poblete, "Beavers imported from Canada are threatening the primeval forests of Patagonia," *Los Angeles Times* January 6, 2017.

had no natural predators. Thriving in the Patagonian environment, the beavers increased to over 200,000 in number. Their activities constructing dams caused floods and threatened forests, severely damaging the environment.)

Arguments against 'de-extinction' projects based on our ignorance of their safety cannot be rescued using strategies extracted from Douglas Walton's 2006 paper. To argue for danger in the context of de-extinction we do not need to employ arguments from ignorance because other arguments are available. To use arguments for danger, based on our failure to know that there is safety in that context, would be to fallaciously appeal to ignorance. Fortunately, for those seeking to argue against de-extinction on the grounds of dangers, a more straightforward — and non-fallacious — strategy is available.

Turning the Tables: Up- and Downgrading of Evaluative Terms in Public Controversies

Jan Albert van Laar and Erik C. W. Krabbe
Faculty of Philosophy, University of Groningen, Netherlands.
j.a.van.laar@rug.nl, e.c.w.krabbe@rug.nl

Abstract

We discuss the strategy of using dyslogistic terms in a novel, laudatory manner, or eulogistic terms pejoratively. By such up- and downgrading of evaluative terms the proponent of a standpoint may attempt to turn the tables in a public controversy: what formerly looked like a bad argument comes to be regarded as a strong one, or vice versa. Is this a licit strategy? We take our lead from Macagno and Walton who examined evaluative words from an argumentative stance.

1 Introduction

Public controversies are more often than not replete with emotion, particularly so where arguers try to get their audiences to agree with them on evaluative points of view. When trying to find common ground they must not only take into account that the audience may cherish values that are at odds with their own preferences, but also that the words one uses to persuade the audience may be differently appreciated. Terms that the arguer intends as expressing recommendable features could come across as expressing what is objectionable and vice versa. When phrasing one's position on issues of evaluation, it is hard to avoid "loaded" terms and to use only formulations that are neutral and not question-begging, i.e. not associated with

---

We acknowledge the intellectual support of COST Action CA17132 'European Network for Argumentation and Public Policy Analysis.' Further, we thank the audiences of presentations of this paper at the conference "Reasons, Citizens and Institutions," organized in this network (4-6 March 2020, University of Wrocław, Poland) and at the Faculty of Philosophy (14 October 2020, University of Groningen, The Netherlands). We thank John Woods for reading our text and commenting on it in its latest stage.

either a negative evaluation of one's antagonist's views, or with a positive one of one's own views. On the one hand, an arguer may have a preference for relatively neutral terms, but on the other hand it is common, expected, and considered tolerable, that any arguer will, within the bounds of reason, further his or her cause by making an opportune selection from the available vocabulary (cf. [22], on strategic maneuvering). And often the selection will include "emotive words" [12, p. 230], which help the arguer to persuade his or her audience to accept certain positive or negative evaluations.

Some evaluative terms, such as "good" and "bad," have an evaluative core meaning that any adequate lexical definition needs to take into account. Others, such as "generosity" and "avarice," have an evaluative meaning in addition to their descriptive meaning. Such terms were discussed by Jeremy Bentham, who spoke of *eulogistic* (laudatory) and *dyslogistic* (pejorative) *appellatives* [2]. They have a core *descriptive* lexical meaning, and an *evaluative* meaning, where either kind of meaning may be shared or not shared by speaker and listener. For instance, discussants often have very different ideas about the core descriptive meaning as well as the evaluative meanings of terms like "liberal," "conservative," "radical," and "popular."

People not only differ in their meaning assignments, they may also be persuaded or manipulated to change them, which makes it possible to design strategies to achieve such an effect. In this paper we discuss one couple of such argumentative strategies linked to evaluative terms, namely the strategies of using, in a positive way, a term that is normally negatively loaded (which we call "upgrading"), or the other way round (which we call "downgrading"). Examples of terms that were recently used in such strategies are: "deplorable" and "multicultural."

These strategies can be used on a local level, effecting a change of meaning within a specific speech event, but also on a global level, gradually changing the speech habits of a community by repeatedly using a term while reversing its customary evaluative meaning.

First, in Section 2, we further explain the concepts of upgrading and downgrading and discuss their relation to other concepts, such as "persuasive definition." We characterize up- and downgrading as argumentative techniques, and show how they can turn the tables: what once generally looked like a worthless argument ("deplorable therefore good" or "multicultural therefore bad"), may no longer do so — thus turning unreason into reason, and what once generally looked like a worth while argument ("deplorable therefore bad" or "multicultural therefore good"), no longer does — turning reason into unreason.

Second, in Section 3, we argue against a complete banishment of all upgrading and downgrading and discuss how up- and downgrading may at times play a positive role. For one, proposing semantic innovations may free issues from old and rusty fashions. Views that have long been taken for granted may be "politicized," and opened up for discussion.

Third, in Section 4, we discuss the possible drawbacks of up- and downgrading: such argumentative tactics may deliver no more than "question begging appellatives" that smuggle controversial starting points into the discussion. Further, they may breed confusion by introducing ambiguities. And also they may just turn the social pressure that a term conventionally generates into a new and reversed kind of social pressure.

Fourth, in Section 5, we conclude that from a normative but not moralistic point of view a ban on up- and downgrading would pose an undue restriction on discussion. We add some remarks about how a normative theory can avoid being overly moralistic.

2 Up- and downgrading as an argumentative strategy

In this section, we discuss the argumentative strategy of upgrading or downgrading evaluative terms. We start by characterizing evaluative terms, proceed by discussing argumentative functions of using evaluative terms and the kind of meaning change involved in up- and downgrading, so as to end with examining the strategic significance of such up- and downgrading in argumentative exchanges.

2.1 Evaluative terms

Evaluative terms are those expressions that, in a context of use, convey a positive or negative evaluation. They are emotive terms in the sense that they "are not simply used to describe reality by modifying the cognitive response of the interlocutor [...], but more importantly to affect the interlocutor's attitudes towards a state of affairs and suggest a course of action" [12, p. 230]. Though it might be possible to convey an evaluation in a purely cognitive way, without aspiring to affect the attitudes of the interlocutor, that would be quite exceptional: generally evaluative terms will be emotive. On the other hand, emotive terms seem always to have an evaluative aspect. In this paper we choose to speak of evaluative terms, thus including the emotive ones.

In special cases, such as when applying the term "good," the evaluation conveyed exhausts the expression's meaning.[1] In most cases, however, the evaluation is to be seen as only a part of the expression's semantic content that is added to a

---

[1] In this paper, when speaking of an expression's "meaning." we refer to the content it has on a particular occasion of use.

descriptive content, such as in the case of the term "generous."[2] In such cases the evaluation may be part of the expression's standard (lexical) content or in a looser and more contextual way connected to the descriptive meaning of the expression (cf. [11, pp. 38ff]). An example of such a contextual connection can be found in the uses of the expression "maximizes direct sunlight," which when used to describe a room considered for renting, would normally express a positive appraisal but not for artistic painters, who wish to prevent any sunlight from coming into their studio (example from: [24, p. 474]). Among evaluative terms, we distinguish between *thick* and *thin terms*, and between *eulogistic* and *dyslogistic terms*.

Thick terms, such as "generosity," "avarice," or "cowardice," systematically — rather than occasionally — combine a descriptive content, such as the (un)willingness to share of one's own or the unwillingness to confront danger, with an evaluative content, presenting the descriptive content as a good or a bad thing ([23], [5, pp. 363–4]). *Thin terms*, such as "good" and "appropriate," are terms that express no more or hardly more than an evaluation. In this paper, we discuss the argumentative use of reversals of the evaluative content of terms. When terms retain their descriptive content and reverse their evaluative content, one would first think of thick terms, but even thin terms can, within specific conversations, be (non-ironically) reversed. For example, "doing good" expresses a positive evaluation, yet "do-gooder" is typically used to decry people who (pretend to) uphold high moral standards, as in cases where "doing good" (e.g. being lenient towards refugees) is perceived as having adverse social effects.

In speaking of eulogistic and dyslogistic terms, we follow Jeremy Bentham, who in his *Book on Fallacies* distinguishes between: appellatives that are *neutral* by being unaccompanied by any sentiment of approbation or disapprobation, such as "labor"; appellatives that are *eulogistic* (or laudatory) by being habitually attached to approbation, such as "generosity"; and appellatives that are *dyslogistic* (or vituperative) by being habitually attached to disapprobation, such as "avarice" [2, p. 436]. But we don't follow him in so far as he seems to deal exclusively with evaluations that are by habit attached to the use of a term, since what we are interested in are precisely those settings in which speakers diverge from habits, or what they probably regard as, old or wrong habits. For the same reason, we think that, instead of speaking of terms that *are* eulogistic or dyslogistic, it is more appropriate to say that *a term is used in either a eulogistic or a dyslogistic way on a particular occasion*. When we use the plain expressions "eulogistic term" and "dyslogistic term," we refer to a

---

[2]Macagno and Walton [12, pp. 230–31] quote Arnauld and Nicole [1, Part I, Chapter 14], who distinguish, as parts of the meaning of a word, a main idea (*idée principale*) and incidental ideas (*idées accessoires*), which roughly correspond to what we call the descriptive meaning and the evaluative or emotive meaning.

term that is normally or typically used eulogistically or dyslogistically.

An example of a (thick) eulogistic term, which we shall keep returning to, is "multi-cultural." About its usage in Canada in the 1970s and 1980s, we quote Michael Dewing:

Example 1. Multicultural society
"The concept of Canada as a 'multicultural society' can be interpreted in different ways: descriptively (as a sociological fact), prescriptively (as ideology) or politically (as policy). As a sociological fact, multiculturalism refers to the presence of people from diverse racial and ethnic backgrounds. Ideologically, multiculturalism consists of a relatively coherent set of ideas and ideals pertaining to the celebration of Canada's cultural diversity. At the policy level, multiculturalism refers to the management of diversity through formal initiatives in the federal, provincial, territorial and municipal domains" [8, p. 1].

When the Canadian Government in 1971 labels a set of measures, such as the policy "to assist cultural groups to retain and foster their identity ... [and] to overcome barriers to their full participation in Canadian society," as the *Multicultural Policy* [8, p. 3], it is most likely that "multicultural" is being used as a eulogistic term — at least in contexts of governmental communication. Further outward signs to this effect are: that in 1972 a junior "minister for multiculturalism" was appointed; that in 1988 the "Canadian Multiculturalism Act" was adopted; that in 1991 the "Department of Multiculturalism and Citizenship" was established; and that from 2002 onwards, Canada celebrates on June 27 "Multiculturalism Day" [8, p. 17].

2.2 The argumentative function of evaluative terms

Before we turn to up- and downgrading, and their argumentative functions, we wish to shortly discuss the more straightforward argumentative function of eulogistic (dyslogistic) terms when they are used in the normal way. That the use of such terms introduces an appeal to emotion doesn't exclude that they yield a valuable contribution to good argument. We agree with Walton when he points out that various types of

> "... arguments that are based on appeals to emotion [...] have a good side and a bad side. These arguments are good when they open up new and valuable lines of argumentation, prompting critical questioning that steers the argument in a constructive direction. From a pragmatic point of view, when used correctly such arguments in a dialogue open up deep feelings about what is important and right and steer the argument along

> by according important presumptions the weight they deserve. On the bad side, accompanying any line of presumptive reasoning is a natural tendency to jump to conclusions that favor one's own interest or bias too quickly. This lack of critical balance needs to be offset by an awareness of fallacies and the tendency to bias." [26, p. 28]

This point of view, we think, generally holds for the use of evaluative or emotive terms, and therefore we can agree with Macagno and Walton that there is more to the use of such terms in argument than their power "to subtly slide over the need to offer support for a claim you're making" and thus present yourself as being in the right [11, p. 2].

For example, Party A favors a broadcasting policy of requiring 60% ethnic programs, whereas Party B is not yet convinced. Then, given that — there and then — "multicultural" is a eulogistic term, the assertion that "this broadcasting policy is multicultural" can be used as a reason to make this policy more acceptable to Party B. But how?

One may understand this from the viewpoint of dialectical argumentation theory by connecting an evaluative term with an argumentation scheme, and the use of the term with an appeal to the acceptability of that scheme (cf. [12, Section 3]). In our example, the general acceptance of "multicultural" as a eulogistic term in an argumentative setting can be seen as implying the implicit acceptance by the participants of the following argumentation scheme:[3] "Policy p is multicultural. Therefore policy p is to be adopted." The idea is not that any argument that instantiates the scheme has premises that, if accepted, necessitates its conclusion, but rather that it has premises that, if accepted, create a presumption in favor of its conclusion. As a consequence, the assertion "this broadcasting policy (of requiring 60% ethnic programs) is multicultural" may enthymematically express an argument that instantiates the scheme just given, namely the argument: "This broadcasting policy is multicultural. Therefore this broadcasting policy is to be adopted." What is more, the opponent can then also be expected to consider the policy's multiculturalism as a good reason for adopting the broadcasting policy, and when she nevertheless wishes to maintain a position of doubt and chooses to continue to critically examine the merits of the policy at hand, it is upon her to provide countervailing considerations, before she can urge the proponent to proceed

---

[3] An argumentation scheme is a pattern of reasoning, with variables, that when instantiated yields a typically defeasible argument for the proponent's standpoint. Argumentation schemes can be very general, and often they are discussed on that level (e.g.: "We should value x positively, because x is p and normally things that are p are to be valued positively"), but also they can be much more specific, as in our example where the scheme revolves around the meaning of one specific term (that is: "multicultural").

in his attempts to discharge his burden of proof. Thus Party A seems to have an easy strategy (using "only a few slick words", [11, p. 2]), whereas Party B will have to struggle to counteract this strategy. But if "multicultural" has indeed been generally accepted as a eulogistic term, it won't be unfair for Party A to make use of the strategy. In this case, as in other cases, the argumentative potential (cf. [12, Section 6]) of such evaluative terms for future arguments is generated by way of plausible and generally acceptable argument schemes characterizing those terms.

There is a connection between this argumentative account of emotive words and Brandom's inferential account of the meaning of expressions of any kind.[4] In Brandom's view, we best understand the conceptual content of an expression as provided by "introduction rules," which provide the conditions that justify one to use the expression, and "elimination rules," which provide the consequences of having committing oneself to an assertion in which that expression is used [4, pp. 61–66].[5] Macagno and Walton [11, 12] focus on the roles of emotive words in conversations, and they detail these roles by means of argument schemes. But these schemes can in many cases be seen as special cases of inferential schemes. For instance, in Macagno and Walton [11], classificatory reasoning (which in a simplified version can be rendered as: "a can be classified as 'F', therefore a is G," where "G" is an evaluative term) seems to function as an introduction rule and some kinds of practical reasoning (for example, and again simplified: "a is G, therefore we should do H," where "G" is an evaluative term) as an elimination rule (see pp. 51ff). We think that Brandom's inferentialism suggests a further specification of Macagno and Walton's argumentative account of evaluative terms: an evaluative term can be characterized jointly by the argument schemes that "introduce" that expression, by means of its occurrence in (only) the scheme's conclusion, and by those argument schemes that "eliminate" that expression, by means of its occurrence in (only) the scheme's premises.[6] The *introduction schemes* may be based on the descriptive content of the

---

[4] We did not find references to Brandom in [11, 12].

[5] In formal logic, introduction and elimination rules are the principal ingredients of systems of natural deduction, which provide a conceptual content (meaning) for logical constants.

[6] Brandom, when discussing the appropriateness or inappropriateness of introduction and elimination rules, uses as an example the French word "boche," a disparaging name for Germans. He objects to the proposal, suggested by Dummett ([4, p. 71]; [7, p. 454]), to require such rules to be conservative, in the sense that the acceptance of a new expression and the accompanying introduction and elimination rules should not legitimize new inferences of conlusions formulated in the original language without that expression. The ground for the objection is that in some settings we accept conceptual changes as conceptual progress (he gives the example of the introduction of new measurement devices for "temperature"). In Brandom's view, we refuse to use a slur like "boche" because we consider the accompanying set of introduction rules (e.g.: "if someone is a German, he or she is a boche") and elimination rules (e.g.: "if someone is a boche, he or she is cruel") as inappropriate or indefensible rather than too newfangled [4, pp. 69–72].

expression, but not exclusively, for the reason that also the evaluative aspect of the term's meaning requires a justification, and the *elimination schemes* may be based on the evaluative content of the expression, yet not exclusively, because a premise with an evaluative key term may provide linguistic support for further descriptive statements.

Macagno and Walton focus on the negative effects of using emotive language: "Emotional judgments are hasty and biased, leading to automatic conclusions of right and wrong" [12, p. 245] and they stress that the use of emotive terms can have a devastating effect on the critical assessment of argumentation (p. 245). On the other hand, they do not qualify the use of emotive terms as fallacious.[7]

We wish to emphasize one problematic aspect of evaluative terms: When the evaluative term at issue is suggestive of social, or moral, or ideological norms that are widely accepted or otherwise dominant in the context of utterance, the evaluative term may put so much social, or moral, or ideological pressure on the addressee as to make her refrain from unraveling and criticizing the implicitly expressed argument for the proponent's position, because she assesses that the argumentative merits of doing so would not weigh up against the reputational or other social costs of thwarting the proponent.

2.3 Upgrading and downgrading

We focus on situations where a speaker uses a term dyslogistically whereas it is normally used in a eulogistic way, or eulogistically whereas it is normally used in a dyslogistic way. Before we get to our characterization of up- and downgrading, and some techniques to up- and downgrade, we discuss a number of related concepts discussed in the literature: "quasi persuasive definition," "reclamation (or reappropriation) of slurs," and "evaluation reversal of thick terms."

2.3.1 Quasi-persuasive definitions

The phenomenon of up- and downgrading of evaluative terms was discussed by Charles Stevenson under the label of "quasi persuasive definition" (Stevenson 1944). Before we turn to this concept, we briefly bring to the fore his concept of "persuasive definition" that he discusses in the context of expressions of specifically moral evaluations, which he refers to as "emotive meanings."

> "A "PERSUASIVE" definition is one which gives a new conceptual meaning to a familiar word without substantially changing its emotive mean-

---

[7] As a matter of fact, the terms "fallacy" and "fallacious" do not occur in their 2019 article.

ing, and which is used with the conscious or unconscious purpose of changing, by this means, the direction of people's interests." [18, p. 331]

A persuasive definition leaves the evaluative ("emotive") meaning of the term at hand more or less untouched, but modifies its descriptive ("conceptual") meaning. For example, the term "cultured" as applied to persons has as its usual descriptive meaning something like "being widely read and acquainted with the arts," and its evaluative meaning is laudatory. The term can then be persuasively redefined as "possessing imaginative creativity," saying or suggesting that this is the really significant or true meaning of "cultured", and thereby dissuading people from admiring wide reading, etc., and inducing them to admire imaginative sensitivity [18, p. 331].[8]

Alternatively, "the emotive meaning may be altered, descriptive meaning remaining roughly constant" [19, p. 277], in which cases Stevenson speaks of "quasi persuasive definitions" (p. 278).[9] Such alterations are part of an attempt to make someone change her positive evaluation of something for a negative one, or the other way round, or — when neutralizing an emotive meaning — to resist persuasion by someone who employs that term. For example: "Culture is only fool's gold; the true metal is imaginative sensitivity and originality" [19, p. 278].

Stevenson stops short of labeling such an up- or downgrading of emotive meaning as a kind of genuine (persuasive) definition, for the reason that the descriptive meaning remains constant. Hence, the hedge "quasi." In our view, however, whether a novel use of a term merits to be qualified as a definition does not depend on whether or not the descriptive meaning diverges from standard usage, and we would be inclined to include some terminological up- and downgrades as being introduced by means of definitions. We speak, permissively, of the act of presenting a *definition* of a term when the speaker or writer somehow communicates the message that this term is used in a specific way, with a certain descriptive meaning in mind, or else with a certain evaluative meaning in mind, or both, and we identify the linguistic (or more generally: communicative) device by means of which this message is conveyed as his or her *definition* (see for an account of definitions as speech acts [25]; see for definitions as the result of goal-directed activities [16]). In our lingo, a definition is the result of a (metalinguistic) speech act of defining.

---

[8]Other examples of persuasive definition can be found among instances of what Jason Stanley would characterize as "undermining propaganda" [17, p. 53] — such as where the term "freedom" is used to refer to the right to exclude others by means of slurs ("freedom of speech"), or to the right of a state to retain slavery, and the term "freedom" is thus used to erode the very ideal of freedom. Stevenson gives a similar example: Socrates first praises justice, and then redefines "justice" as "each of the three classes doing the work of its own class," which amounts to aristocratic propaganda [18, p. 344].

[9]Macagno and Walton [11] use the term "quasi-definition" for the same concept.

A definition can be communicated quite indirectly, for example merely by means of stressing the term at hand ("A piece of *music* must have rhythm, melody, and harmony, but free jazz has not"), or by denying a statement in a way that expresses dissent regarding the way the interlocutor puts a term to use (A: "You failed to read my book!"; B: "This stack of copies is not a book"). The mere fact, however, that someone up- or downgrades a term does not imply that the speaker has defined (or redefined) this term, not even when the speaker is fully aware of his deviant use of the term. After all, he may try to downplay his semantic innovation and try to hide his intention to up- or downgrade, so that this meaning change is not communicated at all. Think for example of commercial or political advertisements, where it can be tried to influence or even manipulate the addressees to associate the advertised product or person with positive feelings and competing products with negative ones. An extreme example, though disputed, can be found in the 2000 G. W. Bush campaign:

Example 2. Rats
"Democratic presidential nominee Al Gore's campaign contacted news organizations about an RNC ad in which the word 'RATS' appears briefly on screen in a spot that criticizes Gore's prescription drug plan. A spokesman for the Texas governor on Tuesday brushed aside suggestions of subliminal advertising as 'bizarre and weird,' while the RNC had no immediate comment. CNN slowed down a copy of the ad, and the word 'RATS' clearly appeared on the screen in large, white letters superimposed over the words 'The Gore Prescription Plan.' In a fraction of a second, the word disappeared, and the words 'BUREAUCRATS DECIDE' showed up in smaller letters. To viewers aware of the presence of the word, it is noticeable when the ad is played at normal speed." [6]

If Gore's campaign is right, the Republican campaign tried to influence viewers to associate a negative evaluation to the expression "The Gore Prescription Plan" and thereby to Gore and his prescription plan, while actively trying to prevent the viewers from even becoming aware of the attempt at persuasion. Thus, if Example 2 exemplifies downgrading (and we think it does), then this downgrading goes without genuine communication and thus without definition.

Yet, because in general novelties catch the eye, upgrading and downgrading typically convey a metalinguistic, definitional act. The speaker of "Culture is only fool's gold; the true metal is imaginative sensitivity and originality" clearly communicates her linguistic innovation, but even if she were not, she might still be seen as trying — in a less than fully communicative way — to downgrade the term "culture."

2.3.2 Reclamation of slurs

There exist a number of linguistic studies on the reclamation (or: appropriation or reappropriation) of slurs, where the initial targets of slurs start to use the terms in a non-derogatory way. According to Bianchi's "echoic" account of reclamation, a reclaimed slur echoes the derogatory content by conveying a critical or mocking attitude towards normal, derogatory uses of the slur. For example, the N-word is used in rap music by dark-skinned people so that it functions: (1) to deprive others from a means of discrimination, or at least to mitigate the effects of discriminatory uses; (2) to create a sense of community, also by being reminded of being the objects of discrimination; (3) to express a critical stance to discriminating uses of the slur; and (4) to subvert discriminating uses [3]. By commenting on normal, derogatory uses, the echoic use of a slur is "metarepresentative."

Such reclamation of slurs, we see as a specific kind of upgrading. In Dutch it is called a "geuzennaam" (literally: name of the "Beggars", but now to be understood as a "slur turned into an honorary title"). The anecdotal etymology of the expression "name of the Beggars," says it all. When in 1566 Margaret of Parma received complaints about the inquisition from a crowd of Dutch gentry, her adviser Charles de Berlaymont whispered in her ear (in French): "N'ayez pas peur Madame, ce ne sont que des gueux" (Don't be afraid, milady; they are but beggars). The revolting gentry used the Dutchification "geuzen" of the French "gueux" as a term of pride, and "geuzen" became a standard label to refer to them. The term "geuzennaam" now stands for any reclamation of a slur.

2.3.3 Evaluation reversal

Cepollaro [5] studies reclamation of slurs in tandem with a phenomenon of wider application, which she calls "evaluation reversal" and which comprises also the use of thick terms that on specific occasions are used with a polarity that is opposite to the polarity that they typically have. As in our account of up- and downgrading, the reversal can be in either direction. In her paper, Cepollaro discusses what kind of linguistic approach best explains the linguistic felicity of such evaluation reversals (see also [23], on the variability of thick terms and concepts). We focus in the upcoming sections on argumentative, rather than linguistic, felicities and infelicities. First, however, we shall more precisely characterize the kind of evaluation reversal we call up- and downgrading, and the strategic significance that comes with it.

2.3.4 Upgrading and downgrading evaluative terms

What kind of activity are upgrading and downgrading? By "upgrading an evaluative term" we refer to the eulogistic usage of a term that is typically or normally used dyslogistically, in an attempt of getting the addressee to regard the term as a eulogistic term, or at least getting her to take the eulogistic usage of the term seriously, and plausibly with an eye to giving broader, societal acceptance, or at least salience, to a eulogistic use of the term at hand. The same definiens applies, *mutatis mutandis*, to "downgrading an evaluative term." Instances of up- and downgrades of a term can often be expected to be echoic in the sense of expressing a critical stance regarding the standard usage of the term at hand [3], for example when reclaiming slurs, or in early instances of a longer series of coordinated attempts to up- or downgrade a thick term. The main argumentative message though derives from the positive (negative) evaluation of the phenomenon to which the up- or downgraded term is applied.

Before we turn to the possible argumentative consequences of successful up- and downgrading, we distinguish between three ways in which a speaker or writer can up- or downgrade a term. We spell this out only for downgrading.

Suppose, term Q is normally and typically used as a eulogistic term.

(i) To downgrade term Q, a discussant may then provide an argument based on a premise that contains Q, such that the only way to make sense of the argument would be to interpret Q as having been downgraded, and thus used in a special, pejorative sense. She can then be seen as using a new, reversely orientated elimination rule for Q. This would be the case if someone argues: "Policy p is Q, therefore p is to be rejected," as in "This proposed policy requiring 60% of the broadcasting to be ethnic is really a multicultural policy. Therefore, we ought to reject this policy." In such cases, the discussant conveys his or her downgrading of Q (here "multicultural") by means of drawing a negative conclusion from the applicability of the term. In the next example, Example 3, we can understand the Conservative MP Burley as arguing: "It was multicultural and leftist, and therefore it was crap," and thereby downgrading both "multicultural" and "leftist."

Example 3. Multicultural crap
"David Cameron will face pressure to remove the Tory whip from the Conservative MP Aidan Burley after he tweeted that the Olympics opening ceremony was 'multicultural crap'. Burley, who was sacked as a ministerial aide last year after he took part in a Nazi-themes stag party in the French Alps, described Danny Boyle's work [Boyle directed the opening] as 'the most leftie opening ceremony I have ever seen'." [27]

(ii) To downgrade term Q, a discussant may also provide an argument that leads to a conclusion that contains Q, and in such a way that its non-typical negatively loaded meaning transpires from the premises. She can then be seen as using a new, reversely orientated introduction rule for Q. This would be the case if someone argues: "Policy p has adverse effects, therefore p is Q," as in "This policy requiring 60% of the broadcasting to be ethnic leads to a serious neglect of our culture. It is, therefore, one of those multicultural policies." In such cases, the discussant is conveying his or her downgrading of a term (here "multicultural") by means of drawing a conclusion containing the term from such premises that it becomes clear that the term is used as a dyslogistic term.

(iii) Instead of downgrading a term Q by advancing an argument, a discussant may do so by merely making one or more assertions. This would be the case if someone asserts: "p is Q", as in "This is a multicultural policy," in such a way that gestures, intonation, mimicry or contextual cues make it clear that Q is used to convey a negative evaluation. Macagno and Walton give an example of Trump's (continual) attempts to downgrade the term "politically correct," in which he asserts that being politically correct is "the big problem this country has" and suggests that bothering about political correctness would be a waste of time ([12, pp. 237–8]; [20]).

In each of these three cases, the metalinguistic message can be (directly or indirectly) communicated, so that some sort of definition is in play. But it is also feasible to employ either of these techniques for the up- or downgrading of terms in a systematic attempt to promote a general change of linguistic usage, yet without *communicating* any specific metalinguistic message. One notorious technique deserves special mention, namely that of repeatedly associating a specific term with negative social contexts. Jason Stanley provides the example of news media that repeatedly connect the term "welfare" with pictures of Blacks that are just hanging around, with the result that "Blacks are lazy" becomes by convention to be associated with the term "welfare," leading to the downgrading of this originally eulogistic term [17, p. 138]. Such practices we would also qualify as kinds of up- or downgrading of terms. But we stop short of including under this heading also linguistic expressions of positive or negative evaluations that are not communicating a metalinguistic message nor part of a deliberate attempt to change linguistic habits. Thus, if someone conveys an evaluation of some phenomenon without redefining any term, and without trying to change any term's semantics, for instance by asserting that "This multicultural policy has led to some serious problems, such as ...", then no up- or downgrading takes places. In practice, it will in many cases be difficult to draw sharp lines between metalinguistic acts of up- and downgrading, and linguistic acts of expressing one's plain evaluation of something. In other cases though, linguistic and contextual clues may provide sufficient clarity.

Up- and downgrading can be globally influential when they have a lasting influence on the cultural perception of the evaluative meaning of the term at hand, or locally when they effect the interlocutors' perception of the subject to which the term has been applied. On the other hand it can also fall dead from the press and have hardly any influence at all. In Anthony Giddens's view, though, the downgrading of "multiculturalism" has — to his regret — been a huge success: "Everyone seems suddenly to be dismissive of the notion" [9].

2.4 Argumentative significance

What argumentative effects can be expected from such up- and downgrading? Why would the proponent of a standpoint engage in such linguistic innovation or deviance? What is the pay-off? Again, by dealing with downgrading we also aim to deal with upgrading.

We focus on the following type of situation. Party A and Party B are engaged in an argumentative exchange about whether or not policy p merits their acceptance. Party A takes the position that p [e.g. this broadcasting policy] can best be rejected; Party B that p is best adopted. The general term Q [e.g. "multicultural"] is typically used as a eulogistic term, and whether or not p is Q is a relevant consideration for deciding on the policy issue. Suppose that Q has a descriptive meaning [e.g. promotes ethnic diversity] such that it is uncontroversial that p is Q. As we have seen in Section 2.2, Party B plausibly has the following *prima facie* cogent argumentation available for justifying her position: "Policy p is Q, therefore policy p merits to be accepted." After all, this argument instantiates an argumentation scheme the acceptance of which derives from Q being a eulogistic term. Now suppose Party A downgrades term Q, using it as a dyslogistic instead of as a eulogistic term. What are the effects Party A hopes for?

One effect may be temporary confusion. Party B may accept that this policy is Q, for the reason that according to any plausible descriptive definition of Q, this policy is in Q's extension. Moreover Party B considers Q to be a eulogistic term. If then, at a later stage of the exchange, Party B is confronted with Party A's use of this concession to support "we should reject this policy," Party B may be at a loss how Party A can draw this conclusion from the accepted premise that this policy is Q, given that Party B never expected that Party A would use term Q as a dyslogistic term. In such circumstances, Party A may profit if Party B is caught off guard and falls silent, so that Party A can go simply galloping ahead and convey the impression, to Party B or to an audience, that he has just made a very convincing case against policy p.

However the most important effect for Party A of a downgrade of Q would consist of poisoning strong arguments for p and breathing life into invalid or weak contra-arguments against p. In so far as Party A considers the existing usage of the term Q as biased against the kind of position it adopts, it regards its downgrading as an attempt at obtaining this argumentative effect and thereby a leveling of an unequal playing field.

More in detail: Suppose that Party B, who is in favor of policy p, had previously available, or even advanced, a serious and cogent argument that was partly based on Q's being a eulogistic term: "Policy p is Q, which forms a reason in favor of accepting p." Then the reverse argument — "Policy p is Q, which forms a reason in favor of rejecting p" — would at that point of the discussion count as a non-starter. In so far as Party A succeeds in downgrading Q, he will turn the tables with regard to these arguments, and possibly to further arguments that hinge on Q, because what was a good argument for Party B has become a bad argument, and what was a bad argument for Party A has become a good argument. Party A's successful downgrading of a term thus invalidates argumentation schemes that could underlie arguments previously advanced or held in storage by Party B. Also it validates new argument schemes that could boost the strength of arguments Party A previously advanced or holds in storage. The same considerations hold for upgrading.

In the following example, David Cameron uses the term "multiculturalism," which is used as a eulogistic term by some of his opponents, as a dyslogistic term by connecting it to the negatively laden expression "community separatism." Next to having the strategic advantage that it may help to "turn the tables", it also brings a risk, as one of Cameron's political associates (Lilico) shows, because it may lead to the wrong impression on the audience's side that Cameron opposes immigration or even that he supports the idea that people who once migrated to the UK should be sent home.

Example 4. Attacking "multiculturalism"
"I think he [David Cameron] wants an end to the threat — perhaps the occasional reality — of community separatism. And in his mind, that threat was best defined [in a speech in Munich, 2013] by the term multiculturalism. (...) Two articles on ConservativeHome, the site for grassroots Tory discussion, caught my eye. (...) In the other, blogger Andrew Lilico concurred [that they should be careful that language is not misinterpreted, and that this is a reason to avoid attacking 'multiculturalism']. 'When laymen hear that such-and-such a commentator or politician has attacked "multiculturalism," they are likely to think that means those commentators are opposed to immigration or perhaps even believe that those that have already arrived from other cultures should be "sent home".' Approach, but cautiously, he said." [14]

The example shows how the strategic significance of using a downgraded term can become a topic of discussion among political associates, and that generally political actors may be highly sensitive to the pros and cons of deploying upgrading and downgrading strategies.

In argumentative discussions, thus, up- and downgrading can be used strategically to try to turn the tables. Can the use of such strategies be seen as dialectically admissible, under certain conditions?

3 The legitimacy of up- and downgrading

As we shall detail in the next section, up- and downgrading easily leads to problems. What is more, there is a reason to see the up- or downgrading of an evaluative term as a violation of a norm for argumentative discussion that requires the participants to formulate their opinions and considerations in a sufficiently clear and unambiguous way (see, for example, the "language use rule," [21, p. 195]. Upgrading or downgrading of evaluative terms may breed confusion about what exactly is expressed, or a misunderstanding, so why not denounce all up- and downgrading as fallacious? Yet, before we come to the problematic side of such tactics, we wish to plead for a tolerating attitude towards up- or downgrading. We approach the matter from two points of view.

3.1 The freedom of the participants

As we saw, up- and downgrading can be used to try to turn the tables by taking advantage of the adversary's means of persuasion for one's own purposes. This is comparable to phenomena in the world of sports.

When in preparation of a sporting event your competitor uses a novel training technique that puts you at a disadvantage, it would only be fair if you were entitled to try to turn the tables, and to make good use of that very same training technique — for your own purposes. Real life examples are: high altitude training in the 1960s, the introduction in the 1990's of the clap skate in speed skating, and recently the introduction of the running shoes with carbon fiber plates in running. Similarly, participants in an argumentative discussion should be entitled to borrow a profitable persuasive device from their interlocutor, and to try to turn it into a profitable device for themselves, such as when one reverses the evaluative content of a term, which the interlocutor was putting to good use. A matter of equity in freedom.

What goes against the analogy is that an argumentative dialogue requires the participants to share a language to enable any progress, so that a divergence from linguistic conventions must be seen as a violation of a rule for argumentative games, rather than as an exercise of the freedom to convince one's interlocutor. What is more, instead of upgrading or downgrading an evaluative term that is already in use, there is always an alternative way to express the evaluative content, without re-using that term.

Yet, we fear that such conservatism regarding the use of language could threaten the freedom of the participants, and turn out to favor the powers that be. After all, the conventional meanings that get associated with key expressions in a public controversy are typically not neutral: When a majority of people (or a majority within a culturally dominant minority) approves (or disapproves) of a type of event or entity E, then expressions that are used to denote exemplars of E get associated with positive (negative) evaluative contents. That a certain term is often used to talk about E may exert some pressure on people to evaluate E in line with such talk. Minorities in public controversies need freedom, not only to express their point of view in language that happens to suit the interests of their opponents, but also to change that language so that it suits their own interests. As a consequence, we plead for some tolerance towards upgraders and downgraders, even if their behavior may affect somewhat the clarity of their use of language. A normative view on discussions should strike a balance between the required freedom for the participants, and the need for clarity in language use, and thus between how to put both the "freedom rule" and the "language use rule" into effect [21, pp. 190 and 195].

The Dutch term "Zwarte Pieten" (Black Petes) is a case in point. Black Petes are impersonated by black painted volunteers and actors who play a part in the feast of St. Nicholas. According to the story, Black Petes assist St. Nicholas in distributing candies and presents to children on his annual visit to the Netherlands. The term is also used as a proper name: "Zwarte Piet" (Black Pete). The feast has been part of Dutch folklore for some centuries, with the first Black Pete appearing in the nineteenth century. Whereas Black Petes formerly often acted as bogeymen, they normally do so no more, and the term "Black Pete" triggers for many quite positive associations. Yet, for years, some people consider the character of Black Pete as a racist element in the festivities, so clearly they don't share the positive evaluative content connected to "Zwarte Piet." Protesters among them have tried to discredit the phenomenon of Black Petes: by exposing the connections with the Dutch colonial past and the Dutch involvement in slavery; by comparing Black Petes with the American racist archetype of blackface; and by repeating the slogan (in translation) "Black Pete is Racism." Though we know of no allegation that the protesters have changed the meaning of "Black Pete," it is most plausible that downgrading the term is also part of what the protesters have been doing.

Whatever we may think of the accusation of racism, it would be unduly conservative and partisan to conceive of attempts to downgrade "Zwarte Piet" as violating a constraint of reasonable discussion, such as the language use rule. Such a ban on downgrading in the name of reasonable dialogue would strongly suggest that the defenders of this part of Dutch folklore are right, and that Black Pete is and should remain a positive concept by definition. As a consequence, a ban would thus well nigh silence the protesters, exclude their opinions from public discussion, and infringe on the freedom of a particular group of participants in the controversy. A similar argument can be made to apply to attempts to discredit multiculturalism and to downgrade the terms "multicultural" and "multiculturalism."

3.2 The merits of the case

Again, we start with an analogy from the realm of athletics. For there to be interesting, good sporting events it is necessary that the rules of sport games allow the parties to compete and to exercise their skills and techniques at the expense of their competitors. The reason is that if the rules of the sporting games are well-designed, and the participants are motivated to play the game and are roughly of equal strength, it is precisely the competitive behavior of the rivals that transforms into the kind of cooperation aimed at finding out who merits to be the winner (cf. for a similar view on competition and cooperation in sports: [15]).

Similarly, in the realm of argumentation, the normative design of an argumentative discussion should assist the participants in their cooperative effort to find out whether or not the thesis (standpoint) at issue merits their acceptance by encouraging each of them to defend or explain their own position and test and challenge the other's position. To promote this dialectical cooperation it may be profitable to allow each participant some room for resisting compliance with possibly biased terminology of the interlocutor and to do so by up- or downgrading.

In an argumentative discussion, the defender of the status quo needs to acknowledge the possibility that what appears to him as something good, really is not so. He should be aware that possibly the interlocutor has a point when downgrading a eulogistic term. Someone who contends that Black Petes are not racist ought, we think, at least acknowledge that the opinion that Black Petes are racist deserves to be discussed, rather than tabooed. The downgrading of a key term in a dispute may come to function as an eye-opener for a participant, or for the spectators of a debate, and instrumental to a well-considered change of mind, or else to at least a better understanding of the merits of the interlocutor's position.

4 Some snares

When someone up- or downgrades a key term, the addressee may object to the linguistic innovation for the reason that it boosts the proponent's side of the controversy, and undermines one's own. An up- or downgrade can be up for discussion, and can be challenged as socially or politically illegitimate, even when its argumentative legitimacy is not at issue. But sometimes the argumentative legitimacy is at issue because upgrading and downgrading of evaluative terms is not without risks to the resolution of disputes, and could be a source of fallacies. In this section we focus on ways in which upgrading or downgrading may be (on the border of becoming) fallacious. We discuss three kinds of risks to see whether they can be reduced if certain dialectical measures are available.

4.1 Question begging appellatives

Bentham, when discussing eulogistic and dyslogistic terms from the viewpoint of begging the question, contends that this fallacy is most effective when employing simply a "question begging appellative":

> "To the proposition of which it is the leading term [predicate], every such eulogistic or dyslogistic appellative, secretly, as it were, and in general insensibly, slips in another proposition of which that same leading term is the subject, and an assertion of approbation or disapprobation the predicate. The person, act, or thing in question, is or deserves to be, or is and deserves to be, an object of general approbation. [Similarly for disapprobation] (...). The proposition thus asserted is commonly a proposition that requires to be proved." [2, p. 436]

In one of Bentham's examples, someone characterizes his policy of spending public money as "liberal," whereas the policy really is depredatory, so that the use of this appellative is question begging by slipping in the proposition that any liberal policy is to be approved of, which in this case is contentious (p. 437): "(Premise 1:) This policy is liberal. (Premise 2:) Any liberal policy is to be approved of. Therefore (Conclusion:), this policy is to be approved of." The mistake then is that a proposition, Premise 2 in our example, that is or should be seen as controversial in the discussion at hand, is in fact treated as if it belongs to the non-controversial common ground, and is thus smuggled into the discussion (see the "starting-point rule," [21, p. 193; see also p. 176]. The same holds, we think, for the use of upgraded or downgraded evaluative terms. The possible abuse clings to the more standard usages and the more innovative (up- or downgrading) ones of eulogistic or dyslogistic appellatives alike.

Evaluative terms can be used in such a question-begging manner, but then they need not. As the preceding section suggests, one may up- or downgrade in a dialectically acceptable way. But the new evaluative content should either be not controversial for the addressees or get accompanied with argumentation in its support or at least be presented transparently as up- or downgrading so that it invites the critical addressees to request such argumentation.

What if your opponent commits the fallacy of begging the question by means of an up- or downgraded term, nonetheless? Among the available dialectical measures should be responses that defuse the (alleged) fallacy and optimize the return to a reasonable, resolution-oriented exchange of arguments. The golden option, then, is to identify the fallacy, explain its dialogical inadmissibility, and ask — possibly anew — for argumentation in support of the controversial proposition: "Whether all liberal spending policies are to be approved of is among the points of contention, thus it's just question begging to label the policy at issue as liberal. What reasons, my dear proponent, can you give me to show that using public money for these purposes is to be approved of as being *liberal*, rather than to be disapproved of as being *depredating*?"[10]

Similarly, one may have reason to counter a downgrading use of "multicultural" as question begging: "whether any broadcasting policy that is "multicultural" is to be disapproved of happens to be among the points of contention, thus it's just question begging to label this policy by means of your disparaging qualification. What reasons, my dear proponent, can you give me to show that requiring 60% of ethnic programs is to be disapproved of as being *multicultural*, as you choose to use the term, rather than to be approved of as being *inclusive and appreciative of cultural diversity*?" Of course, it would be better to convey such messages in slightly less pedantic ways.

4.2 Ambiguity

Changing the meaning of terms possibly breeds confusion or misunderstanding, also when the change concerns the evaluative content of an expression. But, again, not of necessity: One can be quite clear and frank when it comes to the upgrading or downgrading of terms. What if, however, the expression has become confusingly ambiguous? Then, in the interest of the quality of the discussion, things need to be

---

[10]Other responses that can, depending on circumstances, be argumentatively reasonable are: ignoring the fallacy and proceeding the exchange of arguments and criticisms; responding to the fallacy — tit-for-tat — by means of a counterfallacy, hoping that this will make the interlocutor return to reason; shifting to another type of dialogue, such as an eristic dialogue; or, exerting social or legal pressure to return to a fully reasonable argumentative exchange (see [10]).

cleared up; otherwise discussants may end up talking at cross-purposes. Here the golden option is to make a distinction. In the following example, this is precisely what happens: In a setting where the downgrading of "multiculturalism" went together with a shift in the descriptive meaning of the term, Anthony Giddens (2006) sees the need to clear things up by way of a distinction.

Example 5. What multiculturalism has never meant
"There [in Canada] multiculturalism does not mean, and has never meant, different cultural and ethnic groups being left alone to get on with whatever activities they choose. It actually means the opposite. Policy-making in Canada stresses active dialogue between cultural groups, active attempts at creating community cohesion, and the acceptance of overarching Canadian identity" [9].

In Giddens's view, "multiculturalism" is used in one, wrong way with a descriptive meaning of leaving cultural and ethnic groups alone to get on with their activities and an accompanying negative evaluative meaning, and in another, right way with a descriptive meaning of stressing dialogue and cohesion between cultural groups, and an accompanying positive evaluative meaning.

Kenan Malik [13] also distinguishes between two uses of the term "multiculturalism": "multiculturalism" as referring to policies and "multiculturalism" as referring to the lived experience of diversity.

Example 6. Talk of multiculturalism
"The problem of multiculturalism is not one of too much immigration or diversity. It lies, rather, in the impact of the policies enacted to manage diversity. When we talk of 'multiculturalism,' we often conflate the lived experience of diversity with public policies towards minority communities. The failure of those policies has led many to blame diversity itself as the problem" [13].

Thus conflating the lived experience of diversity with failing public policies may result in an unjustified negative evaluation of the former. Example 5 and Example 6 provide examples of mixed cases, where the downgrading of a term went hand in hand with a change in descriptive meaning.

4.3 Social pressure

If an up- or downgraded term gets more widely accepted, this could lead to a softening or neutralizing of the kind of moral, social or ideological pressure that may have characterized the original usage of the term. But it is also possible that a new kind of pressure, in the reverse direction, comes to stick to the up- or downgrading usage of the term. Contemporary uses of "multicultural" may push people to reject whatever is so labeled, especially if the pejorative meaning of the term is presented as obvious and in no need of any further discussion, and may in some cases be regarded as a component of a strategy of exerting pressure on audiences to adopt (or reject) certain policies, such as harsh (lenient) measures regarding refugees. For if people with whom one is acquainted, including family and friends as well as cherished public personalities or other influencers, start using "multicultural" in a certain sense and with a pejorative connotation, there will be a tendency to conform one's judgment and use of language to them.

Some degree of pressure being exerted by the use of evaluative language may be unavoidable. And the same plausibly holds for some degree of back and forth swinging between progressive and conservative pressures. But this does not imply that people can only take part in public controversies as passive recipients of such cultural and linguistic developments. Addressees of arguments can be encouraged to be reluctant when it comes to the strategic usage of linguistic innovations, and to ask for reasons when they identify an alleged case of upgrading or downgrading. Similarly, arguers can be encouraged to explain the reasons for their linguistic innovations, when having been asked to do so or when they anticipate such requests. This enhances mutual understanding. Addressees should not be rushed, however, to linguistic (and substantial) agreement: at no point should arguers pressure audiences unduly by appealing to the need of a shared language and by presenting their use of evaluative terms as the only correct use.

5 Conclusion

We found that participants can follow an argumentative strategy by upgrading a term with a negative evaluative meaning or by downgrading a term with a positive evaluative meaning. We showed how these strategies might turn the tables by converting arguments available for the opposition into useless ones, and by changing previously bad arguments into valuable ones for oneself. Though up- and downgrading can result in a form of question begging, may lead to talking at cross purposes, or produce social pressure, we argued that an outright ban is not a proper response, because each of these problems can be solved or mitigated, and a ban would pose an undue restriction on the discussion, harming both the freedom of the individual discussants and the prospects of developing a resolution of the disagreements at issue on the merits of the case.

To end this paper, we add some explanation of how we think a research program for developing a normative theory of the use of language in argumentation can avoid introducing points of view that would make the theory not just normative but overly moralistic. We may distinguish two senses of being overly moralistic: an ethical sense and a dialectical one. We wish to avoid moralizing in both these senses.

An argumentation theory, we think, would be overly moralistic *in an ethical sense*, if norms included in that theory were motivated, not by dialectical considerations regarding the conversational rights and obligations needed for the verbal resolution of disagreements, but rather by ethical considerations pertaining primarily to non-verbal behavior outside settings of critical discussion. Thus it is not up to *argumentation theories* to plea in support of downgrading "Black Petes" or against the downgrading of "multiculturalism" on account of some ethical position. Ethical and social criticism is different from normative dialectical criticism. From this stance, it could be quite consistent to regard every downgrading of the term "multicultural" as mistaken and wrongheaded from an ethical point of view, and yet to do so without qualifying such downgrading as fallacious, that is, as a breach of the kinds of dialogical norms that provide the confines of critical discussion. Also, though it might not be unlikely that someone who lacks a proper moral sense, or who is led astray by twisted (e.g. egocentric or absolutist) moral ideals might also have perverted normative ideas about how one should engage in a critical discussion, there is no necessary connection (the connection merits empirical investigation but cannot be taken for granted).

An argumentation theory can also be overly moralistic *in a dialectical sense*, if norms included in that theory, though dialectically motivated, are more restrictive than necessary for the verbal resolution of disagreements. Ideally, the norms that govern a discussion are acceptable to all sides of a dispute as applying to discussion and debate more generally. As a result, the dialogue rules that function to define the boundaries of the concept of "critical discussion" should be minimally restrictive and allow the participants maximal freedom compatible with the goal of the resolution of disagreements on the merits of the case. We think that argumentation theorists should encourage discussants to be maximally critical when discussing cases of up- or downgrading, and to challenge such strategies, and if needed to evaluate such cases as wrongheaded from a dialectical point of view, for example on account of the confusion they bring about or their use of loaded (question begging) language. But we also think that we should be careful when it comes to qualifying cases up- or downgrading as fallacious, or to encouraging participants in dialogue to do so. The situation compares to the acceptability of statements: we do not qualify any assertion of a falsehood as a fallacy, but rather choose to see a critical discussion as a setting where discussants can try to sort out for themselves whether or not a statement merits their acceptance. We propose to take a similar attitude to up- and downgrades. Note that there can be exceptions to this stance, so that under specific conditions a case of up- or downgrading does qualify as fallacious: for example, when a participant downgrades a term that applies to one's interlocutor, and turns it into a slur (which happens with expressions such as "homosexual" and "Jew"), he or she would commit an (ad hominem) fallacy, because such a contribution is at odds with genuine dispute resolution and can be expected to be acknowledged as such by any serious discussant.

An argumentation theory that is moralistic in neither of these two ways may all the same provide the critical dialogical tools that enable and encourage discussants to inquire into the virtues and vices of their usage of linguistic means. This task must constitute the core of the research program for developing a normative theory of the use of language in argumentation. By investigating evaluative (emotive) words from an argumentative stance, Macagno and Walton [12] have made a most significant contribution to this research program.

References

[1] Arnauld, Antoine, and Pierre Nicole (1996). *Logic or the Art of Thinking*. Translated and edited by Jill Vance Buroker. UK: Cambridge UP. Original title La logique ou l'art de penser, anonymously first published in 1662.

[2] Bentham, Jeremy (1843). Book of Fallacies. In: John Bowring (Ed.), *The Works of Jeremy Bentham* (pp. 379-487). Edinburgh, W. Tait; London, Simpkin, Marshall.

[3] Bianchi, Claudia (2014). Slurs and appropriation: An echoic account. *Journal of Pragmatics* 66, pp. 35-44.

[4] Brandom, Robert B. (2000). *Articulating Reasons: An Introduction to Inferentialism*. Cambridge, Massachusetts: Harvard University Press.

[5] Cepollaro, Bianca (2018) Negative or Positive? *Croatian Journal of Philosophy* 18 (3), 363-374.

[6] Crowly, Candy (2000). Bush says 'RATS' ad not meant as subliminal message; Gore calls ad 'disappointing development'. CNN.com, September 12, 2000. Retrieved, January 1 2020 from: https://web.archive.org/web/20051125072108/http://archives.cnn.com/2000/ALLPOLITICS/stories/09/12/bush.ad/

[7] Dummett, Michael (1981). *Frege's Philosophy of Language*. (Second edition; original from 1973.) London: Duckworth.

[8] Dewing, Michael (2013). Canadian Multiculturalism. Publication No. 2009-20-E Library of Parliament. Legal and Social Affairs Division Parliamentary Information and Research Service. Retrieved January 3 2020, from: http://www.multiculturalmentalhealth.ca/wp-content/uploads/2014/01/Dewing_Multiculturalism_2009-20-e.pdf

[9] Giddens, Anthony (2006). Misunderstanding multiculturalism. *The Guardian*, October 14, 2006.

[10] Krabbe, Erik C. W., and Jan Albert van Laar. Be reasonable! How to be an optimist in the 'age of unreason'. *Journal of Argumentation in Context*. To appear.

[11] Macagno, Fabrizio, and Douglas Walton (2014). *Emotive Language in Argumentation.* New York: Cambridge University Press.

[12] Macagno, Fabrizio, and Douglas Walton (2019). Emotive meaning in political argumentation. *Informal Logic*, 39, 229-261.

[13] Malik, Kenan, Diversity and immigration are not the problem. Political courage is. *The Guardian*, April 5, 2015.

[14] Muir, Hugh (2013). Multiculturalism: A toxic term for Tories. *The Guardian*, February 11, 2013.

[15] Nguyen, C. Thi (2017). Competition as cooperation, *Journal of the Philosophy of Sport*, 44 (1), 123-137.

[16] Robinson, Richard (1954). *Definition.* Oxford: Clarendon Press.

[17] Stanley, Jason (2015). *How Propaganda Works.* Princeton, NJ: Princeton University Press.

[18] Stevenson, Charles Leslie (1938). Persuasive Definitions. *Mind* 47(187), pp. 331-350

[19] Stevenson, Charles Leslie (1944). *Ethics and Language.* New Haven: Yale University Press.

[20] Tumulty, Karen and Jenna Johnson (2016). Why Trump may be winning the war on 'political correctness'. *The Washington Post*, 4 January.

[21] van Eemeren, Frans H., and Rob Grootendorst (2004). *A Systematic Theory of Argumentation.* Cambridge: Cambridge University Press.

[22] van Eemeren, Frans H., and Peter Houtlosser (2002). Strategic Manoeuvring in Argumentative Discourse: A Delicate Balance. In: Frans H. van Eemeren and Peter Houtlosser, *Dialectic and Rhetoric: The Warp and Woof of Argumentation Analysis*, Dordrecht: Kluwer Academic, pp. 131-159.

[23] Väyrynen, Pekka (2011). Thick concepts and variability. *Philosophers' Imprint*, 11, 1-17.

[24] Väyrynen, Pekka (2014). Essential contestability and evaluation. *Australasian Journal of Philosophy*, 92, 471-488.

[25] Viskil, Erik (1995). Defending Definitions: A Pragma-Dialectical Approach. In: F. H. van Eemeren, R. Grootendorst, J. A. Blair, and C. A. Willard (Ed.), *Perspectives and Approaches: Proceedings of the Third ISSA Conference on Argumentation* (pp. 428-438). Amsterdam Sic Sat

[26] Walton, Douglas N. (1992). *The Place of Emotion in Argument.* University Park, PA: The Pennsylvania University Press.

[27] Watt, Nicholas (2012). Olympic opening ceremony was 'multicultural crap', Tory MP tweets. *The Guardian*, July 28, 2012.

Walton on Ethical Argumentation

James B. Freeman
Hunter College, City University of New York, USA.
jfreeman@hunter.cuny.edu

1 Ethical Argumentation and Layered Maieutic Dialogue

We may model ethical argumentation as a dialogue. Indeed, for Walton this applies to argumentation in general. There will be at least two parties — a proponent and a challenger or respondent. The proponent advocates some ethical judgment or course of action. The challenger may question the acceptability of the judgment or correctness of the course of action. She may also request clarification of the meaning of the claim or she may draw out its logical consequences, perhaps in combination with other statements. If she further advances a contrary judgment of her own, the dialogue will be symmetrical. If she does not advance a viewpoint of her own, the dialogue is asymmetrical. We shall argue that the asymmetrical model is sufficient for our purposes here. The challenger's question may simply ask for direct supporting evidence. By contrast, she may present an objection to the claim. To answer this question satisfactorily, the proponent must counter the objection. Of course, the challenger may challenge further and the proponent may then extend his argument. This challenge and response dialogue (to borrow the title of Wellman's essay [15]) may continue until proponent and challenger come to an agreement. Alternatively, she may find agreement impossible and abandon the dialogue.

If the dialogue ends with agreement, either over of the proponent's original claim or his modification of it in light of the steps of the dialogue, the proponent has presented good reasons (or appropriate modifications which can then be defended with good reasons). But what makes those reasons good reasons lending justification to the viewpoint? Say the proponent endorsed some course of action. The challenger responds not just with the bald assertion that the action is wrong, but that it is an instance of a type of action generally considered wrong, say cheating, stealing, lying, physically or psychologically injuring someone. As Walton points out, these evaluations are generally accepted ethical rules. As such, they are not descriptions, which in principle could be properly supported by observation. How may these rules be justified? Walton's answer is in two parts.

> The first part is the claim that there is a presumption in our everyday practices of interacting with each other that any act that can be classified as "cheating" [Walton's example] is morally wrong (unacceptable behavior). The second point is that this presumption can be tested in an ethical discussion, and can be shown, on balance, to be ethically justifiable at an abstract or general level. Walton [14, pp. 20-21]

But what is the nature of this ethical justification and does it give objective justification to the ethical claim initiating the discussion? Is this justification only subjective, or an attempt at persuasion? This is a central question for Walton's study.

In reply, Walton considers Aristotle's endoxic justification. For Aristotle, *endoxa* are not known to be true but are "accepted by everyone or by the majority or by the wise" (*Topics* 100b 22024, quoted in Walton [14, p. 22]). To use Rescher's terminology in [10, p. 35], *endoxa* are plausible. They should be regarded as truth-candidates, accepted provisionally but defeasible, subject to revision or even withdrawal in light of further evidence. Rescher adds that of rival plausible alternatives, "presumption favors the most plausible" [14, p. 38]. Since *endoxa* need not be universally accepted, disagreements are possible. Walton asks how then can *endoxa* serve to justify ethical claims [14, p. 22]. His answer is that ethical reasoning is dialectical. During the exchange, the proponent may discover reasons for needed refinments to his position. In particular, a simple assertion of the form All S are P may be counterexampled, but the modification All S and R are P avoids the counterexample and still can further his argument.

Justification of an ethical claim may involve recognizing several plausible alternatives to the claim. Assuming each in turn may lead to refuting each of these alternatives but not the claim itself. Alternatively, of several plausible alternatives, the one which best survives this critical examination is deemed the most plausible and supported by the discussion [14, p. 23]. Again the discussion has revealed supportive reasons for this alternative. Walton points out that not every conceivable alternative is an *endoxon*. "To perform the role of an *endoxon* in ethical argumentation, a proposition must be presumed to be the outcome of an intelligent deliberation, so that supportive reasons can be given to back it up" [14, p. 23].

Walton sees this argumentation as maieutic in the sense of the Socratic dialogues, bringing to birth new arguments and insights. In proceeding with a given case, ethical argumentation begins with the facts — genuine or alleged — of the case and then, proceeding from these facts, considers "a set of arguments that brings out more explicitly some central issue posed by the case which could be the basis of a discussion not only of this case, but also of similar cases and general principles involved." [14, p. 40]. The facts themselves present a dilemma: Undertake act A or decline to perform A. This may not be easy to solve. There may be arguments with significant weight on either side.

Argumentation to resolve such a dilemma may proceed on two levels or in two states — deliberation and critical discussion. Deliberation addresses which alternative to choose (there may be more than two). In light of all the available evidence, which alternative is the most prudent? The deliberation may be inconclusive since not all the pertinent evidence may be available and the ethical principles one might take into account are subject to qualifications. Deliberation proceeds from two premises: "A is my goal" and "To bring about A, I need to bring about B" and concludes to "I ought (prudentially) to bring abut B" [14, p. 44]. But, Walton points out, this inference is defeasible. A dilemma may arise because several goals, each legitimate, may conflict. Hence one action may be *prima facie* right because satisfying one goal but *prima facie* wrong by going against another. To use Walton's example, the goal of saving a person's life by performing an operation may conflict with the goal of respecting the person's wishes who will not consent to the operation. Performing the operation and not performing the operation are both prima facie wrong. In this case, there may be no way to avoid an action which is *prima facie* wrong.

Walton holds that use of deliberation in ethical dialogue constitutes the first level of a case [14, p. 49]. A dialogue shift leads to the second level, involving a higher level of generality and a different purpose [14, p. 50]. The shift is from deliberation over which alternative to perform to critical discussion or persuasion concerning justifying the rightness of choosing that alternative. Here again there is a proponent and a challenger. The goal is to engage in resolving this conflict of opinion "by means of rational argumentation" [14, p. 50]. Assumptions advanced are supposed to be supported by arguments. The purpose is "to present a discussion of the issue and principles at stake in the case that would be thought provoking" [14, p. 51]. Being successful at this level does not require a resolution of the conflict of opinion but rather that "the participants both gain insight on the issue that was discussed, in virtue of their having engaged in a well-contested argumentative exchange with an able opponent who has strongly defended the opposed point of view in a partisan but fair exchange" [14, p. 52]. The dialogue then is like a critical discussion but with the object of deepening insight into the issue rather than of resolving a difference of opinion.

Participants begin a persuasion dialogue with certain commitments to certain propositions. Some of these may be explicit, others implicit. Persuasion dialogue makes implicit commitments explicit through argumentation, For example, at one stage of the dialogue, the proponent may be explicitly committed to an unqualified generalization — Doing A is always right or permissible. But the proponent may also implicitly concede certain restrictions on doing A. When the challenger brings this out, the proponent may restore consistency by qualifying his commitment to the proposition. Doing A is right except if conditions $C_1, \ldots, C_n$ hold. Walton points out that these exceptions cannot always be specified in advance of applying a general rule. Rather applying the rule is "like finding the 'mean' or middle way, between two extremes. Becoming able to find the middle way is giving insight into the general principle being applied. Success is measured by how reasonable are the arguments uncovered on each side, their relevance to the issue, how much weight they give to each side, and the extent to which they serve the maieutic function. On the whole, the extent to which the case provokes discussion of both sides of an issue is a measure of its success. To what extent does the discussion of the case get us towards the truth of the matter? [14, p. 64]. The participants may also become aware of the weak points in their position and the fallacies that may have been used in justifying it. This also is a measure of success for the discussion, In addition, the extent to which a participant's understanding of the other side's position is a factor in the success of the critical discussion.

2 The Structure of Ethical Argumentation

Layered maieutic dialogue concerns the stages of ethical argumentation — first deliberation and then persuasion. By contrast, structure concerns the forms of argumentation which may occur at these stages. We may distinguish single or basic arguments from chains of arguments Single argument structure comprises what in [7, pp. xi–xii] we called argument microstructure. Saying that an argument is a categorical syllogism, in particular that it instances the form AAA-1, or that it instances hypothetical syllogism, or *modus ponens*, or statistical syllogism describes it microstructure, how the argument's component premises and conclusions are built up from atomic statements or predicates together with elements of other grammatical categories by sentential connectives, quantifiers, and perhaps further connecting expressions. We discern microstructure by looking "inside" the statements out of which an argument is built and the patterns components of those statements form. Walton holds that simple or basic ethical arguments are syllogistic, i.e. a conclusion is derived from two premises, a general rule premise and a singular or fact premise. Together they yield the conclusion, a singular or factual statement. The form

All S are P

a is an S
Therefore a is a P

is a paradigm example.

In logic, arguments instancing such patterns are called quasi-syllogisms. Syllogisms have exactly two premises which yield exactly one conclusion. Basic ethical arguments, then, are syllogisms, no "quasi" about it. But they are not categorical syllogisms, where all the component statements are categorical. "Quasi-categorical syllogism" better describes their structure.

Not all quasi-categorical syllogisms are deductive arguments.

Most S are P
a is an S
Therefore *ceteris paribus* a is a P

fits the pattern of a quasi-categorical syllogism, regarding "most" as a quantifier. But this argument is defeasible, not deductive. In basic ethical arguments, the major premises assert restricted generalizations. Rescher points out in [10, p. 14] that such statements express qualified or restricted universals.

@(All S are P)

asserts that "*Ceteris paribus* an S is a P."
With such a major premise, the inference is defeasible.

Ethical argumentation consists not just of one argument but at least two arguments chained together. Ethical connotations of classifications for actions, e.g. acts of charity, may be made explicit by a further qualified generalization, e.g. "Most actions of a certain sort, e.g. acts of donating to a charity, are morally right." Where "Alice's donating $100 to disaster relief is an act of charity" is the conclusion of a quasi-categorical syllogism," linking that conclusion with "Most acts of charity which are morally right," produces a syllogism whose conclusion is the evaluation "Alice's donating $100 to charity is morally right." Walton sees this pattern as pervasive in ethical reasoning. We may argue from the fact that someone has performed a certain action, e.g. driven 50 mph in a 25 mph zone and actions of that sort may be classified in a certain way, e.g. violating the Road Traffic Acts, that the person has violated the Road Traffic Acts. But describing the action that way has negative connotations. Such actions *ceteris paribus*, are morally wrong. Making that connotation explicit and linking that statement with the statement that someone has violated the Road Traffic Acts,we may infer that the person has done something morally wrong. The macrostructure of the argumentation looks like Figure 1:

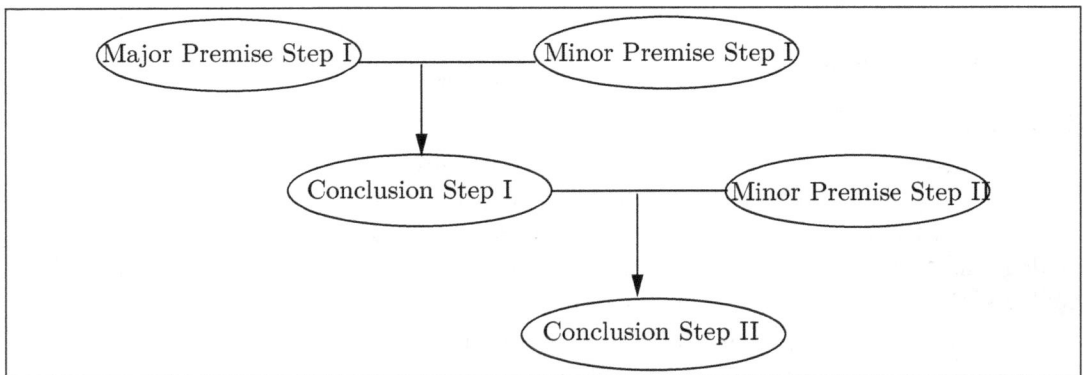

Figure 1

How are the normative premises in such instances of argumentation generated? Walton points out that besides seeing ethical reasoning proceeding top down, from premises to conclusions, we may also see it as working bottom up, from an evaluative conclusion to supporting considerations. Walton sees ethical argumentation moving in this direction. A proponent puts forward an ethical judgment. Although there may be a presumption for that judgment from the proponent's viewpoint, the judgment is open to challenge and the burden of proof is on the proponent. Meeting that challenge involves an abductive type of reasoning, or reasoning to a "best" or "most plausible" explanation of a given set of data. As we shall argue, "inference to the best explanation" may not be the best characterization of the type of reasoning Walton intends here.

Walton holds that terms used to characterize actions may also have emotive connotations which convey moral judgments of good or bad, right or wrong. For example, if one person asserts intentionally a statement he or she knows to be false, the person has lied. From that statement together with the particular statement:

> Smith asserted a statement she knows to be false

we may derive the conclusion

> Smith lied.

But that statement has a morally negative connotation. Walton sees the conclusion as conveying that

> Lying is wrong

Adding to the conclusion the statement that

> All instances of lying are morally wrong,

we may derive the further conclusion

> Smith has done something wrong.

Walton calls the reasoning here a subinference. It is straightforward to see that inference and subinference form a chain.

Both the premises and the conclusion may be challenged. How may the general major premise be justified, if challenged? Walton's answer is central to his concept of ethical argumentation. Describing the premise as a rule, Walton asserts "This kind of rule [of the lying example] ... is connected to the emotive meaning of the term "[lying],""and adds to the ... inference the assumption [premise] of the form

> Anyone who has who has committed [lying] has done something wrong.

Walton then asks "Where do these ethical rules come from?" [14, p. 84]. Alternatively, how can such rules as generalizations be justified? Following Feigl [5, pp. 134-35], Walton points out that such rules can be conclusions of further syllogisms. For example,

> All actions which weaken the general disposition to trust the word of others are morally wrong.
> Lying weakens the general disposition to trust the word of others.
> So lying is morally wrong.

We may then try to justify the rule premise here by appealing to a more general rule yet. This is regressive reasoning. It may continue but will eventually stop somewhere when we reach, in Feigl's words, "Some premise which cannot plausibly or fruitfully be deduced from a still more general or fundamental principle" ([5, p. 135]; quoted in [14, p. 85]).

How does one ascend this ladder? Walton holds that the rule premises are reached by inference to best explanation, selecting the most plausible of the explanations entertained? We shall examine this claim critically in section 4. If we cannot ascend to a higher rung on the ladder of ethical general rules, if we cannot see an even more general explanatory principle supporting this judgment then how is this judgment to be justified? If the parties to the discussion reach agreement, Walton believes that ground may be "the lived ethical experience shared by the two parties to the discussion" [14, p. 86]. We shall examine this claim also in section 4. Walton admits this experience involves feeling, but more is involved. According to Stevenson's emotive theory in [13], "Lying is wrong" means : "I disapprove of lying, do so as well." This explanation does not give a *reason* for "Lying is wrong." The connection is not logical or epistemological or normative, but psychological. But this is precisely the problem. Expressive or emotive meaning simply does not constitute a normative reason or ground for ethical judgment.

By contrast, the chained inference theory "explains precisely how this kind of ethical reasoning to a conclusion on the basis of a good reason to support the ethical claim by reasoning is logically binding. It enables anyone taking part in an ethical discussion to focus on what needs to be supported by evidence, or what needs to be criticized, in order to provide ethical support for a conclusion, or to challenge a claim that such support has been given" [14, p. 87]. Walton sees chained ethical justification as moving backwards to higher and higher principles of justification. The chain will stop somewhere because the interlocutors are human beings with only a finite amount of time. But the chain need not end in agreement. Does failure to reach agreement show that the whole enterprise of maieutic uncovering of ever deeper principles simply constitutes the elaboration of contrasting viewpoints? Does the maieutic process in any way deepen our understanding of the ethical issues surrounding the initial claim? Walton believes that even if the dialogue does not end in resolving the disagreement, it can be deemed successful "to the extent that the fundamental ethical commitments, the ultimate principles and premises on both sides, are revealed maieutically" [14, p. 89]. Do the parties come away with a deeper understanding of the positions on both sides? To the extent that they do, the dialogue has been successful. Although there may be deep disagreement, there may be argumentation on both sides which gives plausibility to each side.

One may ask at this point, looking at the initial inference in a maieutic dialogue, just where does the major premise, a general statement, come from? The minor premise is a description, resting on experience for its justification. The major premise states an evaluation. How much justification does its source give it? Is the proponent justified in accepting it? Walton answers that we should not regard the major premises as totally abstract and independent of the historical context in which they are recognized. Rather, their source lies in generally accepted ethical rules and opinions. "Layered justification should begin by allowing as given data premises that cite what is arguably the generally accepted body of moral opinions in a given case. ... [But] they should not be seen as beyond doubt in disputation" [14, p. 102]. Walton's point is completely in line with plausibility theory. The credibility of sources lends a statement a degree of plausibility, and common knowledge (or opinion) is one source which may vouch for a claim. Walton concedes that this picture makes ethical justification culturally relative. The initial major premises of ethical argumentation are endoxic. But they are the first premises, not endpoints. Ethical argumentation need not end relative to culture. Indeed, it may end with a dissent from the moral opinions of the time. What may bring about moral reform are new situations which the current morality cannot deal with. "The need for a new approach to ethics arises from the fallibility of reasoning based on generally accepted opinions, especially when confronted with a new situation" [14, p. 102]. We may add that the inputs to reform may come when a moral leader shows a contradiction or anomaly with the old systematising from how that system is practiced. Remember Martin Luther King's call for America to rise up and affirm its creed. Maieutic dialogue's stance toward *endoxa* is critical, in no way dogmatic, which may lead to reformation or revision, indeed significant revision, of the prevalent cultural *endoxa*. "Citing generally accepted practice or opinion can be the basis of a reason for holding that something is right or wrong. But it is a basis that should always be treated as subject to further challenge, and to weighing in light of a larger body of relevant evidence in a given case" [14, pp. 104–5].

3 What Ethical Argumentation Needs to Have — The Probative Function

Distinguishing the deliberative and persuasive steps in ethical argumentation and indicating how inferences may chain together pertain to argument structure, specifically macrostructure.[1] But should one discern these features in a given case, the question still remains — Is the argumentation good? Does it give the challenger a good reason to accept the conclusion in light of the premises? Walton is specifically concerned with two types of macrostructure in particular which raise questions about connection adequacy — circular argument and infinite regress arguments. In circular arguments, the conclusion or a statement semantically equivalent to the conclusion, or a statement which presupposes it is introduced as a premise. In an infinite regress argument, a challenger who asks for a reason then asks for a reason for that reason, and no matter how far back in the regression the questioning continues, continues to ask for a reason for the preceding reason, thus embarking on an infinite regress. In both of these forms, not the adequacy of the connection but the acceptability of each premise introduced is always called into question. With neither form can the argument ever establish its conclusion. Yet in both types of argument, the steps from premise to conclusion can be deductively valid. Indeed, p therefore p is deductively valid. Walton responds by characterizing the probative function of argument, using "premises that are evident or have probative weight to transfer that weight to the argument's conclusion, removing or lessening doubt about that conclusion" [14, p. 200]. Every step in "begging the question" or "infinite regress" arguments may be deductively valid, but they seem not to justify

---

[1]For a characterization of macrostructure, see Freeman [7, pp. xi–xiii].

their conclusions. But why? Clearly if one questions that p, one's confidence in p should not be increased by a circular argument which simply takes p as a premise supporting itself. But, Walton asks, why is proving the conclusion the point of an argument? He considers Sanford's view that an argument is supposed to show a conclusion worthy of belief. The problem with circular arguments is epistemic. To increase the belief-worthiness of a conclusion, a premise must be more worthy of belief than the conclusion and the argument must have a structure to transfer that worthiness to the conclusion Clearly, a conclusion cannot be more worthy of belief than itself.

The issue of infinite regress arises when whatever is presented as a premise needs further supporting evidence. But should a premise give that evidence, it will require further evidence from a further premise. The chain apparently will be never ending. The whole chain is based on dubious premises, at least from the challenger's viewpoint. If the challenger's doubts are legitimate, i.e. the challenger is not being simply obstreperous, the proponent has failed to find "a premise or premises that have sufficient probative weight to support the conclusion." [14, p. 212].

If circular and infinite regress arguments fail to fulfill the probative function, what is required to fulfill it? Walton asks us to consider asymmetrical persuasion dialogues.[2] The challenger only asks questions and does not put forward any theses of her own. The proponent's contributions to the dialogue serve the probative function by reducing the "capability of the respondent to ask appropriate critical questions" [14, p. 216].The goal is to reduce this capability until the challenger has no recourse but to concede the proponent's thesis. How this comes about depends on the type of argument involved. Walton identifies three types — deductive, inductive, and abductive. Walton sees basic ethical arguments of the form

>   Most S are P.
> a is an S.
>   So a is a P

as abductive. But while our information about the frequency of As which are Bs allows us to specify that frequency with some degree of precision, for an abductive argument we may be unable to specify it more precisely than just "most." Our evidence will not justify asserting a more specific percentage. With a deductively valid argument, if the challenger accepts the premises, she has no option but to accept the conclusion also. With an inductively strong argument, the challenger still has options to avoid the conclusion if she accepts the premises, but her opportunities are limited, although "only by probability" [14, p. 216]. An abductive argument

---

[2]In [7], we used "challenger" the way Walton uses "respondent" here.

again narrows down the options. Walton here does not indicate how the narrowing down is different in the inductive and abductive cases. It is not difficult to see how one might understand their narrowing. A precise percentage, e.g. 0.5, rules out all other percentages. A range of percentages between 0.4 and 0.6 rules out all percentages outside that range. But a quantifier like "most," although vague, would seem to rule out much more.

The deductive/inductive/abductive distinction has been presented here for one step arguments, from the basic premises to the conclusion. But, as we have seen, ethical arguments may move from basic premises to the final conclusion in a chain of steps involving serial structure. Walton regards this structure as "crucially important for grasping how the probative function works" [14, p. .219]. With a serial argument, the probative function moves forward across the chain [14, p. 219]. To assess the probative function in a serial argument, one not only has to assess each single step but also the whole move from initial premises to final conclusion. Assessing the whole involves assessing each single transition from premises to conclusion going step by step. Walton illustrates what this might mean by asking us to consider the following diagram:

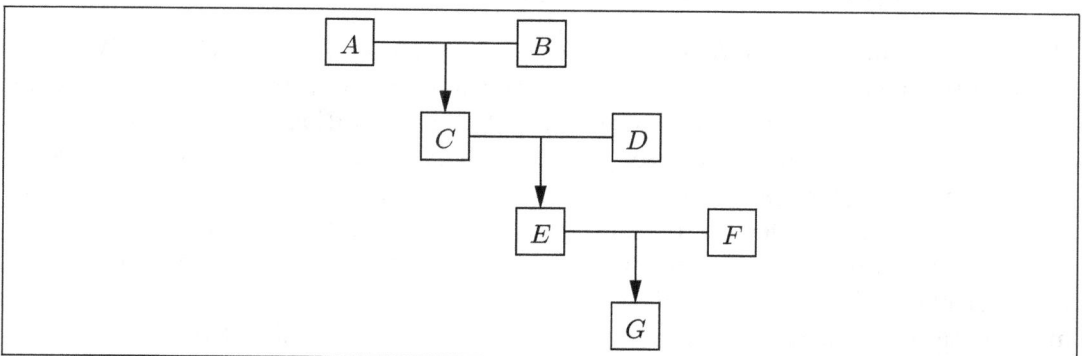

Figure 2

(We are changing Walton's diagraming method somewhat here. He represents liked structure by separate arrows from each premise at a given level, all supporting the same conclusion, to that conclusion itself. In many texts, this configuration represents convergent structure. Our revision of Walton's method brings the representation more in line with the standard practice for representing linked structure.)

At each level, the premises are linked. Assume that prior to assessing probative weight, each statement in the argument has a degree of plausibility. Walton suggests we consider five plausibility weights ranging upwards from 1 to 5. For purposes of illustrating how probative weight moves down a chain, we need not consider here how these probative weights are determined. Walton gives one rule for transmitting weight: In a linked argument, the probative weight of the conclusion will be at least as great as the weight of the least plausible premise [14, p. 216]. He indicates that this is called Theophrastus' rule: He asserts immediately that "the probative function upgrades the plausibility of the conclusion to the level of the least plausible premise" [14, p. 216]. Hence should A, B both have weight 4, and C considered by itself weight 2, the support of A, B linked raises the plausibility of C to 4. If D has weight 3, while E without further evidence has weight 2, its plausibility is raised to 3. If F has plausibility 5, G will have plausibility 3.

We have now presented an account of Walton's view of ethical argument structure as involving layered maieutic dialogue and chained argument. We have sketched how the strength of the premises in ethical argument may be transmitted to conclusions down the chain. We can raise critical questions for Walton's view as presented here. We do this in the next section.

4 Walton's System of Layered Justification for Ethical Arguments: A Critical Question

Walton's system is open to one primary critical question: Are dialectical and rhetorical questions enough for a full account of ethical argumentation? As Wenzel points out in [16], three disciplines are concerned with argumentation.[3] Dialectics concerns coming to agreement over a matter of dispute between two interlocutors by following prescribed rules of procedure. (In [3, 4], van Eemeren and Grootendorst present an example of such rules for reaching agreement.) Conducting the dialogue in a rule governed way renders its persuasion rational persuasion. Rhetoric concerns the persuasiveness of messages addressed to an audience. The issue is not whether the argumentation respects prescribed rules, but whether it "delivers the goods," the audience's accepting "some proposition that they previously doubted or disagreed with" [16, p. 186]. Logic, by contrast, includes discovering the structure in argumentation to assess whether the steps in an argument are epistemically justified. Do the premises at each step give good reasons for the conclusion at that step? The question is both logical and epistemic. Apart from deductive logic, assessing the transfer of acceptability involves considerations on what evidence is available and how weighty is that evidence. Walton gives little consideration to logic even including epistemic concerns.

It is precisely his rejection of the epistemic component of ethical argumentation

---

[3]Given developments since 1979, computer science now also counts as a discipline concerned with argumentation.

that I want to question. Philosophers who are ethical intuitionists allow a cognitive ability of moral intuition as a source of moral judgments. Walton explicitly characterizes intuition in discussing Ross's understanding of right and good. We may judge that certain acts are right or wrong, or certain states of affairs good or bad because we judge they have certain right-making or good-making properties. We may concede that under certain conditions these judgments are overridden by other considerations. Telling the truth may be generally right, but what about telling the truth when it puts someone's life in danger? Leisure is good because pleasurable, but what about the pleasures of sadism? These considerations motivate Ross's distinguishing *prima facie* right from right overriding or without qualification, and likewise *prima facie* good from good without qualification. If two duties conflict, which takes precedence? Ross sees no general rules constantly available to answer these questions. However, he holds that in particular circumstances, we can recognize our duty. "This sense of our partial duty in particular circumstances , preceded and informed by the fullest reflection we can bestow on the act in all its bearings, is highly fallible, but the only guide we have to our duty" [12, p. 42]. Here we are making a judgment about a particular situation. By contrast, for judgments of the *prima facie* rightness or wrongness of acts of a certain kind, for example the rightness of performing act acts of charity, Ross is confident of the correctness of our judgments. For Ross, they are self-evident [12, p. 21, footnote].

Walton objects right here. Since we do not have general principles to weigh the stringency of *prima facie* duties, judgments of priority are subjective. There are hard cases to judge whether an act is right or not, and there is no mechanical rule to decide such questions. In a dispute over a particular case, there seems to be no "dispute resolution method" [14, p. 11]. Walton regards ethical intuition as subjective. To be objective "at least the guidelines or rules for evaluating the arguments on both sides must be objective enough so that they cannot be regarded as based on some kind of unethical bias or favoritism, under a guise of objective assessment" [12, p. 19]. We ask "Is it necessary that there be a *method* for weighing stringency or deciding hard cases? Is having a method the only way to avoid subjectivism in such circumstances?" Answering this question requires us to consider how we come to make moral judgments. Is Walton relying on an adequate picture of moral intuition and of how we come to have moral beliefs, either particular or general?

I propose dividing this question into two parts. Notice that should one say

1. *A* is an act of stealing,

although "act of stealing" may have negative connotations, the assertion is not normative. Given a characterization of "stealing," we can empirically investigate whether an act constitutes theft. By contrast, should I say

2. *A is morally wrong because A is an act of stealing*

I am not only making a moral judgment but also saying that the wrongness of A supervenes on its being an act of stealing. If we were to express (2) as an inference rule,

> From x is an act of stealing
> To infer x is *prima facie* morally wrong

that (2) incorporates a principle of inference is obvious. What belief-generating mechanism "sees" the inferential connection between premise and conclusion here? We hold that it is moral intuition. We are understanding "intuition" here as Cohen understands it in [2], our ability to recognize "what should be inferred, judged or meant in such and such a context, ... about what counts as a reason for what" [2, p. 73]. Cohen allows that the word "intuition" has other meanings as used by some philosophers, as "some rather superior mode of acquiring knowledge, like Plato's 'eye of the soul' or Spinoza's intuitive science" [2, p. 74]. I suspect it is intuition in something like this sense which is how Walton understands the term. But it need not be taken in this sense.

Is intuition in Cohen's sense objective, at least to some degree? He characterizes it as "an inclination of mind that is taken to originate from the existence of a system of tacitly acknowledged rules for making judgments about relevant topics... Each intuition may be presumed to originate from such a system of rules" [2, p. 76]. In this sense, is the notion of intuition obscurantist? Is it totally subjective or the product of learning and assimilating rules which may have objective justification? If we judge that A is a reason for B, have we not judged something objective about the relation between A and B? Given this tacit awareness of the rules, is our judgment in no way objective?

Admitting intuition in Cohen's sense, recognizing what non-moral properties are reasons for moral properties, has significant implications for Walton's account of ethical argumentation. First, concerning the initial level of a chained inference, the major premise need not be endoxic. Consider Walton's example:

> Anyone who has committed theft has done something wrong [14, p. 82].

To appreciate that this statement is acceptable, does one need to turn to a moral principle accepted by everyone, or the majority, or the wise, the probative sources in Aristotles's *Topica*? True, the voucher of these sources may enhance the plausibility of claims. But must one depend on such external sources to generate one's general moral principles? Cannot humans "see" a connection between recognizing that an item belongs to someone else and that one has an obligation not to appropriate that object without the owner's permission? Is this recognition not "on all fours" with Ross's recognition of a connection between one's making a promise and one's being bound to keep it? Does not this hold *mutatis mutandis* for the other *prima facie* duties?

We do not deny that moral intuition is fallible. To the proponent's assertion that all actions satisfying a certain description are morally right or morally wrong, a challenger could reply with a counterexample. But this does not show that moral intuition fails to be reliable. The counterexample satisfies the antecedent of the intuited general principle, but it also satisfies some further property. The proponent can then amend the initial statement of moral obligation, a statement of the form

> *Ceteris paribus* some act which satisfies the description D is morally right (or wrong) to the qualified
> *Ceteris paribus* some act which satisfies the description D and not-E is morally right (or wrong)

Clearly, the proponent may respond with further qualifications should the challenger bring forward further counterexamples. In the end, the process should produce a general principle against which the challenger can bring no further counterexamples.

Walton might still object. We recognize these obligations not through intuiting these connections but because we have internalized popular moral beliefs of our culture. At bottom, the recognition is endoxic. I see two replies to Walton here. First, we see no need to deny that one's culture influences one's moral principles contingently shaping in part our moral intuition. But what of our ability to recognize counterexamples to these principles and to refine our principles in their light? Is the recognition of all counterexamples endoxic?

Walton might object that our presentation of his view is incomplete. He also discusses abductive inference or inference to the best explanation, Here one "goes from given data to a hypothesis that best explains that data" [14, p. 95]. Walton does not stipulate that this hypothesis be disclosed by the culture. Presumably, we can recognize suggestions from the facts themselves without being told by the culture what these facts suggest. By abduction, we come to recognize the major premise of the inference.

I do not understand Walton's point here. An explanation is an interpretation, not an evaluation. If I say

1. The spark occurred because two bare live electrical wires touched,

I am giving an explanation but expressing no value judgment. I am relating the spark to an antecedent event which is connected in a law-like way with that event.

Relating nomically to some wider whole is the hallmark of an interpretation. I am not saying that the spark is in any way good or bad. By contrast, if I say

2. William is valorous because he carried the colors into battle after the flag bearer had been shot

I am expressing an evaluation. I am saying that such an action is praiseworthy. Walton has not shown how *evaluations* are the conclusions of inferences to best explanation. However, we shall return to the point shortly.

According to Walton's view, recognition of a basic moral generalization, in particular the major premise at the first step of a chain of ethical argumentation, is endoxic, taken over from one's culture. The principle may then be refined by critical discussion with a challenger. But is one's acceptance of the major premise always culturally relative? This view rejects moral objectivism at least at the first step of a maieutic dialogue. What can be said for ethical objectivism? The case can be set out straightforwardly.

As Rachels points out in [11], some constraints on human behavior are necessary conditions for human life to exist, at least at a level we recognize as human. Not every specifiable condition need be necessary for every individual. What is necessary may be a disjunction of such conditions. Evidence for such existentially necessary conditions for life is not hard to come by. In the fall of 2020, some commentators on American society noted that Americans do not know just what to believe of what officials are telling them. That is, their trust in the word of their leaders has been seriously damaged. Should there be any question of why this has happened? When leaders repeatedly lie, when they frequently contradict each other, is there any mystery that their trustworthiness is diminished? But what is the existential consequence? Can a *society* really be a society if trust is lacking? Trust among members of society is existentially necessary and actions which threaten trust are existential threats.

Let us now go from interpretation to evaluation. That satisfying an existentially necessary condition is good, objectively good, albeit a *prima facie* good seems as self-evident as making a promise morally binds the person making the promise is self-evident, at least for Ross. If we understand "good" here to mean "good for a living organism," that a condition which fosters the flourishing of living organisms is good, at least for those organisms, seems undeniable. What reason might gainsay it? One might object that what fosters the flourishing of a murderer is certainly not good. But this objection overlooks that the goodness here is *prima facie* goodness. The life of a murderer might be a great good for the murderer, but not an overriding or *ultima facie* good for human life itself. The goodness of a state lies in its having positive existential consequences.[4]

Further examples illustrate that intrinsic goodness or rightness supervene on what is existentially necessary. These examples also motivate that cultural differences over what is right or good do not show that moral objectivism is not a viable position. Moral judgments may supervene on interpretations. Two individuals may interpret the same action differently because of the different rules they have learned. If two persons interpret what they have observed differently, their recognizing different moral properties supervening on what they have "seen" should be no surprise. The controversy over statues of Confederate military leaders in the American Civil War illustrates this point nicely. Some may interpret these larger than life figures on horseback raised above the surrounding area on pedestals as expressing loyalty to the heritage of those memorialized. If loyalty is a virtue, they judge morally good what is represented. By contrast, if they interpret these images as expressions of an ideal of racial dominance and they regard such dominance as morally bad, they judge what is represented as bad. Is there a disagreement over *value* here or disagreement in interpretation?

If the evaluative general premise at a step in ethical argumentation cannot be justified by an inference to best explanation argument and the premise appears not self-evident, how might one argue for it? Consider the highly general

> Any condition which contributes to or sustains what is existentially necessary is intrinsically good.

Can we think of this generalization as a hypothesis or analogous to a hypothesis, a moral hypothesis, not an explanatory hypothesis, about what falls under the extension of "good"? Could we argue for this statement analogously to how we might argue for an explanatory hypothesis by inference to best explanation? If so, we have traded in something obscure in Walton for what we might call an inference to (best) evaluation. How might we construct such an argument and may it have any cogency? Consider:

3. Any contribution to creating or transmitting knowledge is intrinsically good.

We may consider this as a hypothesis about what is included in the denotation of "intrinsically good." We may also subsume this hypothesis under the more general

4. Any condition which contributes to or sustains what is existentially necessary is intrinsically good

---

[4]In proposing existential consequentialism as a theory of what is objectively intrinsically good, we are following Boyd's lead in [1]. He proposes homeostatic consequentialism as a theory of objective goodness. To develop this lead here it beyond the scope of this paper.

as long as we admit that knowledge is existentially necessary for human life. We may think of (4) as a hypothesis — a moral hypothesis, not a physical explanatory hypothesis — about the denotative meaning of "intrinsically good." Could we argue for (4) analogously to how we might argue that a physical hypothesis is the best explanation for the occurrence of some event or condition?

The standard schema for inference to the best explanation or abductive inference has three premises and a conclusion:

(i) (Surprising) event E has occurred or a collection of data has been gathered.

(ii) Hypothesis H explains E.

(iii) H is the best explanation for E. Therefore

(iv) Ceteris paribus H.

Where H is an ethical hypothesis, can we construct an argument for it instancing this pattern? The datum to be explained is the first-level hypothesis itself. The hypothesis argued for is the general second-level hypothesis.

(P1) The creation or transmission of knowledge is intrinsically good.

(P2) That possession and transmission of knowledge are existentially necessary for human life as we know it explains why creation and transmission of knowledge are intrinsically good.

(P3) That possession of knowledge is existentially necessary for human life as we know it is the best reason we can offer for the premise that the creation and transmission of knowledge us intrinsically good. Therefore

(C) *Ceteris paribus* the creation and transmission of knowledge are intrinsically good because existentially necessary.

When the question is the best explanation why some event or condition has come about in the physical world, the first premise is ordinarily a description for which we can seek perceptual verification. That a given hypothesis explains an event can be seen through our background knowledge of the nomic connections between events in the world. Accepting (P1) and (P2) may not then require an argument. (P3) is different, Why should a hypothesis be the best explanation of what needs explaining? As is well known, three criteria are standard for evaluating hypotheses: fruitfulness, consistency with previous results, and simplicity. The hypothesis which comes out best on these three criteria is the best hypothesis.

What criteria are appropriate for seeing a higher-order moral principle as the best reason for a first-order moral generalization? The question has a quick answer and a developed answer. The quick answer is the higher-order principle which comes out best of available higher-order principles under reflective equilibrium. This answer requires some unpacking. First, what makes a theory available? That a higher-order principle is available means its being recognized as a candidate theory, in the case of intrinsic value, for what is intrinsically good. This first answer is completely straightforward. Second, how can reflective equilibrium be a way of assessing the comparative strength of reasons for a given first-order ethical principle?

Cohen's discussion of Sidgwick on intuitionism suggests how intuition is involved. Sidgwick distinguishes three kinds of intuitionism and thus for our purposes here three kinds of intuition — perceptual, dogmatic, and philosophical. Perceptual intuition "appeals to the dictates of a persons conscience in response to particular quandaries on particular occasions" [2, p. 80]. Conscience makes judgments about what particular states of affairs are good or bad, actions right or wrong. Percpetual intuition is what we called in [8] the moral sense. Dogmatic intuition is concerned with general rules rather then particular cases. It thus includes moral intuition. Dogmatic intuition grasps basic level moral principles. Philosophical intuition concerns "fundamental principles that are not evident to ordinary people, in order to explain, justify, or even rectify the morality of common sense" [2, p. 81]. Instead of general principles of right or wrong, intrinsically good or intrinsically bad, philosophical intuition concerns a philosophy of rightness, goodness, virtue, or moral value. Philosophical intuition is concerned with higher-level principles, such as Mill's principle of utility, Hart's harm principle, or Kant's categorical imperative. (See our [9] for our further discussion of these principles.)

We can now appreciate how reflective equilibrium can enter into discerning why a reason for a first-level ethical judgment is the best reason. Consider again (P2) above — Possession of knowledge is existentially necessary for human knowledge as we know it. Why is this the best reason for (P1) the claim that the creation and transmission of knowledge is intrinsically good? We are asking for why existential consequentialism gives the best reason for justifying an ascription of intrinsic value.

What other theories are on the table? Philosophers have offered several answers. In [6], Frankena distinguishes two classes of theories about what is intrinsically good: hedonist and non-hedonist [6, p. 83]. A full discussion of these theories lies far beyond the scope of this paper. Rather, we ask whether any offer better reasons for why the creation and transmission of knowledge is intrinsically good than the hypothesis of existential consequentialism. At best, we can only give a sketch of how an argument would proceed.

Following Frankena, we may summarize hedonism in two theses:

1. All pleasures are intrinsically good and only pleasures are intrinsically good.

2. Being pleasurable is the criterion of intrinsic goodness. It makes things good in themselves [6, p. 84].

If moral intuition testifies that making a promise *ceteris paribus* morally binds one to keep it, does it not also testify that risking one's life to rescue another from danger and the valor such action shows not only exemplify a morally good character but their presence in the world makes the world an intrinsically better place? But is risking one's life, especially given the consequences that it may involve, a pleasurable activity? By contrast, it is existentially necessary, at least in situations necessitating such sacrifice. Hedonism fails to account for such instances of intrinsic goodness, while existential consequentialism explains it. A hedonist who disagrees incurs the burden of proof to show that sacrificing oneself is pleasurable. With this, we conclude that *prima facie* hedonism as a theory does not give a better reason for intrinsic goodness than existential consequentialism.

Can one find a better alternative to existential consequentialism among the rivals of hedonism? We cannot answer that question here because of the multiplicity of non-hedonistic theories proposed. However, we must note that in addition to pleasure, excellence or self-perfection has been claimed to be a good in itself. But are not excellence or self-perfection, properly understood, existentially necessary? It seems plausible that these other theories are equivalent to existential consequentialism or special cases of the view.

What are the consequences of these considerations for appraising Walton's theory of ethical argumentation? First, we have seen that that the major premise — the universal premise — at the first "link" in a chain of ethical argumentation need not be endoxic. A corollary that we are not totally dependent on culture for these initial premises follows. Our moral intuition lets us grasp connections between good or right making properties and being good or right, though our grasp may be defeasible. Generalizations standardly accepted in a given culture may give us points of departure, but they are not the only points of departure. Basic ethical generalizations include more than what is inherited from the culture. Again Walton sees the structure of ethical argumentation as dialectical. One may recognize conflicts between accepted ethical principles, especially between accepted principles and cultural practices. A society may condone slavery yet embrace ethical principles which militate against it. Is there nothing more to recognizing these principles than their endoxic origin? Cannot some of them be recognized as superior moral principles by moral insight? If there is a conflict of endoxic ethical principles, can that conflict only be resolved by some further endoxic principle? May we allow that we may recognize conflicts between ethical principles only by recognizing that they logically conflict and not any other way? If so, what leads us to recognize one of these principles as morally superior? How can that be understood apart from our having some ethical insight? Indeed, if we traverse a chain of argumentation, how do we know we are going up the chain? Walton's essay leaves us with these questions.

References

[1] Boyd, Richard N. How to be a Moral Realist. 1988. In Geoffrey Sayre-McCord, (Ed.) *Essays on Moral Realism*. Ithica, NY USA: Cornell University Press, 181-228.

[2] Cohen, L. Jonathan. 1986. *The Dialogue of Reason*. Oxford: Clarendon Press.

[3] Eemeren, Frans H and Rob Grootendorst. 1984. *Speech Acts in Argumentative Discussions*. Dordrecht-Holland and Cinnaminson-U.S.A. Foris Publications.

[4] Eemeren, Frans H. and Rob Grootendorst. *Argumentation, Communication, and Fallacies*. 1992. Hillsdale, NJ., U.S.A. Lawrence Erlbaum Associates, Publishers.

[5] Feigl, Herbert. 1963. De Princpiis Non Disputandum: On the Meaning and Limits of Justification. In Max Black, (Ed.) *Philosophical Analysis: A Collection of Essays*. Englewood Cliffs, NJ: Prentice Hall.

[6] Frankena, William, K. 1973. *Ethics* Second Edition. 1983. Englewood Cliffs, NJ: Prentice-Hall, Inc.

[7] Freeman, James B. *Dialectics and the Macrostructure of Arguments*. 1991. Berlin and New York: Foris Publications.

[8] Freeman, James B. 2005. *Acceptable Premises*. Cambridge: Cambridge University Press.

[9] Freeman, James B. 2009. Higher Level Moral Principles in Argumentation. In *Argument Cultures*, Juho Ritola, (Ed). Windsor, ON: Ontario Society for the Study of Argumentation, (CD Rom).

[10] Rescher, Nicholas. 1977. *Dialectics*. Albany, NY: State University of New York Press.

[11] Rachels, James. 1986. *The Elements of Moral Philosophy*. New York: McGraw-Hill, Inc.

[12] Ross, W. D. 1930. *The Right and the Good*. Oxford: Oxford University Press.

[13] Stevenson, Charles L. 1944. *Ethics and Language*. New Haven, CT: Yale University Press.

[14] Walton, Douglas. 2003. *Ethical Argumentation*. Lanham, MD: Lexington Books.

[15] Wellman, Carl. 1971. *Challenge and Response: Justification in Ethics*. Carbondale and Edwardsville: Southern Illinois University Press.

[16] Wenzel, Joseph W. 1979. Jürgen Habermas and the Dialectical Perspective on Argumentation. *Journal of the American Forensic Association* 16, 83-94.

Argument is Moral.
Using Walton's Dialectical Tools to Evaluate Argumentation from a Moral Perspective

Katharina Stevens
*University of Lethbridge, A812K University Hall, 4401 University Drive,
Lethbriege, Alberta T1K 3M4, Canada.*
`katharina.stevens@uleth.ca`

1 Introduction

Douglas Walton's dialectical theory of argumentation, developed in a swath of papers and several monographs, most central of which are *The New Dialectic* and, co-authored with Eric Krabbe, *Commitment in Dialogue*, is one of the most thoroughly developed, detailed and fruitful theories of argumentation available (see, e.g. [25, 26, 27, 29, 30, 36]). But Walton's dialectical theory of argumentation is valuable not only as a comprehensive framework. It also contains important insights fit to illuminate and answer questions that arise outside of his framework and that he may or may not have intended to address. Here, I will sketch how Walton's concepts of dialogue types and dialogue shifts, which allowed him to build a highly original fallacy theory, can be used for constructing a method of argument-evaluation from a moral perspective.

Walton has the goal of making it possible to evaluate argumentation without needing to attempt a determination of the arguer's intentions, which he thinks would require too much psychological speculation (see, e.g. [27, pp. 244–249]). It may seem that the moral evaluation of argumentative behavior requires exactly this. However, I argue that Walton's method of grounding argument evaluation in the goals of argumentative dialogues can be adapted by identifying moral goals of argumentation associated with harm avoidance and respect for dignity. This makes it possible to determine whether argumentative behavior is desirable or undesirable from a moral point of view without needing to inquire into an arguer's intention. Crossing that line is only necessary if we want to determine whether moral blame should be put on the arguer.

2 Dialogue Types and Dialogue Shifts as Tools for Evaluating Argumentative Behavior Beyond Fallacies

The concepts of dialogue type and dialogue shift are central in Walton's account of fallacies. However, these concepts do more than aid in the determination of an arguments-as-product's cogency (following O'Keefe [18]: argument1). They are also tools for evaluating behaviors during arguments-as-processes (argument2) that might otherwise have escaped attention. Here I will provide a selective summary of Walton's use of dialogue types and dialogue shifts for argument-evaluation, highlighting and extending the different ways in which Walton uses them for the evaluation of argumentative behavior. This will prepare the ground for sketching how they can also be used to evaluate argument from a moral perspective in the next section.

2.1 Dialogue Types

Walton (e.g. [27, 29, 27]) argues that individual arguments1 cannot be evaluated independently of their dialectical context as being parts of arguments2. One reason for this is that we use argumentation in pursuit of different goals, such as for example the goal of rationally resolving a conflict of opinions, the goal of jointly determining a best course of action, or the goal of coming to an equitable compromise. These goals determine how the argument2 should be carried out, which in turn determines what types of argument1 are appropriate. It is acceptable to appeal to a party's interest by making a tit-for-tat offer in an argument2 aimed at determining a compromise, but the same is fallacious where the goal is to determine the most just decision.

In order to capture these differences in a systematic way that lends itself to an ordered method of argument evaluation, Walton (sometimes with Krabbe) defines six normative dialogue types: the persuasion, the inquiry, the negotiation, the information seeking, the deliberation and the eristic dialogue.[1] These dialogue types are meant to do two things at once: 1) They are meant to be descriptive abstractions that, while unlikely ever to be realized fully in real life, approximate real argument practices closely enough that they can, 2) serve as meaningful normative ideals fit to provide the basis for the context-sensitive evaluation of real arguments.[2]

---

[1] In 2010, Walton recognized one additional dialogue type, the discovery dialogue, but this dialogue type is not discussed in many of his works and it is not especially interesting for the purposes of this paper [33].

[2] It should be noted that Walton acknowledged that many arguments2 in the real world would have features of more than one of these dialogue types and pursue more than one goal, so he allowed for the possibility that there might be mixed dialogues. For example, a dialogue between a doctor and her patient who seek the most prudent course of treatment together might be a mixture between an information seeking dialogue and a deliberation. Dialogues may also have flavors of

Each dialogue is defined over the initial situation in which the argument arises, the goals of the argument's participants and the goal that the argument should accomplish in order to be a good instance of the dialogue-type. For most of the dialogue types, Walton provides more detailed profiles (by himself, with colleagues, or he approves those provided by others).[3] However, for my purposes, the argumentative burdens (like, e.g., the burden of proof) that rest on participants are most important, so I added those in the table below as I understood them from Walton's explanations, e.g. in Walton [27, 33]:

Dialogue Type	Initial Situation	Participant Goals	Argumentative Burdens	Dialogue Goal
Persuasion	Conflict of opinions	Persuade other party	Burden of proof on the proponent to show that a claim is acceptable based on the premises accepted by the opponent, burden of production (of evidence) on each party when they assert a claim.	Resolve or Clarify Issue
Inquiry	Need to have proof	Find and verify evidence	Joint (high) burden of proof on the inquiring parties together to either conclusively prove or disprove the hypothesis.	Prove (Disprove) Hypothesis
Negotiation	Conflict of Interests	Get What you Most Want	Burden to make and counter offers/threats and burden to respond to them.	Reasonable Settlement that Both Can Live With

---

another dialogue. For example, a persuasion dialogue might over time become more and more adversarial, adopting an eristic flavor [29].

[3]The persuasion dialogue gets an extensive treatment in the central books of the theory [29, 36]. For the others, see: Inquiry [35]; negotiation [24, 38]; information seeking [2, 28]; deliberation [37]. There does not seem to be too much on eristic dialogues in the Waltonian framework.

Information Seeking	Need information	Acquire of Give Information	(Sometimes) burden to justify the request for information on the side of the asker, (sometimes) burden to provide information and show the appropriateness of the information on side of the informant.	Exchange Information
Deliberation	Dilemma or practical choice	Co-ordinate Goals and Actions	Joint burden to provide proposals and reasons. Once a proposal is made, a burden of justification may be assigned to the one making it.	Decide Best Available Course of Action
Eristic	Personal conflict	Verbally Hit out at Opponent	No obvious burdens	Reveal Deeper Basis of Conflict

Which dialogue type arguers will engage in is determined during the opening stage of the argument. According to Walton [32], arguments arise when there is a problem, difference of opinion or question that must be resolved for at least one of the arguers. Each argumentative dialogue has three stages, the opening stage, the argumentation stage, and the closing stage. During the opening stage, arguers determine what kind of dialogue they will engage in, but also how they will approach the issue, what rules they will follow and who will play which argumentative roles. Even though Walton associates different initial situations with different dialogue types, this is not necessarily fully pre-determined by the context in which the argument arises. For example, if arguers have to tackle the question which dog-breed they should select for the family dog, this could be determined via a persuasion dialogue (or several) or via a deliberation. Even a negotiation is possible. During the argumentation stage, arguers exchange their arguments and attempt to meet their respective argumentative burdens. The closing stage is reached when the dialogue's goal has been accomplished and/or the participants agree to end the argument.

Walton's normative model has not been uncontroversial. For example, the idea that arguers go through an opening stage (shared with the pragma-dialectical school of argumentation (see, e.g. [22])) has been criticized as unrealistic by Jacobs [14], Krabbe [16] and myself [20]. Indeed, I do not believe that there are many arguments that begin with explicit agreements of this kind. And I agree with others that few actual argumentative dialogues will resemble any one of Walton's dialogue types exactly, especially with respect to the more elaborate versions he or others have provided. Nonetheless, I believe that the dialogue types identify kinds of arguments that we are socialized to tell apart, even if the boundaries are blurrier and the associated norms less clearly determined than Walton and others describe them. For example, I know that there is a difference between participating in an argument in which my interlocutor is my opponent and one in which she works with me to find the best possible decision. And I have a feeling for what is acceptable behavior in each situation. I also believe that over the course of the first moves in an argument, arguers reach a mutual understanding about what it is they are doing. Afterwards they rely on this understanding to determine what is expected of them, what they can expect of their interlocutors and what will be met with disapproval.

Further, while it is important to create a good descriptive account of how these things are determined in actual arguments,[4] such an account is too complex to be a useful for all but the most involved argument evaluation. The abstraction Walton makes with his three-stage model of the argumentative dialogue and his six different types allows for a method of argument evaluation that is medium-fine-grained: Context-sensitive enough to avoid gross errors of oversimplification while simple enough not to invite errors that arise from over-complexity where information and resources might be sparse. For example, his tools allow for a context-sensitive but nonetheless methodological evaluation of arguments1. According to Walton, the acceptability of using a certain argument-type depends on whether the argument1 contributes to accomplishing the goal of the dialogue-type that the argument2 can be identified as belonging to. *Ad hominem* attacks, for example, are perfectly acceptable during an eristic dialogue because they can promote the goal of understanding the basis of a conflict better by airing grievances. But they are inappropriate in an inquiry dialogue because they will not contribute to the joint determination of whether a hypothesis is true and can even create a barrier to this. Using an argument type that is inappropriate to the present dialogue type means committing a fallacy. This approach to fallacy evaluation is useful even if the real argument2 does not fully embody a specific dialogue type (or mixed dialogue). Even a rough identification with a dialogue type makes it possible to use the type's goal as the normative basis for evaluation.

2.2 Illicit Dialogue Shifts

A dialogue shift happens when arguers who have so far argued in one dialogue type start arguing according to the norms of another. Some shifts amount to fallacies because they consist of the arguer using an argument appropriate only in a dialogue-type other than the one that the arguers are engaged in and hinder the dialogue's goal. Such shifts are illicit because they obstruct the goals of the original dialogue. They are especially problematic if an arguer performs a shift unnoticed by other arguers, so that the outcome of the new dialogue is mistaken as the result of the old one (compare Walton [27, p. 237]). For example:

---

[4]For a detailed account of how I believe this happens, see Stevens (2019). In this, I follow design-theorists of argumentation [8, 13, 14].

Deliberation	John: We still have to decide which dog would fit best into our family. Jane: I think a dog that can handle a long walk would be good. We love taking hikes. John: I also like to be lazy sometimes, but I agree that we like to hike, and if this is a high priority for you, I think it is best to look for a dog with at least medium energy. Jane: And I really do not want a dog that will shed a lot. This will double my cleaning time and take away from our family time. John: Oh, but long-haired dogs are very beautiful. Jane: True, and I know how important aesthetics are to you. Aren't there other beautiful dogs?
Unnoticed Negotiation	John: Well, if you cannot deal with the long hair, then I want a lazy dog! I should not be the only one who has to make concessions. Jane: I am sorry. John: Clearly, a long haired, medium energy dog is best for all of us then. Jane: I guess.

Here, John commits the fallacy of bargaining during a deliberation dialogue by using an argument type that would only be appropriate during a negotiation. Jane does not notice the shift but takes the argument to make a valid point in the context of the original deliberation. As a result, she mistakenly accepts the outcome of getting a long haired medium energy dog for the best course of action available (the goal of a deliberation), when the argument2's outcome could at most show that it might be an acceptable compromise (the goal of a negotiation), but most likely not even that because Jane did not get a chance to negotiate back.

Importantly for my purposes, not all illicit dialogue shifts coincide with the use of a fallacy. As Walton puts it, fallacies are babtizable types of (defeasible) argument1 used in the wrong way [27]. But Walton's concept of illicit dialogue shifts can also be used to understand why argumentative behavior that cannot be identified with the use of a fallacy can be unacceptable. For example, illicit dialogue shifts can also happen if an arguer successfully assigns a burden of providing argumentative content to their interlocutor when, according to the norms of the dialogue, the interlocutor should not have that burden and this obstructs the goal of the dialogue. Atkinson, Bench-Capon, and Walton [1] describe how an arguer in a deliberation dialogue might try to covertly adopt the role of an opponent as it exists in persuasion dialogues. Walton et al. explain why such behavior is problematic: One main difference between a deliberation and a persuasion is that in a persuasion dialogue, the proponent has to *persuade* the opponent by using premises that the opponent already accepts, including premises about goals. This means that in a persuasion dialogue, a proponent would have the burden of proof — she has to show the opponent that a certain course of action would meet the opponent's goals. In a deliberation dialogue, there is only the burden to provide proposals and reasons as arguers come up with them. The goals of all participants enter the deliberation equally and are equally relevant to what will count as the best course of action. An arguer in a deliberation who is successful in illicitly adopting the opponent-role may thereby manage to push her interlocutors into accepting burdens that are associated with the proponent-role in a persuasion dialogue. This will indirectly give her own goals heightened importance. If that goes unnoticed, the interlocutors might still believe to be in a deliberation and treat the outcome of the now *de facto* persuasion as the outcome of a deliberation. This would mean that they behaved as if the delinquent arguer's goals reflected the goals of the whole group instead of treating them as goals that only need to be considered as much as anyone else's. As a result, the deliberation would now be less likely to lead to the best course of action for everyone involved and more likely to lead to a course of action that was best for the offending arguer (potentially without anyone noticing).

Importantly, all this might happen without the arguer using any inappropriate argument types, simply through a slight shift in her argumentative behavior. She might, for example, stop acknowledging the importance of the goals of any other arguers and insist that others show how her goals will be met:

Deliberation	John: We still have to decide which dog would fit best into our family. There are Collie puppies, Jack Russel puppies and Schnauzer puppies available. Jane: I think a dog that can handle a long walk would be good. We love taking hikes. John: I also like to be lazy sometimes, but I agree that we like to hike, and if this is a high priority for you, I think it is best to decide between the Schnauzers and the Collies. Jane: And I do not want a dog that sheds. John: Oh, but Collies are so beautiful and have medium energy, and they would be good for the kid, they are so gentle.
Shift into Persuasion	Jane: Can you show me that Collies do not shed? John: I admit that they shed a little. But I really appreciate a beautiful dog, and the kid needs a really gentle dog. Jane: Well, if they shed, then they are out, they are not the dog for us. John: I guess so.

By taking the role of an opponent and successfully assigning burdens to other arguers that they should not have, Jane effects an illicit dialogue shift that cannot be associated with a fallacy. However, it does threaten to prevent reaching the original dialogue's goal, namely identifying the dog that would be best for the family as a whole.

2.3 Required Dialogue Shifts

Not all dialogue shifts are illicit. Arguers engaged in a deliberation may, for example, shift to an information seeking dialogue because they need additional information to meet the goal of the deliberation, which is to decide on the best course of action. In this case, the shift to the information seeking dialogue does not obstruct the goals of the deliberation dialogue but helps advance them. This is so because when the arguers have completed the information-seeking dialogue, they can return to the deliberation dialogue with additional information that will help them identify the best course of action for everyone. The information seeking dialogue is then *embedded* in the deliberation dialogue, propelling it forward (see, e.g. [27, 29, 36]). For example:

Deliberation	John: We still have to decide which kind of dog we should get. Jane: It is important that the dog is very kid-friendly. After all, we have a toddler and a baby on the way. John: I agree. I think Collies are kid friendly, Golden Retrievers and Poodles are too. Jane: I also do not think I can deal with a dog with too much energy. John: It is important to me that our dog will be playful though. So we do not want a couch-potato either.
Information Seeking	Jane: Hmm. Difficult. I do not know whether there are any dogs among the kid-friendly ones that are medium energy like this. You know a lot about dogs, Sam. What are Golden Retrievers like? Sam (a family friend): Golden Retrievers have very high energy. So have Poodles, to be honest. Jane: What about Collies? Sam: They are not low energy, but if you give them a good walk every day, they are calm in the house and like to take a nap.
Deliberation	Jane: So a Collie is going to be kid-friendly, and will like to play outside but be calm inside. But won't a Collie shed a lot with all that hair? John: Yes, but given all the other perks, can't we live with that? Jane: I guess we can..

The possibility that dialogue shifts may advance the goals of a dialogue suggests that in addition to licit and illicit dialogue shifts, there may also be *required* dialogue shifts. Sometimes, a dialogue, like the deliberation in this case, may reach an impasse, where the dialogue's goal can only be reached if the arguers embed another type of dialogue, like an information seeking dialogue. Imagine that Jane and John have been deliberating about the dog for the family for a while, and the energy levels of the acceptable dog-breeds have been recognized as the point on which things turn. In this case, they might find themselves unable to decide unless they gather more information about dog-energy-levels: The deliberation will fail *unless* an information seeking dialogue is embedded.

To my knowledge, Walton does not devote a detailed discussion to required dialogue shifts, and limits himself to discussing licit and illicit shifts in, e.g. *The New Dialectic* (1998), *Commitment in Dialogue* (1995) and *A Pragmatic Theory of Fallacy* (1995). But this does not mean that the possibility is entirely unacknowledged in his work. In fact, in his co-authored paper "Missing Phases for Deliberation Dialogue for Real Applications", he endorses something similar for deliberation dialogues [37]. In this paper, Walton et al. discuss the model of deliberation dialogues that was developed by McBurney, Hitchcock and Parsons [17] in order to expand it to deal with some problems they identify. One of these problems is the need to be able to integrate new information into the deliberation, and

Walton et. al. suggest that the knowledge base on which the deliberation proceeds must be kept open. In this context, they concur with McBurney, Hitchcock and Parsons that shifts to information seeking dialogues can become necessary.

It seems a small step from acknowledging that shifts can be necessary to the idea that an arguer who obstructs such a shift may sometimes be criticizable just like an arguer who causes an illicit shift. Imagine, for example, that John wants a Golden Retriever independent of whether this will be the best dog for the family. He might have an inkling that Golden Retrievers are high-energy dogs and therefore be reluctant to allow an embedded information seeking dialogue about energy-levels. If he then tries to disrupt or stop the information seeking dialogue with Sam, he would be guilty of intentionally hindering the deliberation from reaching its goal through the illicit refusal to shift dialogues. He might even be guilty of this if he merely knows that Sam has information about the energy-levels of dog breeds and fails to initiate the necessary shift to an embedded information seeking dialogue.

3 Dialogue Types and Dialogue Shifts as Tools For Evaluating Argumentative Behavior from a Moral Perspective

So far, I have discussed how Walton's concepts of dialogue types and dialogue shifts can be used not only for identifying fallacies, but also for evaluating more subtle argumentative behaviors. I have drawn on Walton's description of illicit and licit shifts and expanded on his acknowledgement that shifts may be needed in deliberations to add the idea of necessary shifts and illicit refusals to shift dialogues. It is now time to provide a preliminary sketch of how Walton's concepts of dialogue types and dialogue shifts can be developed into tools for the evaluation of argumentative behavior from a moral perspective. Developing these tools fully would require considerably more space than available in a single paper. However, I think that it is possible to provide a general outline.

Walton himself does not talk about the normativity of argumentation as a moral normativity much. This is presumably because he wants to create a method for the evaluation of argument without the need for inquiring into the arguer's intentions. Doing this, to him, would have turned argument evaluation into too much of a "psychological task" (see, e.g. [27, pp. 244–249]). Walton treats the original choice of the dialogue type mainly as a given, relying on the assumption that arguers have freely agreed to it in the opening stage.[5] Fallacies and illicit dialogue shifts are fallacious and illicit because they obstruct the goals of the dialogue that the arguers are *de facto* engaged in, not because arguers use fallacies and

---

[5]It should be remarked that he does offer the typical conditions under which the choice for a dialogue is made, and the way he talks about dialogue goals shows that some situations call for some kinds of dialogues. He also establishes a sort of hierarchy of dialogues and describes shifts from a deliberation/inquiry, to a persuasion, to a debate/negotiation and finally to a quarrel as a downward cascade that diminishes the quality of the dialogue [36]. But by that he likely means that the avowed goals of these dialogues are less and less grounded in a pursuit of the truth. None of this gives the normativity of argumentation he envisions a moral flavor.

shift dialogues with nefarious intent. I therefore take it to be my task not only to show that Walton's tools for argument evaluation can be useful to the analyst who adopts a moral perspective. But also that such evaluation from a moral perspective is possible, at least to some degree, without having to inquire into the arguer's intentions.[6]

3.1 Are Intentions Needed? Example: Fallacies

Admittedly, the easiest way in which to use Walton's theory for the evaluation of argumentative behavior from a moral perspective is by adding an inquiry into the arguer's intentions to Walton's method for fallacy-identification. For example Blair [3] claims that it is "unethical deliberately to invite the interlocutor to commit or be deceived by what one *believes* to be a fallacy — that is, is it unethical to offer an argument one believes to be fallacious as if it were legitimate or to make a fallacious argumentative move *on purpose*. [italics added by me]" [3, p. 24]. Blair goes on to explain that using fallacies intentionally is a kind of dishonesty or deception, and that the intentional use of specific fallacies, like versions of the *ad hominem* can amount to bullying. This shows that to Blair, the intentions of the arguers determine whether the judgement that an argument is fallacious should go hand in hand with moral blame. It might seem as if using Walton's method for argument-evaluation from a moral perspective requires an additional inquiry into the arguer's intend and therefore the kind of psychological speculation Walton wanted to avoid.[7]

However, while a determination of intention may be necessary to assign blame, concentrating on the moral blameworthiness of individual arguers is not the only way we can evaluate argumentative behavior from a moral perspective. We can instead follow Walton in anchoring the evaluation of argumentation on the goals of an argument. We can identify certain morally relevant goals that should be accomplished during an argument and ask how argumentative behavior impacts these goals. Doing so will allow argumentation analysts to determine whether the arugment2 they are studying was morally problematic. It will also help arguers to meet their moral responsibilities by allowing them to screen the arguments they are engaged in for signs that changes are needed to avoid a morally undesirable development of the argument.

For example, Blair [3] argues that we can legitimately ask moral questions about argumentation because argumentation and its outcomes have effects on people's wellbeing: Acquiring true beliefs or making just decisions on the basis of arguments that reach correct conclusions *prima facie* leads to morally relevant goods. And the normative goals Walton ascribes to his various dialogue types are goals that may be considered *prima facie* morally desirable: The reasonable resolution of a difference of opinion is desirable because it generates valuable epistemic good, as does the transference of reliable information and the

---

[6] I should point out that I do not share Walton's worry about integrating intention into theorizing about argument. However, I take it to be important to follow his methodological commitments in a paper meant to expand his theory rather than develop my own.

[7] Still, for those who are willing to take intentions into account, Walton's theory of fallacy has eliminated a pitfall in assigning blame and praise to argumentative behavior — namely the temptation to suspect people of being deceptive or bullies on the basis of the argument types they use alone.

accomplishment of definite proofs or disproofs. And if a decision must be made, one that takes all the goals and reasons of the participants into account or that balances their interests equitable is more just than one that does not. Fallacies can then be diagnosed as morally relevant problems in argumentation by concentrating on their consequences: They make it more likely that the outcome of arguments impact the practical and epistemic well-being of arguers (and others) negatively by leading to unjust decisions in deliberations, untrue conclusions in inquiries or imbalanced compromises in negotiations. A reliable method for the detection of fallacies automatically becomes a method for determining, at least *prima facie*, whether the behavior of arguers during an argument is morally desirable or not.

3.2 The Choice of Dialogue Type

I said that Walton's method for the detection of fallacies can only lead to *prima facie* judgements on the moral desirability of argumentative behavior. Obviously, whether it is morally undesirable to commit a fallacy and obstruct a dialogue-goal will depend heavily on the context in which the arguers engage in the dialogue. If arguers engage in deliberation to determine how best to set off an explosion that will eliminate all life on earth, obstructing the dialogue's goal is the morally desirable thing to do.

But it is not only the moral acceptability of the use to which an argument2's outcome will be put that is relevant to its moral evaluation. The choice of dialogue-type, too, is relevant and can be evaluated as desirable or undesirable from a moral point of view by determining its impact on the attainment of morally relevant goals. I will assume that it is morally desirable that arguments will not result in avoidable epistemic or unjust practical losses, and that the way the argument proceeds is respectful of the arguers' status as reasonable, dignified beings.

Coerced Choices of Dialogue Type

As I said above, Walton [32] places the choice of the dialogue type in the opening stage of argumentative dialogues. He attaches some importance to the idea that arguers should agree to the type of dialogue they will engage in and should be aware of its implications and associated rules. However, it is important for the moral evaluation of arguments to realize that this cannot be taken for granted. It is not possible to simply assume that arguers choose the type of dialogue they engage in with a full awareness of the available options and the argumentative burdens that will be placed on them. Even the assumption that they make any kind of reflective or consensual choice may often be too much. Jacobs [14] points out that such things are usually determined more or less in the flow of the argument, simply through the way in which arguers present themselves. And I [20] argue that arguers determine which dialogue type they will engage in through similar means as they may later use to shift dialogue types: They simply start arguing according to the norms of a particular argumentative role and see if their interlocutor accepts the complimentary role. Indeed, Walton [32, p. 9] is aware that agreements about dialogue types and applicable rules will not always be made explicitly and remarks that they are often simply replaced by normal expectations of custom and politeness. But this is not the same as freely given consent. And power differences between arguers can have a considerable impact on who gets to determine what the argument will look like: If one arguer is positioned in the better social, economic or professional position and chooses an argumentative role (like the opponent role of a persuasion dialogue), her interlocutor might find herself unable to refuse the associated dialogue type and her role in it (compare [20]). This might be because the interlocutor has to fear that questioning the dialogue type could lead to face- or relationship losses. It could also lead to a refusal from the better situated arguer to continue an argument about an issue that the interlocutor desperately needs resolved.

This is itself morally objectionable not only because coercion is *prima facie* objectionable, but also because the dialogue type through which an issue is handled has an impact on the way the issue will be resolved. Whether John and Jane determine the dog-breed of their new pet through a negotiation or a deliberation might well change the outcome. So if the choice of dialogue type is not consensual but, to some degree, coerced, then the result of the argument retains a coerced element, which is at least *prima facie* morally undesirable.

But this is not all. Feminist argumentation theorists have shown how some arguers, in some context, will be unable to give their reasons effect in certain dialogue types because they are unable to gainfully play the associated roles [12, 19, 20]. For example, persuasion dialogues require of their proponents to meet the burden of proof: It is up to them alone to defend their position and persuade their opponent. But arguers who are emotional, not fully and self-reflectively aware of their position, or suffer from hermeneutical lacunae [4] may not be able to shoulder that burden. Forced into a persuasion dialogue, they might find themselves unable to influence the argument's outcome because they cannot play their role.

Such a situation is doubly problematic: It is potentially harmful because the disadvantaged arguer (or others) might suffer epistemic losses or unjust practical disadvantages because of her inability to make her reasons heard. And it is humiliating — that is, it threatens the disadvantaged arguer's dignity. I have argued in [21] that this is so because argument (except maybe for eristic dialogues) is at its basis a cooperative activity centered around the exchange of reasons.[8] Inviting another person to argue about an issue means inviting them at least purportedly into an activity during which the issue will be dealt with through the identification, integration and evaluation of reasons, *their* reasons included. It means at least purporting to address them in their status as a dignified being capable of having, understanding and communicating reasons. After all, good faith argumentation is dignity-affirming exactly because arguing with someone means expressing one's recognition of their dignified status as a reasonable being.[9] However, forcing an arguer into a dialogue type the role of which they cannot effectively play, so that they cannot make their reasons heard, turns this purportedly respectful activity into a humiliating farce. It withdraws the respect that was seemingly offered without, at the same time, withdrawing the associated demands to show the same kind of respect for the other arguer. The resulting humiliation also has a psychologically and emotionally harmful effect because it makes the affected ar-

---

[8] In this I follow Govier [9] and am in agreement with Walton e.g. in [27].

[9] Compare, for example, literature arguing for the dignity affirming effects of deliberative instead of exclusively voting-based representative democracy (e.g. [10, 11]).

guers realize that they are not being treated as valuable individuals. Readers might be able to recall the feeling of frustration and exclusion when they noticed that the *way* in which an argument proceeded made it impossible for them to be heard.

Consensual Choices of Dialogue Type

Even where there is no reason to suspect that the choice of dialogue was coerced, the choices associated with the opening stage can be subject to moral evaluation. The reason is that they can have an impact on the argument's outcome and thereby on the well-being of the arguers and others.

Take again John and Jane's decision about a family dog. As I mentioned above, the decision could be made via a negotiation, a deliberation or a persuasion dialogue in which either Jane or John aims to persuade the respective other of their preferred dog-choice. But depending on the context in which the choice is made, even an equitable compromise between Jane and John — the goal of a negotiation — might not be preferable to the outcome of a well-executed deliberation. The kinds of reasons that will be given weight in the respective dialogues are different: The deliberation aims at taking into account all reasons relevant to the best dog for the family (including, e.g. reasons pertaining to the children and the annoying aunt living with them). By contrast, the negotiation will integrate these reasons only if they can be translated into interests of the negotiating parties.[10] This will have an impact on the argument's outcome and, thereby, an impact on how this outcome will interact with the well-being of the arguers and others.

Additionally, the dialogue type may negatively impact the ability of one or both arguers to make their reasons heard even if they have consented to it. After all, we are not always fully aware of our rhetorical limitations or the boundaries of our understanding of our own position. So arguers may happily enter a dialogue type but find that they cannot fulfill the associated argumentative burdens after all. As a result, the reasons they may have been able to contribute in another dialogue type will not have the appropriate impact. But in some situations, it is important in morally relevant ways that the reasons of all arguers are fully taken into account. For example, Jane's and John's well-being will be impacted by the final decision about the family dog. Choosing a dialogue-type in which one of them cannot effectively play their role runs the risk of undervaluing their reasons, such as Jane's reasons for rejecting long-haired dogs. If this is so, then a persuasion dialogue, for example, may not be appropriate.

Of course, evaluating the choice of dialogue type with an eye to the morally relevant goals of avoiding epistemic and unjust practical losses and humiliation requires a lot: It takes background knowledge about the arguers and the context in which the argument arises to determine whether the choice of dialogue type may be detrimental or coerced. And a close observation of the arguers is necessary to determine whether an arguer is struggling to give her reasons effect. However, this simply continues a familiar theme in Waltonian argument

---

[10]Godden and Casey [7] argue against van Laar and Krabbe [23] that the outcome of a negotiation cannot functionally replace the outcome of another dialogue type like a persuasion or deliberation dialogue.

evaluation — namely that the context of the argument is hugely important for its evaluation. It does not diminish that Walton has provided necessary, medium-fine grained tools that can be adapted to determine methodically and express relatively precisely why an argument2, in virtue of the type of dialogue it is and the way in which this type was chosen, may be morally undesirable. He has provided these tools by distinguishing between his six dialogue types, the associated argumentative roles and the argumentative burdens they do or do not come with.

3.3 Dialogue Shifts

If the choice of dialogue type is important from a moral perspective, then it is not surprising that dialogue shifts are important in this way too, and for all the reasons discussed above. As I described above, Atkinson, Bench-Capon and Walton [1] mention that arguers can trigger shifts in dialogue simply by taking a role associated with their target dialogue. Thereby, they can change the argumentative burdens that rest on arguers, the goal which the argumentative sequence can accomplish and the types of argument that are taken to be permissible to use. Alternatively, arguers can use meta-dialogues to propose and agree on a shift in dialogue, potentially simply by asking for help in formulating their reasons or by suggesting to find a compromise.[11] Dialogue shifts can be illicit, permissible or required from a moral point of view. Here I will discuss illicit and required dialogue shifts.

Illicit Dialogue Shifts

As I explain above, I do not believe that arguers often explicitly agree on the dialogue type or the norms that will govern their argument, at least not in normal, every-day arguments. Nor do I believe that arguers are explicitly or implicitly aware of norms that are as specific and complex as the norms Walton and others have attributed to dialogue types in their more detailed treatments.[12] Nonetheless, I do believe that arguers come to an understanding about the general kind of argument they are having during the first few moves of the argument. And I think that the six argument types are a good rough estimation of the kinds of arguments real-life arguers a socialized to tell apart. Everyday arguers associate at least roughly different norms with an argument in which they are bargaining in order to come to a compromise than with an argument in which they all chip in proposals and reasons for or against them in order to solve a common problem.

Once arguers have determined the dialogue type for their argument and distributed the argumentative roles, they can be expected to rely on the assumption that the norms associated with the argument type govern the argument. They invest their cognitive energy and time into fulfilling their argumentative roles and meeting their argumentative burdens under the assumption that the issue of the argument will be resolved via the chosen dialogue type: An arguer who assumes they are negotiating for a compromise might not concentrate much energy on justice-related arguments. The associated reliance reasons speak against

---

[11]See Krabbe [15] and Walton [31] for discussions of meta-dialogues.
[12]See Footnote 3.

shifting dialogues without a justifying reason, and dialogue shifts that are not necessary or at least conducive to the argumentative goal can be presumed to be illicit from a moral point of view.

Relatedly, arguers will also assume that, if they follow the norms of their dialogue type and reach a result, then this result will have the kind of validity that the dialogue can provide it with: An outcome reached through a well-conducted negotiation counts as a mutually acceptable compromise, one reached through a well-conducted inquiry counts as a proof or disproof of the hypothesis in question. Dialogue shifts that go unnoticed can therefore be assumed, at least *prima facie*, to be undesirable from a moral point of view because they lead arguers to accept a kind of validity that is not actually warranted. This creates risks of epistemic and unjust practical losses. Take, for example, this unnoticed shift from a negotiation to a persuasion dialogue:

Negotiation	Jane: I do not want to do all the additional housework associated with a long-haired, shedding dog. I am willing to consider one if you agree to de-hairing the couches and chairs every week and to do any additional sweeping
	John: That is basically all the additional work associated with the shedding. I will do it every second week and I will take the dog for a walk every day if you allow me to make some exceptions if I have something special planned.
	Jane: No deal. I do not want anything to do with those hairs. I am even willing to take over bringing the garbage out if you just take care of the hair.
Shift into Persuasion	John: You clearly want a beautiful dog too, otherwise you would not even think about it. So I do not see what this discussion is even about. You should be looking forward to having a beautiful dog.
	Jane: Shedding will make my life much harder if I have to do so much additional housework.
	John: But you will get the enjoyment of the dog out of it. So I do not see the big deal. If I walk the dog every day, that is a bonus for you.

What John is doing here is trying to move into the opponent position in a persuasion dialogue instead of maintaining the negotiation. He attempts to use Jane's willingness to think about a long-haired dog in exchange for concessions as a commitment to wanting a long-haired dog for the purposes of a persuasion. If he succeeds, the result will be that Jane has to shoulder the burden of persuading him that he should take care of the dog-hair while accepting the premise that they both want a long-haired dog. If she fails to do so, he may try to sell the outcome of him getting his long-haired dog in exchange for doing most of the walks as a fair compromise, which would presumably be an unjust practical outcome.

Apart from these two ways in which dialogue-shifts can be undesirable from a moral point of view, they can also be undesirable for the reasons why an initial dialogue choice can be problematic: The new dialogue type may make it hard for the shifting arguer's interlocutor to give their reasons effect, a problem that is compounded if the shift happens in a coerced way.

Required Dialogue Shifts

Walton (e.g. [29]) claims that dialogue shifts should be evaluated mainly with an eye to whether the new dialogue contributes to accomplishing the goal of the original dialogue. However, I believe that this is only the case if the original choice of dialogue was desirable from a moral point of view. If it was undesirable from the start, or if new information shows that it is undesirable after all, shifts to other dialogues that change the argument2's goal instead of supporting the original goal can be permissible, or even required.[13]

Arguers might, for example, begin a deliberation dialogue to make a practical decision. However, in the course of the argument it might turn out that one of the arguers was wedded to a certain decision from the start. Her important goals outside the argument and her emotional attachment to the decision might make it impossible for her to argue with the degree of cooperativeness required in a deliberation. It might then be necessary to shift to a persuasion dialogue to ensure that all arguers will have equal opportunities to give their reasons effect in the argument.

Think back to Jane's behavior in the example in which she performed an illicit shift to a persuasion dialogue:

Deliberation	John: We still have to decide which dog would fit best into our family. There are Collie puppies, Jack Russel puppies and Schnauzer puppies available. Jane: I think a dog that can handle a long walk would be good. We love taking hikes. John: I also like to be lazy sometimes, but I agree that we like to hike, and if this is a high priority for you, I think it is best to decide between the Schnauzers and the Collies. Jane: And I do not want a dog that sheds. John: Oh, but Collies are so beautiful and have medium energy, and they would be good for the kid, they are so gentle.
Shift into Persuasion	Jane: Can you show me that Collies do not shed? John: I admit that they shed a little. But I really appreciate a beautiful dog, and the kid needs a really gentle dog. Jane: Well, if they shed, then they are out, they are not the dog for us. John: I guess so.

Jane might have acted in this way because she is committed to not being saddled with an even larger share of the housework than she is currently doing. She might be deeply emotionally affected by her perception that she is being unjustly disadvantaged in the distribution of housework. As a result, it might be too much to ask of her to participate in a

---

[13]Compare Godden and Casey [7], who discuss the "retrospective evaluation" criterion and its exceptions.

deliberation about the importance of finding a dog that does not shed. This might become manifest in her body-language or the way she formulates her contributions. Her illicit and covert shift is morally problematic because its result may be that the arguers mistake the result of their argument for the outcome of a deliberation. Nonetheless, it might be similarly morally problematic for the group to continue with a deliberation. The necessary course of action may be to shift to a persuasion dialogue and give up on the idea that the decision can be made based on a deliberation's result.

Importantly, this shift to a persuasion dialogue might or might not be permanent. It is possible that the persuasion dialogue about the importance of shedding in dogs could be embedded into the deliberation dialogue. Walton et al. [37] recognize that deliberation dialogues might sometimes profit from the embedding of persuasion dialogues when arguers find themselves committed to certain proposals for action. In situations like the one described above, such embeddings might be necessary to give the deliberation a chance at reaching its goal while also ensuring that all arguers have an equal opportunity to give their reasons effect. Refusing such shifts may be problematic from a moral perspective because it increases the risk of the argument's outcome generating epistemic or unjust practical losses and because it denies an arguer the chance of being heard.

Above, I introduced the idea of necessary dialogue shifts and the illicit refusal of dialogue shifts by using the example of embedded information dialogues. I described a scenario in which John, secretly preferring a Golden Retriever, refuses a shift to an information seeking dialogue to prevent information about Golden Retrievers' energy levels to influence the family's decision. Such behavior is *prima facie* problematic from a moral point of view for the same reason that fallacies are. It obstructs the accomplishment of the dialogue's goal and does so potentially covertly, thereby increasing the risk of epistemic losses and bad decisions.

However, there is a variant of needing to embed an information seeking dialogue where the information that is needed concerns the commitments, beliefs, goals, feelings or experiences of some of the arguers themselves. This variant, I believe, is even more interesting when evaluating argumentation from a moral perspective, so I will take some time to explore it.

Walton's dialectical theory contains the idea that argumentative dialogues may have a maieutic function because they may bring commitments of the arguers to light that either their interlocutors or even themselves were not aware of (see, e.g. [27]). Walton explains that one of the ways in which argumentative dialogues can create maieutic results is by challenging arguers to justify their reflectively known commitments and prompting them to explore their own systems of commitment: If John challenges Jane's claim that long-haired dogs are repulsive, her attempt to justify herself via her commitment to not wanting to clean more may lead her to realize that she believes that her share of housework is unfair. This is one reason why a shift from a deliberation to a persuasion dialogue might be helpful. But the needed maieutic effects might not always be accomplishable simply by challenging the arguer to justify her claims and then leaving her to explore her position alone.

While Walton is mostly interested in the maieutic function in so far as it can bring commitments of propositional content to light, Gilbert [5, 6] takes this idea further. He replaces the term "dark side commitments" that Walton uses to denote unknown aspects of an arguer's position with "dark side- items". This is so because Gilbert believes that an arguer's position is not only propositional, but also connected to emotional predispositions, strong intuitions and the implications of one's socio-economic standing and the accompanying viewpoints. This makes argumentative positions very complex. Gilbert places a lot of emphasis on the problem that arguers may not always be fully aware of their own goals, or they might be unable to formulate them precisely enough for others to understand. Arguers might also misunderstand themselves or commit to goals that, once they understand their own positions more fully, would not be exactly the goals they would want to commit to. Finally, though Gilbert does not use this term, arguers might suffer from hermeneutical lacunae (whether they are culturally shared or private), so they might not be able to fully grasp their own experiences or feelings. Gilbert suggests that argument is unlikely to be fruitful if arguers do not understand each other. Therefore the first thing arguers should do in an argument is explore and try to understand their own and each other's positions. In other words, Gilbert believes that the success of an argument relies on producing Walton's *maieutic* outcomes of argumentative dialogues. These maieutic outcomes, however, are not only achieved by arguers requiring justification from each other. Instead, Gilbert advocates a cooperative exploration of the arguers' commitments in order to determine how their respective claims are connected to all aspects of their epistemic systems, personalities and experiences.

I do not want to follow Gilbert quite so far as to say that such common exploration is needed in all arguments. But I do think that sometimes it may be necessary for an interlocutor to assist the arguer's self-exploration through questions, suggestions and co-construction of explanations and justifications. In Waltonian terms, I think this would amount to a sub-type of the information seeking dialogue, where the information that the interlocutor is trying to gain is information about the arguer's position. Embedding such information seeking dialogues regarding aspects of an arguer's position may become necessary whenever accomplishing the goal of a dialogue requires knowledge about those aspects of the arguer's position. For example:

Deliberation	John: We still have to decide which dog would fit best into our family. Jane: I really do not want a dog that will shed a lot. I hate hairy beasts. John: Sure, shedding is a minus. But I think the dog's character and personality is more important. Jane: Yes, those are important. But even the nicest shedding dog is a no-go. It is just disgusting to me. John: I do not know whether I understand where your priorities come from. Jane: They are what they are!

Information Seeking	John: Let's talk about this. What is it about hairy dogs that is so bad? Jane: Its just disgusting, their hair everywhere. You will see it on our dark floors. John: Couldn't that be solved by sweeping a little more often? Jane: Of course that is your answer! John: You seem upset at the thought. Jane: Of course I am upset. I am already cleaning way more than my fair share! Getting a hairy dog will just increase my workload. John: Oh, I understand now.
Deliberation	Jane: It is very important to me that the dog does not increase my workload too much. John: Well, how much dogs shed is not only associated with the length of their hair. We should look into this more. (...)

Importantly, such information seeking dialogues may not only be necessary to preserve the dialogue's chances at accomplishing its goals, but also to give an arguer a chance to give their reasons effect in the argument. As such, they might be required to avoid exposing the arguer to humiliation or exclusion. Such shifts may therefore be required from a moral perspective for reasons of harm-avoidance as well as reasons of avoidance of humiliation. And, if such information seeking dialogues exploring an arguer's position can be necessary, then it is also possible to illicitly refuse such a shift.

Importantly, it is surely not acceptable to require arguers to share their innermost feelings in any dialogue that would profit from it. Nor can we expect that arguers will always be able to invest the time and energy that is required to bring the aspects of a position to light that are so far kept in the dark. As before, whether an information seeking dialogue of this kind is required will be very context dependent. Nonetheless, I do think that we can identify instances where the refusal of embedding this kind of information seeking dialogue is not justified by important time- or resource-constraint. It may then be illicit both because it hinders the dialogue in reaching its goal and because it threatens an arguer's ability to give her reasons effect.

4 Conclusion

In this paper, I have argued that the dialectical tools Walton has created can also be used to evaluate arguments from a moral point of view. This can be done without having to inquire into the intentions of arguers. It is enough to determine whether the argumentative behavior on display hinders the morally relevant goals of avoiding epistemic and unjust practical losses and of respecting the dignity of arguers as reasonable beings. However, if the goal is to assign praise or blame, then evaluating the intentions of individual arguers might become necessary. This is an additional step that can be helpful in some circumstances but is not necessary to conduct any moral evaluation of argumentative behavior at all.

References

[1] Atkinson, K., Bench-Capon, T., & Walton, D. (2013). Distinctive features of persuasion and deliberation dialogues. *Argument & Computation*, 4(2), 105-127.

[2] Birrer, F. A. J. (2001). Expert Advice and Argumentation: Some Remarks on the Work of Douglas Walton. *Argumentation*, 15(3), 267-276.

[3] Blair, A. J. (2011). The Moral Normativity of Argumentation. *Cogency*, 3(1), 13-32.

[4] Fricker, M. (2007). *Epistemic injustice: power and the ethics of knowing*. Oxford: Oxford University Press.

[5] Gilbert, M. A. (1995). Coalescent Argumentation. *Argumentation*, 9, 837-852.

[6] Gilbert, M. A. (1997). *Coalescent Argumentation*. Mahwah, New Jersey: Lawrence Erlbaum Associates.

[7] Godden, D., & Casey, J. (2020). No Place for Compromise: Resisting the Shift to Negotiation. *Argumentation*, online first.

[8] Goodwin, J. (2001). One Question, Two Answers. *OSSA Conference Archive*. doi:https://scholar.uwindsor.ca/ossaarchive/OSSA4/papersandcommentaries/40?utm_source=scholar.uwindsor.ca%2Fossaarchive%2FOSSA4%2Fpapersandcommentaries%2F40&utm_medium=PDF&utm_campaign=PDFCoverPages

[9] Govier, T. (1987). A New Approach to Charity. In *Problems in Argument Evaluation and Analysis* (pp. 133-158). Dordrecht: Foris.

[10] Gutmann, A., & Thompson, D. F. (2004). *Why deliberative democracy?* (STU - Student ed.). Princeton: Princeton University Press.

[11] Habermas, J. (1994). Three normative models of democracy. *Constellations*, 1(1), 1-10.

[12] Hundleby, C. (2013). Aggression, Politeness, and Abstract Adversaries. *Informal Logic*, 33(2), 238-262.

[13] Jackson, S. (2015). Design Thinking in Argumentation Theory and Practice. *Argumentation*, 29(3), 243-263.

[14] Jacobs, S. (2017). *On how to do without the opening stage: arguers and argumentation theorists*. Paper presented at the 2nd European Conference on Argumentation: Argumentation and Inference, Fribourg, Switzerland.

[15] Krabbe, E. C. W. (2003). Metadialogues. In F. H. van Eemeren, A. J. Blair, C. A. Willard, & A. F. Snoek Henkemans (Eds.), *Anyone who has a view: Theoretical contributions to the study of argumentation* (pp. 83-90). Dordrecht: Kluwer Academic Publishers.

[16] Krabbe, E. C. W. (2007). On How to Get Beyond the Opening Stage. *Argumentation*, 21(3), 233-242.

[17] McBurney, P., Hitchcock, D., & Parsons, S. (2007). The eightfold way of deliberation dialogue. *International journal of intelligent systems*, 22(1), 95-132.

[18] O'Keefe, D. J. (1977). Two Concepts of Argument. *Argumentation and Advocacy*, 13(3), 121-128.

[19] Rooney, P. (2010). Philosophy, Adversarial Argumentation, and Embattled Reason. *Informal Logic*, 30(3), 203-234.

[20] Stevens, K. (2019). The Roles We Make Others Take: Thoughts on the Ethics of Arguing. *Topoi* 38, 693-709.

[21] Stevens, K. (Forthcoming). Charity for Moral Reasons?

[22] van Eemeren, F. H. (2018). *Argumentation Theory: A Pragma-Dialectical Perspective* (Vol. 33). Cham: Springer.

[23] van Laar, J. A., & Krabbe, E. C. W. (2017). Splitting a Difference of Opinion: The Shift to Negotiation. *Argumentation*, 32(3), 329-350.

[24] van Laar, J. A., & Krabbe, E. C. W. (2018). The Role of Argument in Negotiation. *Argumentation*, 32(4), 549-567.

[25] Walton, D. (1989). Dialogue theory for critical thinking. *Argumentation*, 3(2), 169-184.

[26] Walton, D. (1992). Types of Dialogue, Dialectical Shifts and Fallacies. In F. H. Van Eemeren, R. Grootendorst, A. J. Blair, & C. A. Willard (Eds.), *Argumentation Illuminated* (pp. 133-147). Amsterdam: SICSAT.

[27] Walton, D. (1995). *A pragmatic theory of fallacy*. London [England];Tuscaloosa: University of Alabama Press.

[28] Walton, D. (1997). *Appeal to expert opinion: arguments from authority*. University Park, Pa: Pennsylvania State University Press.

[29] Walton, D. (1998). *The new dialectic - Conversational contexts of argument*. Toronto: University of Toronto Press.

[30] Walton, D. (2000). The Place of Dialogue Theory in Logic, Computer Science and Communication Studies. *Synthese* (Dordrecht), 123(3), 327-346.

[31] Walton, D. (2007). Metadialogues for Resolving Burden of Proof Disputes. *Argumentation*, 21(3), 291-316.

[32] Walton, D. (2008). *Informal logic - A pragmatic approach* (2nd ed.). New York: Cambridge University Press.

[33] Walton, D. (2010). *Types of Dialogue and Burden of Proof*. Paper presented at the Computational Models of Argument Conference, Desenzano del Garda, Italy.

[34] Walton, D. (2015). Profiles of Dialogue: A Method of Argument Fault Diagnosis and Repair. *Argumentation and Advocacy*, 52(2), 91-108.

[35] Walton, D. (2016). *Knowledge and Inquiry. In Argument Evaluation and Evidence*. Cham: Springer.

[36] Walton, D., & Krabbe, E. C. W. (1995). *Commitment in dialogue - Basic concepts of interpersonal reasoning*. Albany: State University of New York Press.

[37] Walton, D., Toniolo, A., & Norman, T. J. (2014). Missing Phases of Deliberation Dialogue for Real Applications. *Proceedings of the 11th International Workshop on Argumentation in Multi-Agent Systems*, 1-20.

[38] Walton, D. & Macagno, F. (2007). The Fallaciousness of Threats: Character and Ad Baculum. *Argumentation*, 21(1), 63-81.

Dialogue Types, Argumentation Schemes, and Mathematical Practice: Douglas Walton and Mathematics

Andrew Aberdein
School of Arts & Communication, Florida Institute of Technology, Melbourne FL
aberdein@fit.edu

Abstract

Douglas Walton's multitudinous contributions to the study of argumentation seldom, if ever, directly engage with argumentation in mathematics. Nonetheless, several of the innovations with which he is most closely associated lend themselves to improving our understanding of mathematical arguments. I concentrate on two such innovations: dialogue types (§1) and argumentation schemes (§2). I argue that both devices are much more applicable to mathematical reasoning than may be commonly supposed.

1 Dialogue Types

Several decades ago, Douglas Walton proposed a classification of dialogue types: different contexts in which argumentation may arise [67, 68]. His elegant presentation of the key differences between the most central types (from joint work with Erik C. W. Krabbe) is summarized in Table 1. Dialogue types are distinguished by two main factors: the *initial situation* or circumstances in which the interlocutors find themselves and their *main goal* in pursuing a dialogue. Some situations admit more goals than others: if the situation is strongly adversarial, the disputants may be seeking a full determination of the matter at hand, requiring one to *persuade* the other; or they may need to decide on a course of action and *negotiate* a practical consensus; or they may have little intent beyond airing their respective positions, however quarrelsome or *eristic* the exchange. Whereas, if the interlocutors are addressing an open problem where neither has any prior commitments, the last of these goals would be incoherent, but the disputants may still *inquire* into the problem or *deliberate* on how best to act. And if the interaction arises simply because one party has knowledge the other lacks, only the first sort of outcome makes any sense: an

	Initial Situation		
Main Goal	Conflict	Open problem	Unsatisfactory spread of information
Stable agreement/ resolution	*Persuasion*	*Inquiry*	*Information Seeking*
Practical settlement/ decision (not) to act	*Negotiation*	*Deliberation*	N/A
Reaching a (provisional) accommodation	*Eristic*	N/A	N/A

Table 1: Walton and Krabbe's systematic survey of dialogue types [73, p. 80]

information-seeking dialogue. Thus we arrive at six principal dialogue types. However, some of these types may be further subdivided or combined [72, p. 31], and the classification is not intended to be exhaustive.

Walton itemizes the goals of the interlocutors, individual or collective, and the potential benefits that may accrue from dialogues of each of the main types in Table 2 (taken from [71, p. 605]; see also [68, p. 413], [73, p. 66]). Different patterns of argument may be appropriate in different dialogue types: what is reasonable in a negotiation would be improper in a persuasion dialogue; almost anything goes in a quarrel but well-conducted inquiries require respect for procedure, and so forth. Another important feature of the picture that Walton and Krabbe present is the *dialectical shift*: in the course of a dialogue its type may change [73, pp. 100 ff.]. This can be a positive development—as a conversation unfolds, its participants can productively shift their attention to different ends. But dialectical shifts can also be troublesome, especially if they go unnoticed by one or more of the participants, leading to the use of argumentative tactics that are now contextually inappropriate.

Walton does not discuss mathematical dialogues but, in other work, his collaborator Krabbe observes that proofs may occur in several different contexts:

1. thinking up a proof to convince oneself of the truth of some theorem;
2. thinking up a proof in dialogue with other people (inquiry dialogue...);
3. presenting a proof to one's fellow discussants in an inquiry dialogue (persuasion dialogue embedded in inquiry dialogue...);
4. presenting a proof to other mathematicians, e.g. by publishing it

Type of Dialogue	Initial Situation	Individual Goals of Participants	Collective Goal of Dialogue	Benefits
Persuasion	Difference of opinion	Persuade other party	Resolve difference of opinion	Understand positions
Inquiry	Ignorance	Contribute findings	Prove or disprove conjecture	Obtain knowledge
Deliberation	Contemplation of future consequences	Promote personal goals	Act on a thoughtful basis	Formulate personal priorities
Negotiation	Conflict of interest	Maximize gains (self-interest)	Settlement (without undue inequity)	Harmony
Information-Seeking	One party lacks information	Obtain information	Transfer of knowledge	Help in goal activity
Quarrel (Eristic)	Personal conflict	Verbally hit out at and humiliate opponent	Reveal deeper conflict	Vent emotions
Debate	Adversarial	Persuade third party	Air strongest arguments for both sides	Spread information
Pedagogical	Ignorance of one party	Teaching and learning	Transfer of knowledge	Reserve transfer

Table 2: Walton's types of dialogue [71, p. 605]

in a journal (persuasion dialogue...);
5. presenting a proof when teaching (information-seeking and persuasion dialogue) [44, p. 457].

The primary, if not exclusive, concern of Krabbe's account (and of my own earlier application of dialogue types to mathematics [2]) is with proof. This may reflect what has been called in another context "proof chauvinism"—a tendency in philosophers of mathematics to privilege proof over other aspects of mathematical practice [19]. Nonetheless, proof is the aspect of mathematical practice where the applicability of informal logic is most unexpected. Hence I shall again begin with proofs.

Are proofs always dialogues or can they be monologues? The conception of the mathematician as isolated genius has a firm grip on the popular imagination [37]. It is true that mathematicians coauthor papers less than most other scientists, and there are some celebrated examples of solitary endeavour, such as Srinavasa Ramanujan labouring in obscurity or Andrew Wiles's years of solo work prior to his surprise announcement of a proof of Fermat's Last Theorem. Nonetheless, this impression is incomplete at best: Ramanujan only began to fulfil his potential after travelling to Cambridge to collaborate with Hardy and Littlewood [42]; Wiles discovered gaps in his solo work which were eventually bridged by a collaboration with Richard Taylor [53]. On the other hand, as Paul Ernest suggests, there are many ways in which mathematics is underpinned by "symbolically mediated exchanges between persons"—conversations or dialogues:

> First, the ancient origins as well as various modern systems of proof use dialectical or dialogical reasoning, involving the persuasion of others [see also [23].] ... Second, mathematics is a symbolic activity using written inscriptions and language; it inevitably addresses a reader, real or imagined, so mathematical knowledge representations are conversational. Third, many mathematical concepts [such as epsilon-delta definitions of limit in analysis and hypothesis testing in statistics] have an internal conversational structure. Fourth, the epistemological foundations of mathematical knowledge, including the nature and mechanisms of mathematical knowledge genesis and warranting, utilise the deployment of conversation in an explicitly and constitutively dialectical way. Fifth, ... mathematical facts stand on the basis of collective agreement and are part of institutional reality ... built on interpersonal communicative interactions, that is, through conversation [25, p. 74].

Once we agree that mathematical proof is dialogical, we may ask in what dialogue type it characteristically arises. As Krabbe indicates, the proving process, at least at its inception, might best be thought of as an inquiry dialogue: a collaborative exchange between mathematicians with the shared goal of settling an open question, which neither of them has prejudged. Certainly such exchanges can be found in mathematics, at least in the context of discovery of mathematical results (see, for example, [66]). However, there is also an unavoidable element of adversariality in the epistemology of mathematical proof: mathematicians only trust proofs that have gained wide assent from the mathematical audience [14]; the value of that assent lies in the assumption that the proofs have been sufficiently challenged. Catarina Dutilh Novaes has sought to capture this idea in terms of prover/sceptic dialogues [21, 22, 23]. Prover and sceptic are (idealizations of) two complementary roles in the process that leads to the eventual acceptance of a proof by the mathematical public: the prover presents a putative proof; the sceptic responds with searching but fair questions; through their successive exchanges the proof is improved where necessary and eventually comes to be generally accepted (or is exposed as unsound). Prover/sceptic dialogues are persuasion dialogues because the parties start from a difference of opinion, as required by their contrasting roles.

Journal referees can play the sceptic role, at least if they are sufficiently thorough in their scrutiny [10]. But so can collaborators, at an earlier stage in the development of a proof, or other mathematicians, at a later stage, who expand or refine the published proof in their own work. Imre Lakatos's celebrated imaginative reconstruction of the development of a proof of the Descartes–Euler conjecture (linking the numbers of vertices, edges, and faces of polyhedra) takes the form of a dialogue between characters loosely representing various nineteenth-century mathematicians [47]. Lakatos identifies a range of dialectical manoeuvres whereby mathematicians either present apparent counterexamples of various kinds to a working conjecture or respond to such apparent counterexamples. (Alison Pease and colleagues have shown how these Lakatosian manoeuvres can be captured in terms of dialogue games [55].)

But perhaps the dialogue type in which proofs are most frequently presented is neither inquiry nor persuasion, but pedagogical information-seeking. Proofs are presented in countless classrooms at school and university level and even research mathematicians will attend essentially didactic presentations of novel but settled results. Such exchanges are best understood as information-seeking dialogues. Hence, as Krabbe notes, the development of a proof may be seen as a sequence of dialectical shifts, from an initial inquiry phase, to a more verification-focussed persuasion dialogue, and eventually, if the proof survives these earlier stages, to a dissemination phase characterized by information-seeking dialogues. Of course, the progress of most significant proofs is seldom this smooth, so the dialectical shifts are likely to be more numerous, as failed attempts at verification send mathematicians back to more open-ended inquiry, or at least open up subsidiary discussions of how localized problems may be addressed. Michael Barany and Donald MacKenzie, in an ethnographic treatment of mathematical research, describe how some of these processes can work:

> When a suitable partial result is obtained and researchers are confident in the theoretical soundness of their work, they transition to "writing up".

Only then do most of the formalisms associated with official mathematics emerge, often with frustrating difficulty. Every researcher interviewed had stories about conclusions that either had come apart in the attempt to formalize them or had been found in error even after the paper had been drafted, submitted, or accepted. Most saw writing-up as a process of verification as much as of presentation, even though they viewed the mathematical effort of writing-up as predominantly "technical", and thus implicitly not an obstacle to the result's ultimate correctness or insightfulness [15, pp. 111 f.].

Although inquiry, persuasion, and information-seeking dialogues are perhaps the dialogue types most hospitable to proof, they do not exhaust the range of dialogue types in which mathematical argumentation may occur. By analogy with the device used to indicate problematic sporting records, I have elsewhere used "proof*" to refer to "species of alleged 'proof', where there is either no consensus that the method provides proof, or there is broad consensus that it doesn't, but a vocal minority or an historical precedent which points the other way" [3, p. 2]. Amongst the proofs* I included "proofs* predating modern standards of rigour, picture proofs*, probabilistic proofs*, computer-assisted proofs*, textbook proofs* which are didactically useful but would not satisfy an expert practitioner, and proofs* from neighbouring disciplines with different standards". Each of these cases can be seen two ways: either as a (perhaps very) disputed form of mathematical proof or as an undisputed form of mathematical reasoning that ought to be characterized as something other than proof. Hence, if our goal is to repudiate proof chauvinism and characterize mathematical reasoning in general, then we must pay attention to proofs*.

Table 3, adapted from [2, p. 148], summarizes the principal mathematical dialogue types discussed so far: inquiry, persuasion, and pedagogical information-seeking. It also lists three dialogue types in which proof* is likely to be more at home than proof: deliberation, negotiation, and a non-pedagogical form of information-seeking. Deliberation and negotiation abandon the goal of stable resolution that we would normally expect of proof whereas oracular information-seeking pursues that goal in an unconventional manner. In one of his foundational papers on computability, Alan Turing briefly considers the case of a machine "supplied with some unspecified means of solving number-theoretic problems; a kind of oracle as it were" [64, p. 172]. Subsequent authors expanded this remark into a theory of relative computability [60]. There is nothing necessarily supernatural about an oracle machine: a laptop with access to an online database would meet the broad definition (if we ignore Turing's statement that the oracle "cannot be a machine" [64, p. 173]). However, an oracle is by definition a "black box": its inner workings are inscrutable to

Dialogue Type	Initial Situation	Main Goal	Goal of Proponent	Goal of Respondent
Inquiry	Open-mindedness	Prove or disprove conjecture	Contribute to main goal	Obtain knowledge
Persuasion	Difference of opinion	Resolve difference of opinion with rigour	Persuade respondent	Persuade proponent
Pedagogical Information-Seeking	Respondent lacks information	Transfer of knowledge	Disseminate knowledge of results and methods	Obtain knowledge
Oracular Information-Seeking	Proponent lacks information	Transfer of knowledge	Obtain information	Inscrutable
Deliberation	Open-mindedness	Reach a provisional conclusion	Contribute to main goal	Obtain warranted belief
Negotiation	Difference of opinion	Exchange resources for a provisional conclusion	Contribute to main goal	Maximize value of exchange

Table 3: Some mathematical dialogue types

the local machine; in principle, they could be inscrutable to any analysis. Sceptics of the proof status of unsurveyably large computer-assisted proofs, such as Thomas Tymoczko, have suggested that the appeal such proofs make to a computer should be seen in similar terms [65]. Analogously, Yehuda Rav proposes as a thought experiment a computer that could answer any mathematical question with certainty but without proof. For Rav, such a machine would be "a death blow to mathematics, for we would cease having ideas and candidates for conjectures" [57, p. 6]. Tymoczko and Rav are both concerned about fallacious appeal to authority in mathematical proof, an issue I will return to below.

The combinatorialist Edward Swart was also concerned with "lengthy proofs (whether achieved by hand or on a computer)". He coined the term "agnograms" to refer to the resulting "theoremlike statements" since we are, at least for the immediate future, required to be agnostic about their truth value, as they "have been neither adequately formalized nor adequately surveyed and are suggestive rather than definitive", due to the limitations of our available resources [62, p. 705]. Establishing an agnogram is thus more of a practical settlement than a stable resolution, suggesting that the dialogue in which it results may better be seen as deliberation or even negotiation, rather than inquiry or persuasion. Likewise, in a widely discussed polemical proposal, Arthur Jaffe and Frank Quinn sought a clear demarcation between "speculative and intuitive work" in "theoretical mathematics" and a "proof-oriented phase" of "rigorous mathematics" [41, p. 2]. Of course, speculative and intuitive work is characteristic of the earlier, inquiry phase of a proof dialogue. However, Jaffe and Quinn anticipate an outlet for responsibly labelled speculation; since this is provisional in character, the process by which it is derived might be seen as deliberation. Something similar might also be said about conjectures, particularly the wide-ranging, fruitful conjectures that comprise the framework of mathematical research programmes, sometimes called "architectural conjectures" [49, p. 198]. Even more speculative is the suggestion of Doron Zeilberger that in the not so distant future "semi-rigorous mathematics" may essentially assign price tickets to proofs, indicating the quantity of computational resources needed for certainty, thereby situating mathematical proof within a negotiation dialogue [78]. This proposal has not generally been well received [12]. Nonetheless, in a weaker form it reflects a truism: even the purest of mathematicians cannot ignore issues of funding, even if the link to their work is not as intimate as Zeilberger suggests. Lastly, even eristic dialogues have had some role to play in mathematical reasoning, as witnessed by such celebrated quarrels as that between the early modern mathematicians Girolamo Cardano and Niccolò Tartaglia [59]. The salient detail is not the asperity of their exchange, which ultimately turned on an accusation of theft of intellectual property, but the adversarial strategy mathematicians of that era adopted to convince the mathematical public of their successes. Rival mathematicians would keep their methods (in this case of solving cubic equations) secret but challenge each other to public contests, each solving problems set by the other until the winner posed a problem the loser could not solve.

2 Argumentation Schemes

An argumentation scheme is a stereotypical pattern of reasoning. In recent decades, the study and classification of argumentation schemes has been the most influential aspect of Douglas Walton's work [70, 75]. Although antecedents of the argumentation scheme can be traced back millennia to the tradition of loci or topoi, Walton's work set it on a new foundation of rigour and clarity. Building on that foundation, Hans Hansen has proposed the following definition of argumentation scheme: "(i) a pattern of argument, (ii) made of a sequence of sentential forms with variables, of which (iii) at least one of the sentential forms contains a use of a schematic constant or a use of a schematic quantifier, and (iv) the last sentential form is introduced by a conclusion indicator like 'so' or 'therefore' " [32, p. 349]. Schemes also generally include 'critical questions', which itemize possible lines of response. The critical questions are key to the evaluation of defeasible schemes: whether the argument should be judged to have succeeded or whether it has been defeated will turn on whether the questions can receive a satisfactory answer.

Walton argues that in principle all defeasible argumentation schemes could be understood as special cases of a defeasible version of modus ponens [75, p. 366]:

Argumentation Scheme 1 (Defeasible Modus Ponens)

Data: P.
Warrant: As a rule, if P, then Q.
    Therefore, ...
Qualifier: presumably, ...
Claim: ... Q.

Critical Questions:

1. *Backing:* What reason is there to accept that, as a rule, if P, then Q?
2. *Rebuttal:* Is the present case an exception to the rule that if P, then Q?

I have reconstructed Walton's scheme for defeasible modus ponens so as to bring out its resemblance to another very general model of defeasible reasoning, the Toulmin layout [7, p. 829]. (On the relationship of schemes to layouts, see also [54, pp. 22 ff.]; for a contrasting view, see [38].) This is not an accidental choice: Toulmin layouts have lately found widespread employment in the analysis of mathematical argumentation (for recent surveys, see [43, 46]). Although Walton emphasized defeasible schemes, deductive rules of inference can also be seen as argumentation schemes: the schemes framework is "illatively neutral" [32, p. 355]. This means that argumentation schemes can provide a unified treatment of a wide range of arguments employed in mathematics. Indeed, a number of authors have applied argumentation schemes to mathematical reasoning [1, 4, 5, 6, 7, 20, 50, 51, 54].

The illative neutrality of the schemes framework licences scepticism about the "standard view" [13] of mathematical argumentation as purely comprised of derivations, that is arguments in which every step instantiates a deductive inference rule.

To that end, I have elsewhere proposed a threefold distinction between A-, B-, and C-schemes:

- **A-schemes** correspond directly to derivation rules. (Equivalently, we could think in terms of a single A-scheme, the 'pointing scheme' which picks out a derivation whose premises and conclusion are formal counterparts of its data and claim.)

- **B-schemes** are exclusively mathematical arguments: high-level algorithms or macros. Their instantiations correspond to substructures of derivations rather than individual derivations (and they may appeal to additional formally verified propositions).

- **C-schemes** are even looser in their relationship to derivations, since the link between their data and claim need not be deductive. Specific instantiations may still correspond to derivations, but there will be no guarantee that this is so and no procedure that will always yield the required structure even when it exists. Thus, where the qualifier of A- and B-schemes will always indicate deductive certainty, the qualifiers of C-schemes may exhibit more diversity. Indeed, different instantiations of the same scheme may have different qualifiers ([7, p. 829]; cf. [6, pp. 366 f.]).

So the widespread "standard" view of mathematical proof, that it is identical to derivation, could be expressed as denying C-schemes a place in proofs. I have argued against that view [6, p. 375], but even if it were to be conceded, it would still leave room for C-schemes in other forms of mathematical reasoning.

What sort of schemes might C-schemes be? Some of them may be unique to mathematics, but we should expect others to resemble schemes that have been found useful in addressing non-mathematical reasoning. Walton and his collaborators have made a number of attempts to classify such general purpose argumentation schemes. Table 4 is based on a recent classification he developed with Fabrizio Macagno. Walton and Macagno employ a series of binary distinctions: first between source-dependent arguments and source-independent arguments; then subdividing the latter into practical reasoning and epistemic reasoning; which is in turn divided into discovery arguments and arguments applying rules to cases. Each of the four resulting headings are then further subdivided into various thematic groups of individual schemes. However, Walton and Macagno concede that this classification is incomplete, notably omitting some linguistic arguments [74, p. 24].

I have annotated Table 4 with citations to works in which mathematical versions of each scheme are discussed. As may be seen, mathematical arguments have been

Discovery arguments	Applying rules to cases	Practical reasoning	Source-dependent arguments

1. Arguments establishing rules
• Argument from a random sample to a population
• Argument from best explanation
2. Arguments finding entities
• Argument from sign [20, 50, 51]
• Argument from ignorance [5]
	1. Arguments based on cases
	• Argument from an established rule
	• Argument from verbal classification [4, 50, 51]
	• Argument from cause to effect
	2. Defeasible rule-based arguments
	• Argument from example [5, 7, 50, 51]
	• Argument from analogy [6, 51]
	• Argument from precedent [54]
	3. Chained arguments connecting rules and cases
	• Argument from gradualism [5]
	• Precedent slippery slope argument
	• Sorites slippery slope argument
		1. Instrumental argument from practical reasoning
		• Argument from action to motive
		2. Argument from values
		• Argument from fairness
		3. Value-based argument from practical reasoning
		(a) Argument from positive or negative consequences [5, 50, 51]
		• Argument from waste
		• Argument from threat
		• Argument from sunk costs
			1. Arguments from position to know
			(a) Argument from expert opinion [1, 5, 50]
			(b) Argument from position to know [5]
			• Argument from witness testimony
			2. Ad hominem arguments
			(a) Direct ad hominem
			(b) Circumstantial ad hominem
			• Argument from inconsistent commitment
			• Arguments attacking personal credibility
			  i. Arguments from allegation of bias
			  ii. Poisoning the well by alleging group bias
			3. Arguments from popular acceptance
			• Argument from popular opinion [4]
			• Argument from popular practice [4]

Table 4: Walton & Macagno's partial classification of schemes (adapted from [74, p. 22]), with applications to mathematical argumentation indicated.

identified under each of Walton and Macagno's four main headings. In addition, mathematical applications have been found for several schemes that are missing from Table 4 but which are found in the more exhaustive (but less structured) list in [75]. These include linguistic arguments, such as arguments from arbitrariness or vagueness of a verbal classification [54] and argument from definition to verbal classification [5], but also source-dependent arguments, such as ethotic argument [5], practical reasoning arguments, such as argument from positive consequences [5], discovery arguments, such as abductive argument [54] and argument from evidence to a hypothesis [5, 7, 54], and arguments applying rules to cases, such as argument from an exceptional case [54]. Conversely, not all of the individual schemes in Table 4 have yet been found useful in discussing mathematics. While some of these omissions may merely be oversights, others are to be expected. For example, causal reasoning, whether argument from cause to effect or the various kinds of slippery slope, is unlikely to be of direct application to mathematics, since mathematical objects are generally understood to be causally inert. In the remainder of this section, I will discuss the mathematical applications of a sample of schemes, chosen in part to remedy some of the omissions in Table 4.

2.1 Epistemic reasoning

Walton and Macagno subdivide epistemic reasoning into two subcategories, discovery arguments and arguments applying rules to cases. Arguments of both kinds can be readily found in mathematical reasoning. In particular, many discovery arguments are broadly abductive in character and abduction has been proposed as an account of mathematical reasoning in a wide range of situations, including classroom discussion [29, 52]; concept formation in mathematical practice [35]; and the selection and defence of axioms [36]. There are several abductive schemes in Walton's catalogue, including multiple subtypes of abductive argument [75, p. 329] and argument from evidence to (verification of) a hypothesis, which I have discussed elsewhere [5, 7, 54]. Another such scheme is argument from sign:

Argumentation Scheme 2 (Argument from Sign [75, p. 329])

Specific Premise: A (a finding) is true in this situation.
General Premise: B is generally indicated as true when its sign, A, is true.
Conclusion: B is true in this situation.

Critical Questions:

1. What is the strength of the correlation of the sign with the event signified?

2. Are there other events that would more reliably account for the sign?

Argument from sign has been discussed since antiquity, particularly in the context of medical reasoning [9]. The explicit application of Scheme 2 to mathematics is due to Ian Dove, who uses it to analyse a surprising but widely discussed class of proofs*: those employing molecular computation [20]. This consists in encoding a mathematical problem into strands of DNA which are then subject to standard laboratory assays that determine the solution of the problem with high likelihood [8]. Hence the outcome of the assay is a sign of the mathematical problem having a specific solution, and the mathematician infers the latter from the former in accordance with Scheme 2. This, and other less esoteric probabilistic methods, such as the Miller–Rabin primality test, are generally viewed by mathematicians as heuristically useful but falling short of the standards of rigour required for proof. Nonetheless, the intellectual defensibility of this perspective has also been the subject of a debate in philosophy of mathematics [24, 26, 28]. Much of this debate could be understood as offering competing answers to the critical questions for Scheme 2. Dove also suggests that what I have referred to above as oracular information seeking could be analysed as employing the same scheme [20, p. 144].

Argument from sign also illustrates the importance of the illative neutrality of argumentation schemes: not all its instances need be defeasible. We can find deductive instances of Scheme 2. For example, much of twentieth and twenty-first century mathematics employs increasingly complex mathematical infrastructure or tools: that is, mathematical theories designed to help us investigate other areas of mathematics. Mathematical tools, such as Galois theory or K-theory, establish rigorous relationships between outwardly unrelated classes of mathematical objects. As Jean-Pierre Marquis observes, the function of such tools is to "reveal important *properties* of the objects studied, and *only* these properties" [48, p. 264]. In other words, a result in one of the two related areas may be taken as a sign that one of the presumably less tractable objects in the other area has a particular property. However, since the relationship between the areas can be rigorously established, the sign is not merely generally indicative, but infallibly so.

Applying rules to cases is also a very widespread practice in mathematics. Several of the schemes that fall under this heading, such as argument from verbal classification, argument from example, and argument from analogy, have mathematical applications that I have discussed elsewhere [4, 5, 6]. Walton and Macagno also include chained arguments, which comprise a substantial proportion of mathematical reasoning [5, p. 235]. But here I shall focus on a different scheme:

Argumentation Scheme 3 (Argument from an Established Rule [75, p. 343])

Major Premise: If carrying out types of actions including A is the established rule for x, then (unless the case is an exception), x must carry out A.
Minor Premise: Carrying out types of actions including A is the established rule for a.
Conclusion: Therefore, a must carry out A.

Critical Questions:

1. Does the rule require carrying out types of actions that include A as an instance?
2. Are there other established rules that might conflict with or override this one?
3. Is this case an exceptional one, that is, could there be extenuating circumstances or an excuse for noncompliance?

Scheme 3 is framed in terms of actions to be carried out. That might initially appear to be an obstacle to its application to mathematics. However, as Wilfrid Hodges has observed, informal mathematical arguments include not only the "object sentences", in which some mathematical content is explicitly given, and "stated or implied justifications for putting the object sentences in the places where they appear" but also "instructions to do certain things which are needed for the proof" [39, p. 6]. The last of these, carrying out actions, has perhaps received least attention from logicians, but it is ubiquitous and important. Many proofs instruct us to " 'Suppose C', 'Draw the following picture, and consider the circles D and E', 'Define F as follows' " and so forth [39, p. 6]. Language of this sort is phrased conventionally as an instruction to the reader, but it is also a description of the actions undertaken by the deviser of the proof. But how did the proof's author know which actions to carry out? At least in some cases, by application of Scheme 3.

Carrying out a rule has also been the focus of a significant debate in the philosophy of mathematics, inspired by the work of Ludwig Wittgenstein. Wittgenstein considers the case of a pupil who learns to follow a rule whereby he writes down a series of natural numbers each greater than its predecessor by 2. However, after he gets to 1000 he increments the numbers by 4, instead of 2, but takes himself still to be following the same rule [76, §185]. What ought we to make of such behaviour? It has been suggested that Wittgenstein's intent was to suggest a general scepticism about rule-following [45]. If that were to be the case, Critical Question 2 in Scheme 3 would always receive an affirmative answer: there would always be another rule which might override any rule we may consider. Less radically, we could read Wittgenstein as counselling against a platonist interpretation of rules as existing independently of the practices they govern [77, p. 91]. Rather we should understand rules as implicit within our practice but nonetheless as carrying normative force. The ontological status of rules is the subject of a difficult and important debate. Fortunately, Scheme 3, and related rule-establishing and applying schemes, are neutral as to the outcome of that debate.

2.2 Practical reasoning

Practical reasoning is an inevitable component of resource-sensitive mathematical deliberation dialogues whether limited by time, money, or processor capacity. If numerical approximation methods are easy and cheap and an exact answer would be expensive and slow, we may settle for the former. More broadly, a dialogue can shift to addressing this sort of question within reasoning about a problem whenever a choice of methods arises. So practical reasoning is not just a project management phase to be completed before the real work begins, but potentially a recurrent phenomenon throughout the research process. For example, James Franklin points out one context in which practical reasoning occurs in the career of almost every research mathematician: choice of Ph.D. topic. A Ph.D. thesis is expected to address an open question which must also be "tractable, that is, probably solvable, or at least partially solvable, by three years' work at the Ph.D. level" [30, p. 2]. Determining whether a problem is tractable is not something which can be established with certainty. But it is critical to the success of the Ph.D. Elsewhere I have addressed some special cases of practical reasoning, including argument from positive consequences: if important results would follow from a conjecture, that at least provides good reason to investigate it more thoroughly than similar, but less consequential conjectures [5, pp. 235 f.]. However, I have not directly discussed the most general practical reasoning scheme in Walton's taxonomy (see also [69, p. 131]):

Argumentation Scheme 4 (Practical Inference [75, p. 323])

Major Premise: I have a goal G.
Minor Premise: Carrying out this action A is a means to realise G.
Conclusion: Therefore, I ought (practically speaking) to carry out this action A.

Critical Questions:

1. What other goals that I have that might conflict with G should be considered?
2. What alternative actions to my bringing about A that would also bring about G should be considered?
3. Among bringing about A and these alternative actions, which is arguably the most efficient?

4. What grounds are there for arguing it is practically possible for me to bring about A?
5. What consequences of my bringing about A should also be taken into account?

Scheme 4 could be used to analyse much of the embedded negotiations about resource allocation discussed above. However, it can also play a more direct role in mathematical reasoning: Yacin Hamami and Rebecca Morris have proposed an account of plans and planning in proving in terms of intentions and practical reasoning, building on Michael Bratman's work in the philosophy of action [16]. The process of finding a proof, at least if the proof is of any complexity, may involve the construction and execution of a carefully devised proof plan, which Hamami and Morris define as "an ordered tree whose nodes are proving intentions, whose root is the proving intention corresponding to the theorem at hand, and where each set of ordered children consists of a subplan obtained from the parent node through an instance of practical reasoning" [31]. The plan is not the proof, any more than the map is the journey. But, they suggest, the plan is essential not only to successfully finding the proof, but also to subsequently understanding the proof.

2.3 Source-dependent arguments

Mathematicians have an ambivalent attitude to authority: there is "a schism in the mathematical community ... [between those] who think that one should never use a result without having understood its entire proof ... [and those who] don't share that view" (anonymous mathematician, interviewed in [11]). Unlike the empirical sciences, where replication of experiments can require substantial resources, it is in principle always possible for mathematicians to work through every step of every proof they use. But, for many mathematicians, a division of labour is unavoidable and even welcome. Hence there is a place in mathematics for one of the most discussed argumentation schemes, that for argument from expert opinion [1, 5, 50]. Notably, there are several disputes over the role of testimony in mathematical reasoning that may be understood in terms of the critical questions of this scheme. "Folk theorems" are one such troublesome case. These are results which are widely used and accepted despite lacking a clear source in the literature. In a pioneering study drawing attention to their prevalence, the theoretical computer scientist David Harel suggests that "popularity, anonymous authorship, and age ... seem to be necessary and sufficient for a theorem to be folklore, [although] the ways in which they appear and can be established are by no means clear-cut" [33, p. 379 f.]. Don Fallis raises the concern that the citation of folk theorems may represent "universally untraversed gaps" in mathematical reasoning since "everyone is convinced that these theorems are provable, but no one has bothered to work through all the details of a proof" [27, p. 62]. And, of course, everyone could be wrong. Colin Rittberg and colleagues raise a different problem: the ambiguous status of folk theorems, neither rigorously proved nor strictly open problems, presents a hazard to young researchers [58]. Actually proving a folk theorem can be an unrewarding project, since referees may reject such work as unoriginal—despite being unable to cite any prior proof. A possible resolution to this and related problems lies in the work of Kenny Easwaran, who has posited a property of "transferability" that may distinguish the proofs which safely support argument from expert opinion from those that do not. Transferable proofs are those that "rely only on premises that the competent reader can be assumed to antecedently believe, and only make inferences that the competent reader would be expected to accept on her own consideration" [24, p. 354]. Hence folk theorems lack transferable proofs, unless there is a proof simple enough for any competent mathematician to reconstruct. Many other proofs* would be untransferable too, including unsurveyably long proofs, probabilistic proofs, and proofs that rely on empirical procedures. This suggests a revision, or precisification, of the critical questions of the expert opinion scheme.

Argument from expert opinion is not the only source-dependent argument relevant to mathematics. Walton draws an important distinction between argument from expert opinion and argument from position to know. The distinction is familiar from legal practice, as that between expert and fact witnesses. I have suggested elsewhere that argument from position to know provides a model for appeals to intuition in mathematics [5, p. 240 f.]. In this I follow philosophers, such as Elijah Chudnoff, for whom intuition is analogous to perception [17, 18]. Reports on (reliable) perceptions support cogent instances of argument from position to know, so if intuition may be treated analogously, then the same scheme should apply. Arguments from popular opinion and popular practice also have important applications to mathematics [4, p. 283 ff.]. However, here I will focus on yet another source-dependent argument:

Argumentation Scheme 5 (Ethotic Argument [75, p. 336])

Major Premise: If x is a person of good (bad) moral character, then what x says should be accepted as more plausible (rejected as less plausible).
Minor Premise: a is a person of good (bad) moral character.
Conclusion: Therefore, what a says should be accepted as more plausible (rejected as less plausible).

Critical Questions:

1. Is a a person of good (bad) moral character?

2. Is character relevant in the dialogue?
3. Is the weight of presumption claimed strongly enough warranted by the evidence given?

Superficially, this scheme may appear to be a poor fit for mathematical argument, but there is empirical research that suggests the applicability of something much like it. Matthew Inglis and Juan Pablo Mejía-Ramos gave an informal mathematical argument for the presence of one million sevens in the decimal expansion of π to samples of undergraduates and research mathematicians and asked them to rate how persuasive they found it [40]. Participants in both groups for whom the argument was correctly attributed to the prominent mathematician Tim Gowers ranked it as more persuasive than those for whom it was presented anonymously, significantly so for the researchers (who were presumably more likely to have heard of Gowers). Of course, it is not Gowers's (doubtless exemplary) moral conduct which leads us to trust his arguments, but rather his demonstrably high standards as a working mathematician. As one of the research subjects in this study comments, "We are told the argument is made by a reputable mathematician, so we implicitly assume that he would tell us if he knew of any evidence or convincing arguments to the contrary" [40, p. 42]. This suggests that we should localize Scheme 5 to mathematics, replacing instances of "moral" with "mathematical":

Argumentation Scheme 6 (Ethotic Mathematical Argument)

Major Premise: If x is a person of good (bad) mathematical character, then what x says should be accepted as more plausible (rejected as less plausible).
Minor Premise: a is a person of good (bad) mathematical character.
Conclusion: Therefore, what a says should be accepted as more plausible (rejected as less plausible).

Critical Questions:

1. Is a a person of good (bad) mathematical character?
2. Is mathematical character relevant in the dialogue?
3. Is the weight of presumption claimed strongly enough warranted by the evidence given?

A (presumably implicit) invocation of Scheme 6 would explain Inglis and Mejía-Ramos's finding, but it raises other questions, most centrally: what is mathematical character? This is a question which recent work applying virtue epistemology to mathematical practice has sought to answer [63]. Mathematicians have also employed virtue talk to describe their activities. For example, George Pólya asserts that the following "moral qualities" are required of a mathematician:

- First, we should be ready to revise any one of our beliefs.
- Second, we should change a belief when there is a compelling reason to change it.
- Third, we should not change a belief wantonly, without some good reason [56, vol. 1, p. 8].

He goes on to expand on these points in explicitly virtue-theoretic terms, telling us that intellectual courage is required for the first, intellectual honesty for the second, and wise restraint for the last. The first of these is a widely discussed intellectual virtue and the second is, at least in this context, closely related to the even more widely discussed intellectual humility. Wise restraint is perhaps more familiar as prudence or practical wisdom. More recent mathematicians who have discussed character virtues relevant to their profession include Michael Harris [34] and Francis Su [61]. Of course, these mathematicians would not necessarily endorse every application of Scheme 6. Indeed, as we saw above, some mathematicians consider it to be a virtue to never take mathematical results on trust and insist on convincing themselves of the proof of any result they cite. Nonetheless, many other mathematicians, especially when reasoning speculatively rather than writing up proofs, rely on a division of labour which makes essential use of the informal arguments of their peers, and may be expected to take those arguments more seriously when they have more reason to trust their authors.

3 Conclusion

Recent work in the philosophy of mathematical practice has drawn attention to mathematical reasoning in contexts other than proof and challenged the traditional conception that mathematical proof is essentially reducible to formal derivation. This leaves a conspicuous lacuna in our understanding of how mathematics works. Formal logic is an excellent tool for the analysis of formal derivations, but it is less well adapted to the analysis of informal reasoning. However, the tools developed by informal logicians such as Douglas Walton are a rich source for remedying this deficit. In particular, as we have seen, dialogue types help to contextualize the different levels of rigour that mathematical argument can exhibit and argumentation schemes provide a valuable taxonomy of the steps that comprise such arguments.

References

[1] Andrew Aberdein. Fallacies in mathematics. *Proceedings of the British Society for Research into Learning Mathematics*, 27(3):1–6, 2007.

[2] Andrew Aberdein. The informal logic of mathematical proof. In Bart Van Kerkhove and Jean Paul Van Bendegem, editors, *Perspectives on Mathematical Practices: Bringing Together Philosophy of Mathematics, Sociology of Mathematics, and Mathematics Education*, pages 135–151. Springer, Dordrecht, 2007.

[3] Andrew Aberdein. Mathematics and argumentation. *Foundations of Science*, 14(1–2):1–8, 2009.

[4] Andrew Aberdein. Observations on sick mathematics. In Bart Van Kerkhove, Jean Paul Van Bendegem, and Jonas De Vuyst, editors, *Philosophical Perspectives on Mathematical Practice*, pages 269–300. College Publications, London, 2010.

[5] Andrew Aberdein. Mathematical wit and mathematical cognition. *Topics in Cognitive Science*, 5(2):231–250, 2013.

[6] Andrew Aberdein. The parallel structure of mathematical reasoning. In Andrew Aberdein and Ian J. Dove, editors, *The Argument of Mathematics*, pages 361–380. Springer, Dordrecht, 2013.

[7] Andrew Aberdein. Evidence, proofs, and derivations. *ZDM*, 51(5):825–834, 2019.

[8] Leonard M. Adleman. Molecular computation of solutions to combinatorial problems. *Science*, 266:1021–1024, 1994.

[9] James Allen. *Inference from Signs: Ancient Debates about the Nature of Evidence*. Clarendon, Oxford, 2001.

[10] Line Edslev Andersen. Acceptable gaps in mathematical proofs. *Synthese*, 197(1):233–247, 2020.

[11] Line Edslev Andersen, Hanne Andersen, and Henrik Kragh Sørensen. The role of testimony in mathematics. *Synthese*, 2021. Forthcoming.

[12] George E. Andrews. The death of proof? Semi-rigorous mathematics? You've got to be kidding! *The Mathematical Intelligencer*, 16(4):16–18, 1994.

[13] Marianna Antonutti Marfori. Informal proofs and mathematical rigour. *Studia Logica*, 96:261–272, 2010.

[14] Zoe Ashton. Audience role in mathematical proof development. *Synthese*, 2020. Forthcoming.

[15] Michael J. Barany and Donald MacKenzie. Chalk: Materials and concepts in mathematics research. In Catelijne Coopmans, Janet Vertesi, Michael E. Lynch, and Steve Woolgar, editors, *Representation in Scientific Practice Revisited*, pages 107–129. MIT Press, Cambridge, MA, 2014.

[16] Michael Bratman. *Intention, Plans, and Practical Reason*. Harvard University Press, Cambridge, MA, 1987.

[17] Elijah Chudnoff. *Intuition*. Oxford University Press, Oxford, 2013.

[18] Elijah Chudnoff. Intuition in mathematics. In Barbara Held and Lisa Osbeck, editors, *Rational Intuition: Philosophical Roots, Scientific Investigations*, pages 174–191. Cambridge University Press, Cambridge, 2014.

[19] William D'Alessandro. Mathematical explanation beyond explanatory proof. *British Journal for the Philosophy of Science*, 71(2):581–603, 2020.

[20] Ian J. Dove. Towards a theory of mathematical argument. *Foundations of Science*, 14(1–2):137–152, 2009.

[21] Catarina Dutilh Novaes. Reductio ad absurdum from a dialogical perspective. *Philosophical Studies*, 173:2605–2628, 2016.

[22] Catarina Dutilh Novaes. A dialogical conception of explanation in mathematical proofs. In Paul Ernest, editor, *The Philosophy of Mathematics Education Today*, pages 81–98. Springer, Cham, 2018.

[23] Catarina Dutilh Novaes. *The Dialogical Roots of Deduction: Historical, Cognitive, and Philosophical Perspectives on Reasoning*. Cambridge University Press, Cambridge, 2021.

[24] Kenny Easwaran. Probabilistic proofs and transferability. *Philosophia Mathematica*, 17:341–362, 2009.

[25] Paul Ernest. A dialogue on the ethics of mathematics. *The Mathematical Intelligencer*, 38(3):69–77, 2016.

[26] Don Fallis. The epistemic status of probabilistic proof. *Journal of Philosophy*, 94(4):165–186, 1997.

[27] Don Fallis. Intentional gaps in mathematical proofs. *Synthese*, 134:45–69, 2003.

[28] Don Fallis. Probabilistic proofs and the collective epistemic goals of mathematicians. In Hans Bernard Schmid, Marcel Weber, and Daniel Sirtes, editors, *Collective Epistemology*, pages 157–175. Ontos Verlag, Frankfurt, 2011.

[29] Elisabetta Ferrando. *Abductive Processes in Conjecturing and Proving*. PhD thesis, Purdue University, 2005.

[30] James Franklin. Non-deductive logic in mathematics. *British Journal for the Philosophy of Science*, 38(1):1–18, 1987.

[31] Yacin Hamami and Rebecca Morris. Plans and planning in mathematical proofs. *Review of Symbolic Logic*, 2020. Forthcoming.

[32] Hans V. Hansen. Argument scheme theory. In Catarina Dutilh Novaes, Henrike Jansen, Jan Albert Van Laar, and Bart Verheij, editors, *Reason to Dissent: Proceedings of the 3rd European Conference on Argumentation, Groningen 2019*, volume 2, pages 341–356. College Publications, London, 2020.

[33] David Harel. On folk theorems. *Communications of the ACM*, 23(7):379–389, 1980.

[34] Michael Harris. *Mathematics without Apologies: Portrait of a Problematic Vocation*. Princeton University Press, Princeton, NJ, 2015.

[35] Albrecht Heeffer. Abduction as a strategy for concept formation in mathematics: Cardano postulating a negative. In Olga Pombo and Alexander Gerner, editors, *Abduction and the Process of Scientific Discovery*, pages 179–194. Centro de Filosofia das Ciências da Universidade de Lisboa, Lisbon, 2007.

[36] John Heron. Set-theoretic justification and the theoretical virtues. *Synthese*, 2021. Forthcoming.

[37] Reuben Hersh and Vera John-Steiner. *Loving and Hating Mathematics: Challenging the Myths of Mathematical Life*. Princeton University Press, Princeton, NJ, 2011.

[38] David Hitchcock. Toulmin's warrants. In F. H. van Eemeren, J. Blair, C. Willard, and A. F. Snoeck-Henkemans, editors, *Anyone Who Has a View: Theoretical Contributions to the Study of Argumentation*, pages 69–82. Kluwer, Dordrecht, 2003.

[39] Wilfrid Hodges. An editor recalls some hopeless papers. *Bulletin of Symbolic Logic*, 4(1):1–16, 1998.

[40] Matthew Inglis and Juan Pablo Mejía-Ramos. The effect of authority on the persuasiveness of mathematical arguments. *Cognition and Instruction*, 27(1):25–50, 2009.

[41] Arthur Jaffe and Frank Quinn. "Theoretical mathematics": Toward a cultural synthesis of mathematics and theoretical physics. *Bulletin of the American Mathematical Society*, 29(1):1–13, 1993.

[42] Robert Kanigel. *The Man Who Knew Infinity: A Life of the Genius Ramanujan*. Charles Scribner's Sons, New York, NY, 1991.

[43] Christine Knipping and David A. Reid. Argumentation analysis for early career researchers. In Gabriele Kaiser and Norma Presmeg, editors, *Compendium for Early Career Researchers in Mathematics Education*, pages 3–31. Springer, Cham, 2019.

[44] Erik C. W. Krabbe. Strategic maneuvering in mathematical proofs. *Argumentation*, 22(3):453–468, 2008.

[45] Saul A. Kripke. *Wittgenstein on Rules and Private Language: An Elementary Exposition*. Blackwell, Oxford, 1982.

[46] Götz Krummheuer. Methods for reconstructing processes of argumentation and participation in primary mathematics classroom interaction. In Angelika Bikner-Ahsbahs, Christine Knipping, and Norma Presmeg, editors, *Approaches to Qualitative Research in Mathematics Education: Examples of Methodology and Methods*, pages 51–74. Springer, Dordrecht, 2015.

[47] Imre Lakatos. *Proofs and Refutations: The Logic of Mathematical Discovery*. Cambridge University Press, Cambridge, 1976. Edited by J. Worrall and E. Zahar.

[48] Jean-Pierre Marquis. Abstract mathematical tools and machines for mathematics. *Philosophia Mathematica*, 5(3):250–272, 1997.

[49] Barry Mazur. Conjecture. *Synthese*, 111(2):197–210, 1997.

[50] Nikolaos Metaxas. Mathematical argumentation of students participating in a mathematics–information technology project. *International Research in Education*, 3(1):82–92, 2015.

[51] Nikolaos Metaxas, Despina Potari, and Theodossios Zachariades. Analysis of a teacher's pedagogical arguments using Toulmin's model and argumentation schemes. *Educational Studies in Mathematics*, 93(3):383–397, 2016.

[52] Michael Meyer. Abduction—a logical view for investigating and initiating processes of discovering mathematical coherences. *Educational Studies in Mathematics*, 74(2):185–205, 2010.

[53] C. J. Mozzochi. *The Fermat Diary*. American Mathematical Society, Providence RI, 2000.

[54] Alison Pease and Andrew Aberdein. Five theories of reasoning: Interconnections and

applications to mathematics. *Logic and Logical Philosophy*, 20(1-2):7–57, 2011.

[55] Alison Pease, John Lawrence, Katarzyna Budzynska, Joseph Corneli, and Chris Reed. Lakatos-style collaborative mathematics through dialectical, structured and abstract argumentation. *Artificial Intelligence*, 246:181–219, 2017.

[56] George Pólya. *Mathematics and Plausible Reasoning*. Two Volumes. Princeton University Press, Princeton, N.J., 1954.

[57] Yehuda Rav. Why do we prove theorems? *Philosophia Mathematica*, 7(3):5–41, 1999.

[58] Colin Jakob Rittberg, Fenner Stanley Tanswell, and Jean Paul Van Bendegem. Epistemic injustice in mathematics. *Synthese*, 197(9):3875–3904, 2020.

[59] Tony Rothman. Cardano v. Tartaglia: The great feud goes supernatural. *The Mathematical Intelligencer*, 36(4):59–66, 2014.

[60] Robert I. Soare. Turing oracle machines, online computing, and three displacements in computability theory. *Annals of Pure and Applied Logic*, 160(3):368–399, 2009.

[61] Francis Su. *Mathematics for Human Flourishing*. Yale University Press, New Haven, CT, 2020.

[62] Edward Swart. The philosophical implications of the four-color problem. *The American Mathematical Monthly*, 87:697–707, 1980.

[63] Fenner Stanley Tanswell and Ian James Kidd. Mathematical practice and epistemic virtues and vices. *Synthese*, 2021. Forthcoming.

[64] Alan Mathison Turing. Systems of logic based on ordinals. *Proceedings of the London Mathematical Society*, 45:161–228, 1939.

[65] Thomas Tymoczko. The four-color problem and its philosophical significance. *Journal of Philosophy*, 76:57–83, 1979.

[66] Cédric Villani. *Birth of a Theorem: A Mathematical Adventure*. Macmillan, London, 2015.

[67] Douglas N. Walton. *Informal logic: A Handbook for Critical Argumentation*. Cambridge University Press, Cambridge, 1989.

[68] Douglas N. Walton. What is reasoning? What is argument? *Journal of Philosophy*, 87:399–419, 1990.

[69] Douglas N. Walton. *A Pragmatic Theory of Fallacy*. University of Alabama Press, Tuscaloosa, AL, 1995.

[70] Douglas N. Walton. *Argumentation Schemes for Presumptive Reasoning*. Lawrence Erlbaum Associates, Mahwah, NJ, 1996.

[71] Douglas N. Walton. How can logic best be applied to arguments? *Logic Journal of the IGPL*, 5:603–614, 1997.

[72] Douglas N. Walton. *The New Dialectic: Conversational Contexts of Argument*. University of Toronto Press, Toronto, 1998.

[73] Douglas N. Walton and Erik C. W. Krabbe. *Commitment in Dialogue: Basic Concepts of Interpersonal Reasoning*. State University of New York Press, Albany, NY, 1995.

[74] Douglas N. Walton and Fabrizio Macagno. A classification system for argumentation

schemes. *Argument & Computation*, 6(3):219–245, 2015.

[75] Douglas N. Walton, Chris Reed, and Fabrizio Macagno. *Argumentation Schemes*. Cambridge University Press, Cambridge, 2008.

[76] Ludwig Wittgenstein. *Philosophical Investigations*. Blackwell, Oxford, 1953.

[77] Crispin Wright. Wittgenstein on mathematical proof. *Royal Institute of Philosophy Supplements*, 28:79–99, 1990.

[78] Doron Zeilberger. Theorems for a price: Tomorrow's semi-rigorous mathematical culture. *Notices of the American Mathematical Society*, 46:978–981, 1993.

Walton on Argument, Arguments, and Argumentation

Harvey Siegel
University of Miami

John Biro
University of Florida

Douglas Walton's contributions to logic (both formal and informal) and the theory of argumentation are legion and well known. Hugely prolific and highly influential, the scope of his work on argument schemes, fallacies, and related topics is unmatched. It is an honor to contribute to this special issue of the *Journal of Applied Logics* celebrating his legacy. In what follows, we examine Walton's treatment of two related notions fundamental to argumentation theory: *argument* and *argumentation*. We suggest that that treatment, though undeniably insightful, is flawed in several respects. In particular, it mischaracterizes the distinction between argument and argumentation, overvalues dialogicality, mistakenly regards argument evaluation as inherently instrumental, and downplays the centrality of epistemic criteria in assessing the quality of arguments.

1 A Thought Experiment

You walk into an empty classroom ready to teach your next Philosophy 101 class. On the board, you see the following:

1. God is the being greater than which none can be conceived.
2. Existence in reality is greater than existence in the understanding alone.
3. Therefore, a being that exists in the understanding alone is not God.
4. Therefore, God exists in reality.

Thanks to John Woods for his advice and encouragement.

You immediately recognize these lines as a canonical statement of one version of Saint Anselm's ontological argument, one of philosophy's most famous and controversial arguments. In the few minutes before your class begins you consider once again Anselm's argument. Although it was left on the board by an instructor in an earlier class and is not what you intend to cover today ("Why don't my thoughtless colleagues erase the board when their class is over?", you silently fume), as your students trickle in, you engage them in casual conversation concerning its strengths and weaknesses, and perhaps mention Gaunilo's and Kant's criticisms and possible replies, until finally a sufficient number of your students take their seats and you erase the board and get on with today's topic.

What should we make of this example? We will return to this question below. First, let us consider Walton's view of argument and argumentation.

2 Argument v. Argumentation

Walton defines 'argument' as "a social, interactive, goal-directed tool of persuasion." (p. 401)[1] One good thing about this definition is that it understands that tool to be composed of items, things or actions — usually speech acts[2] — that are used in social, interactive activities. Those activities are usually understood as instances or episodes of *argumentation*, an activity people engage in when they attempt to persuade[3] others by using *arguments*. Arguments are the tools, and argumentation is the social practice in which they are deployed. Unfortunately, Walton rejects the distinction as just articulated, and urges that 'argument' take on the meaning just ascribed to 'argumentation'.[4] We will consider this conflation further below.

Walton offers what he calls "the dialectical conception of an argument" (p. 409),

---

[1] All untethered page references in the text are to Walton [31]. This paper is insightful and impressive in several respects (as is Walton's work more generally), especially concerning reasoning, argument, and dialogue types, as well as Aristotle's treatment of argument, and the radical reconceiving of logic. We regret that we have no space to treat these topics here.

[2] Here Walton credits van Eemeren and Grootendorst (p. 400).

[3] People of course use arguments to do many things other than to persuade [10, pp. 30–32]. More on this below.

[4] Whether this is in fact what 'argumentation' means may be questioned. The activity and practice in question is, we think, better described as one of *arguing*. ('Arguing', of course, itself has other senses: disagreeing, disputing, debating, even fighting, which are not relevant here.) We sometimes speak of an author's or a paper's argumentation as being weak or wanting in some respect. When we do so, we are saying something about the strength, relevance or formulation of the arguments offered, or of the argumentative strategy employed. Thus we think that the very expression 'argumentation theory' commonly used to denote the study of arguing and arguments is off the mark. However, the use is so entrenched that there is no point in resisting it beyond merely noting its impropriety.

according to which "an argument is more than just a set of propositions. It comprises many kinds of speech acts, evaluated in a goal-directed, normative model of dialogue" (p. 400), as an alternative to the more traditional 'set of propositions' conception. He credits Frans van Eemeren and Rob Grootendorst [13] here for their account of argument as, in Walton's words, "a rule-governed kind of discussion entered into by two or more parties in order to resolve a conflict of opinions." (p. 400)[5] It is true that English speakers sometimes use 'argument' in this way, in sentences such as "Trump's argument with his Democratic opponents concerning immigration has descended into a chaotic mess of *ad hominem* irrelevancies and has failed to bring about a viable political resolution" and "The argument among Boston, Los Angeles, and Chicago fans concerning the best NBA team of all time has gone on for decades", thus taking 'argument' in this 'discussion' sense of the term: an entire discussion can in this sense be considered an argument. But it is clearly a different sense of the term from that in 'ontological argument', and should be treated as such.[6] There are several reasons for this, but the most important reason is also the most straightforward: the hoped-for resolution of a conflict of opinion by an argument (in the 'discussion' sense of the term) is accomplished, at least ideally, by the use of good arguments (in the 'set of propositions' sense of the term). If not, there is something normatively suspect about the resolution: if the parties resolve their difference of opinion not on the basis of arguments that secure the epistemic propriety of the resolution, there is no reason to regard such a resolution as preferable to any other possible resolution.[7]

3 Argument/Argumentation as Intrinsically Dialogical or Dialectical

According to Irving Copi, "An argument, in the logician's sense, is any group of propositions of which one is claimed to follow from the others, which are regarded as providing support or grounds for the truth of that one." ([12, p. 6], quoted

---

[5] As noted above, he also defines 'argument' as "a social, interactive, goal-directed tool of persuasion." (p. 401)

[6] In Biro and Siegel ([8, p. 92] and elsewhere), we distinguish between arguments as abstract objects — usually structured strings of sentences or propositions — and arguments as sequences of events, usually acts of arguing. (Cf. also Biro and Siegel [10, p. 30].) The distinction is fundamental, especially to an understanding of the proper object of study of argumentation theory. Walton, following van Eemeren and Grootendorst, urge the second; in Biro and Siegel [9], we argue that both are proper objects of study, so long as they are "clearly distinguished and their relation... properly understood." [8, p. 92].

[7] Cf. Biro and Siegel [8, p. 94].

by Walton, pp. 408-9) In discussing Copi's definition, Walton suggests that the expressions 'claimed' and 'regarded as providing support' in Copi's definition

> refer to a kind of stance or attitude taken up or conveyed by the proponent of the argument. To claim that a proposition is true and can be supported is to assert that proposition and commit oneself to its truth, implying a commitment to defending its truth, as alleged, against attacks or undermining of it by any potential opponent. In this sense, the term 'claim' tacitly presupposes an interactive (dialectical) framework of a proponent upholding a point of view and an opponent questioning that point of view. A claim is an upholding of some particular proposition that is potentially open to question. (p. 409)

This is one strand in Walton's case for regarding argument as intrinsically dialogical. (pp. 411 ff.) Walton is right, we think, that insofar as Copi takes the 'claimed' and 'regarded as providing support' to be essential to the definition, he seems to be committed to the intrinsic dialogicality of 'argument'. But why should we follow Copi on this? Consider the ontological argument again. It may be true that Anselm so claimed (of the conclusion) and so regarded (the premises), and that 'proponents of the argument' must as well, but must we? Surely a critic of the argument will demur, and claim (and so regard) no such thing. What about the 101 students who confront the argument for the first time, or the seasoned scholar who reconsiders it in the course of writing a monograph, tentatively titled *Arguments for God's Existence: A Survey and Evaluation*? It seems better to say simply that Copi's definition conflates two distinct things: the argument, and the attitude of its proponent. The argument is simply the string of propositions, or the sentences with which they are expressed, in question.

Does this imply that any such string is an argument? That seems a stretch. Consider the following string:

1. The sky is blue.

2. Snow is white.

3. Therefore, the moon is made of green cheese.

Is this an argument? We could say that it is, though an exceedingly bad one, for the premises offer no support whatsoever to the conclusion. Alternatively, we could say that it is not, because for it to count as an argument the premises must provide at least a modicum of support to its conclusion. Or we could require at least that the sentences in question be linguistically meaningful in order to be eligible for inclusion in any string deserving of 'argument' status.[8] We think that it does not much matter which way we go here. Our inclination is to err on the side of inclusion, and regard any such string as an argument, however bad — even a terrible argument is an argument. But we see the force of the objection that doing so will force us to regard even sequences of meaningless or absurd sentences as arguments, and surely

1. The president is made of orange cheese.

2. The mountain wishes it were a river.

3. Therefore, colorless green ideas sleep furiously.

strains our intuitive conception of an argument. So, let us leave the question open for now. The important point is that we should not follow Copi in thinking that, as Walton puts it, the "kind of stance or attitude taken up or conveyed by the proponent of the argument" is a proper part of any given argument. The argument is one thing; a proponent's stance or attitude to it quite another. If so, the case for the intrinsic dialogicality of arguments is not yet made.[9]

But Walton offers an additional reason for thinking that arguments are intrinsically dialogical or dialectical. This reason is noted above: the term 'argument' "has a broader meaning", at least "[a]mong those not corrupted by logic courses". (p. 409) In this broader meaning, 'argument' is used to refer to the larger dialectical event among interlocutors in which reasons are offered, considered, and evaluated in the context of interactive dialectical exchanges: "a long thread or fabric that runs through and holds together an extended discourse or argumentative text." (p. 410) Walton notes that this broader meaning of 'argument' is "more like what van Eemeren and Grootendorst call *argumentation*, a goal-directed form of interactional (communicative) activities wherein two parties attempt to resolve a conflict of opinions." (p. 410, emphasis Walton's) For this reason Walton endorses

---

[8]We might go further still and require that the premises be *relevant* to the conclusion, as relevance is regarded as an obvious criterion of argument quality, along with premise acceptability and sufficiency, by many informal logicians. (The three criteria were originally proposed in Johnson and Blair [21]; for a systematic survey of and acute commentary on central topics in informal logic, including this one, see Hitchcock [20].) But relevance seems to us to be an idle wheel, since it is subsumed by sufficiency, which is itself best understood in terms of the central epistemic criterion of *evidential support*: premises may be relevant but offer little or no (or even negative) support to their purported conclusion; while if they offer support, that fact itself renders them relevant. For a brief case that informal logic, no less than argumentation theory, rests upon an epistemological account of argument quality, see Biro and Siegel [7, pp. 97–98].

[9]Thanks here to Dan Cohen for enlightening and enjoyable conversation.

a new way of defining 'argument' to make it *coextensive with argumentation*. The only difference between the two is one of connotation. 'Argumentation' refers to the global process of defending and criticizing a thesis (point of view) which spans the whole context of discussion. The term 'argument' can also have this meaning, but is often used for practical purposes to refer to a local segment of a chain of argument, comprising specifically designated premises and conclusions. According to this usage, the term 'argument' can be used in a restricted way somewhat reminiscent of the logicians' truncated definition. Accordingly, a new definition of 'argument' is now proposed.[10]

Argument is a social and verbal means of trying to resolve, or at least to contend with, a conflict or difference that has arisen or exists between two (or more) parties. An argument necessarily involves a claim that is advanced by at least one of the parties. (pp. 410-411, first emphasis added)

Walton here explicitly rejects the distinction, drawn above, between 'argument' and what he and others call (however improperly) 'argumentation'. This is a mistake, since the distinction is both clear and essential for properly delineating the relevant objects of study. (Biro and Siegel [9]) Most importantly, maintaining the distinction enables us to study, independently, both the epistemic strengths and weaknesses of particular claims and their alleged supporting reasons (e.g., how much, if any, support do the reasons offered afford the claim?), and the rhetorical and persuasive effectiveness of particular dialectical moves (e.g., how much, if at all, does this or that particular remark or strategy move the audience to embrace the hoped-for resolution?). These are both well worth extended study, both practically and epistemically. Because it is obvious that people can be and are regularly persuaded by bad arguments, and equally obvious that they can fail to be persuaded by good ones, both sorts of study are worthwhile. But they are not remotely identical. Argument quality is one thing; argumentative effectiveness quite another. (Biro and Siegel [8, p. 93])[11,12]

---

[10][Note added:] Here we have a slightly different case of the misuse of 'argumentation' noted earlier. In defending a thesis one does not offer an argumentation; one offers an argument or arguments. The global/local distinction is a red herring.

[11]Walton clearly acknowledges the distinction as it is ascribed to reasoning (pp. 403-404), noting that "the logical point of view on reasoning is distinctively different from the psychological point of view." (p. 404) It is curious, to say the least, that he is apparently unwilling to ascribe it to either argument or argumentation.

[12]John Woods similarly rejects Walton's claim that all arguments are inherently dialectical. Following Aristotle, Woods contrasts "arguments in the broad sense with **arguments in the narrow**

4 Argument Evaluation as Instrumental

Walton argues that arguments should be evaluated *instrumentally*, in terms of their success in realizing the pragmatic goal of the dialogue in question: persuasion, the resolution of a difference of opinion (*a la* van Eemeren and Grootendorst and their pragma-dialectical fellow travelers), or some other goal set by the context of the dialogue. They should be "evaluated in a goal-directed, normative model of dialogue" (p. 400) because "argument is essentially goal-directed." (p. 411) Walton offers a list of eight different types of argumentative dialogues, each with its own argument-type goal.[13] (p. 413) He emphasizes that "argumentation, as a field" essentially concerns "the uses of argumentation schemes in a context of dialogue, an essentially pragmatic undertaking." (p. 418)

The pragmatic, instrumental, goal-directed sort of evaluation that Walton urges[14] is indeed one dimension along which arguments can be evaluated. But, as we have urged, this is not the most important dimension of argument evaluation, since (among other reasons) bad arguments can and often do succeed at persuasion, and locutions of the sort 'He was persuaded by that bad argument' are not only perfectly sensible but, alas, often accurate: as the psychology of reasoning literature amply demonstrates, people are in fact often persuaded by bad arguments and fail to be persuaded by good ones. If a given argument can succeed at persuasion but still be bad, that badness cannot be a function of its lack of persuasive power. And if its badness is not a matter of its failure to persuade, what is it that makes it bad? According to the view we favor, the badness is an epistemic matter: its premises — i.e., the reasons adduced in support of its conclusion – fail to provide adequate

---

sense. Arguments in the broad sense are social exchanges between parties who hold conflicting positions about some expressly or contextually advanced thesis. Arguments in the narrow sense are abstract sequences of categorical propositions, of which the terminal member is its conclusion and the remaining ones its premises... There is *nothing* dialectical or social or interactive about arguments in the narrow sense." ([33, p. 10], emphasis in original) Woods' 'narrow/broad' distinction parallels the 'argument/argumentation' distinction drawn above. See also Woods [34].

[13]Walton is a pioneer in this area and his work on dialogue types has been highly influential. Among many of his contributions, see his *Informal Logic: A Handbook of Critical Argumentation* [30], and his and Erik Krabbe's *Commitment in Dialogue* [32].

[14]Here Walton follows van Eemeren and Grootendorst. For critical discussion of the pragma-dialecticalist's instrumental view of argument evaluation, see Siegel and Biro [28, pp. 192–3].

support for that conclusion, [7, 8, 9, 27, 28, 29].[15,16]

5 The Purposes of Argumentation, and the Centrality of Epistemic Criteria of Argument Quality

Why do we argue? Sometimes we argue in order to persuade others. But persuasion is only one possible goal of argumentation. We also argue to tease, to annoy, to antagonize, to amuse, to waste time, to entertain, to score points, to defeat one's opponent in a quarrel [10, pp. 30–32], and, as Walton, van Eemeren and Grootendorst emphasize, to resolve differences of opinion. Nevertheless, the fundamental purpose of argumentation is that of helping us determine what we ought to believe, epistemically speaking. Insofar, it is *epistemic improvement*, rather than the pragmatic furthering of our practical ends, that is the key to argument quality.

Given the speech-act model Walton espouses, it is surprising that he does not notice that his account fails to respect the distinction between illocutions and perlocutions, Austin [1]. We are all familiar with the abstract or introductory paragraph that begins "In this paper I argue that..." Maintaining, as Walton does, that what one announces oneself as doing is setting out to persuade the reader is to take what Austin called the perlocutionary object of a speech act, in this case, persuading one's interlocutor, as defining what speech act is being performed. But this is a mistake. For one thing, the same act of arguing can be performed with different perlocutionary objects. One may give the same argument to two people, aiming to persuade one and provoke or reassure the other. Announcing oneself as giving an argument does not in and of itself tell us which, if any, perlocutionary object one has beyond performing the illocutionary act of arguing. *That* act must therefore be understood independently of whatever perlocutionary aim one may have in performing it.

It should also be noted that the perlocutionary object — the effect one aims at — and what Austin called the perlocutionary sequel — the effect actually resulting — sometimes fail to coincide. In giving someone an argument, one may aim to persuade and succeed only in amusing, irritating, or even causing skepticism. The general

---

[15] We should note that instrumentality and epistemic propriety are not contraries; an argument can be both instrumentally efficacious and epistemically forceful. Our point is simply that the latter is not a function of the former. The sort of instrumental evaluation of argument quality Walton favors is neither the only nor the most important sort of argument evaluation; how far an argument furthers the arguer's interests is one thing; how good it is *qua* argument quite another. Thanks here to John Woods for helpful conversation.

[16] An underlying philosophical issue concerns the appropriateness of understanding epistemic rationality itself as a matter of instrumental goal-satisfaction. For a survey of recent literature and two arguments against an instrumental conception of epistemic rationality, see Siegel [26].

point is that the connection between illocutionary acts and both perlocutionary objects and sequels — between arguing and seeking to persuade and succeeding (or not) in persuading — is contingent. By contrast, the connection between the illocutionary act of arguing and the aim of providing reasons for belief is a conceptual one. To mean by 'argument' what Walton suggests we should mean would be to change the subject. It is not that words cannot change their meaning over time, of course. And Walton's proposal could take root, with 'argument' eventually coming to mean what he wants to talk about. But then we would need a new word to allow us to talk about what the word means now.

Rather than conceiving of argument evaluation in instrumental, pragmatic terms, the epistemic view we favor understands such evaluation of both the speech act of arguing and the quality of the argument deployed in it in terms of the latter's ability to *justify, or render rational*, the belief (in the conclusion) of someone who considers it. To give an argument *just is* to provide reasons for belief. Thus judgments about the quality of an argument must employ standard epistemic criteria: How good are the reasons/premises offered? Do they support the claim/conclusion under consideration? Are there other considerations that count against it? If the reasons offer sufficient support for the claim under consideration, and so enhance the knowledge or justified belief of the person in question, the argument is a good one; if not, not.[17]

It should be evident that the same considerations apply to the pragma-dialectician's claim that the aim of argumentation is conflict-resolution. As with Walton, the attempt to define the act of arguing in terms of a non-epistemic goal misfires because such goals are not conceptually connected with arguing, as reason-giving is. They are external to the practice, whereas the latter is internal to it: if one is not giving reasons for belief, one is not arguing.

6 The Ontological Argument Reconsidered

To return to our earlier question: What should we make of the opening thought experiment? It is hard to deny that the ontological argument is an argument. If it is an argument, how does it fare on Walton's account of argument? It certainly seems to be "a kind of abstract structure"[18] — namely, a structure composed of

---

[17]Biro and Siegel [7, 8, 9, 10]; Siegel and Biro [27, 28, 29]. Fellow epistemic theorists (albeit with varying emphases and account details) include Mark Battersby [5, 6]; Sharon Bailin and Battersby [2, 3]; J. Anthony Blair and Ralph Johnson [11]; Richard Feldman [14, 15, 16]; Alvin Goldman [17, 18]; David Godden [19]; Christoph Lumer [22, 23, 24]; and Robert Pinto [25].

[18]Walton actually attributes this abstract structure to *reasoning*, not argument: He defines reasoning as "a kind of abstract structure, which can nevertheless be dynamic and interactive in some cases, as well as static and solitary in other cases", and is "characteristically used in argument,

abstract objects, propositions (or the sentences that express them), organized in premise-conclusion form. Is it composed of "speech acts evaluated in a goal-directed, normative model of dialogue"? Is it evaluated in terms of "a goal-directed, normative model of dialogue"? In the usual case, no: it is, rather, evaluated in terms of the ability of its premises to warrant belief in its conclusion. This is because, as we argue (in Biro and Siegel [7, p. 92]; Siegel and Biro [27, p. 290]), it is a conceptual truth that the central purpose of arguments, understood as abstract objects that can be articulated in oral or written form and deployed in episodes of argumentation, is to yield knowledge or reasonable belief. As we say, "an argument succeeds to the extent that it *renders belief rational*" [7, p. 96], [27, p. 278] (italics in originals); as we also put it, "arguments are good when their reasons/premises increase the knowability or rational believability of their conclusions" [10, p. 29]; likewise, "the intrinsic merit of an argument must be judged relative to the intrinsic goal, the *raison d'etre*, of arguments: to provide good reasons for belief." [8, p. 94][19] If these claims are sound, Walton's new definition of 'argument' must be rejected.

References

[1] Austin, J. L. (1962): How to Do Things with Words. Cambridge, Mass: Harvard University Press.

[2] Bailin, Sharon, and Mark Battersby (2016): Reason in the Balance: An Inquiry Approach to Critical Thinking, 2nd ed. Indianapolis: Hackett Publishing Company.

[3] Bailin, Sharon, and Mark Battersby (in press): 'Is There a Role for Adversariality in Critical Thinking?' Topoi.

---

but it can be used in other pragmatic contexts as well." (p. 401) At first blush this definition appears problematic, since reasoning is an activity – something we *do*, something we *engage in* — hence not an 'abstract structure' of any obvious kind. But since that abstract structure can, on Walton's account, be "*used* in argument...[and] in other pragmatic contexts as well" (*ibid.*, our italics), charity suggests we interpret Walton to be suggesting that reasoning involves the *use* of that abstract structure. Philosophers usually take reasoning to involve *transitions between mental states or attitudes*, and *good* reasoning to involve transitions that result in later states or attitudes that constitute *gains in knowledge, justified belief, and/or rational belief, attitude, or decision.* (Walton characterizes this sort of account of reasoning as 'internalist', in contrast with 'externalist' accounts of reasoning as "a process of linguistic interaction that appears to be more sociological than psychological" (p. 400).) The 'transitions between mental states' characterization reflects our advocacy of the epistemic theory of argumentation. (We speak explicitly of rational belief transitions in several of our papers, e.g., [7, pp. 87–88ff], [27, pp. 278–80]; the idea is echoed in later papers.) An important new collection of essays on philosophical issues concerning reasoning is Balcerak Jackson and Balcerak Jackson [4]).

[19]The last quote is altered by the deletion of an embedded note and a pair of parentheses.

[4] Balcerak Jackson, Magdalena, and Brendan Balcerak Jackson, eds., (2019): Reasoning: New Essays on Theoretical and Practical Thinking. Oxford: Oxford University Press.

[5] Battersby, Mark E. (1989): 'Critical Thinking as Applied Epistemology: Relocating Critical Thinking in the Philosophical Landscape.' Informal Logic 11.2: 91-100.

[6] Battersby, Mark E.(2016): 'Enhancing Rationality: Heuristics, Biases, and the Critical Thinking Project'. Informal Logic 36.2: 99-120.

[7] Biro, John, and Harvey Siegel (1992): 'Normativity, Argumentation, and an Epistemic Theory of Fallacies'. In F. van Eemeren, et. al., eds., Argumentation Illuminated: Selected Papers from the 1990 International Conference on Argumentation. Dordrecht: Foris, pp. 85-103.

[8] Biro, John, and Harvey Siegel (2006a): 'In Defense of the Objective Epistemic Approach to Argumentation'. Informal Logic 25.3: 91-101.

[9] Biro, John, and Harvey Siegel (2006b): 'Pragma-Dialectic versus Epistemic Theories of Arguing and Arguments: Rivals or Partners?' In P. Houtlosser and A. van Rees, eds., Considering Pragma-Dialectics: A Festshrift for Frans H. van Eemeren on the Occasion of his 60th Birthday. Mahuah, NJ: Erlbaum, 2006, pp. 1-10.

[10] Biro, John, and Harvey Siegel (2015): "Argument and Context". Cogency 7.2: 27-41.

[11] Blair, J. Anthony, and Ralph H. Johnson (1993): 'Dissent in Fallacyland, Part 1: Problems with van Eemeren and Grootendorst'. In R. E. McKerrow (ed.), Argument and the Postmodern Challenge: Proceedings of the Eighth SCA/AFA Conference on Argumentation. Speech Communication Association, Annandale VA, pp. 188-190.

[12] Copi, Irving (1986): Introduction to Logic. New York: Macmillan, 7th ed.

[13] Eemeren, Frans H. van, and Rob Grootendorst (1984): Speech Acts in Argumentative Discussions. Dordrecht: Foris.

[14] Feldman, Richard (1994): 'Good Arguments'. In F. F. Schmitt (ed.), Socializing Epistemology: The Social Dimensions of Knowledge. NY: Rowman and Littlefield, pp. 159-188.

[15] Feldman, Richard (2005a): 'Deep Disagreement, Rational Resolutions, and Critical Thinking'. Informal Logic: 25.1: 13-23.

[16] Feldman, Richard(2005b): 'Useful Advice and Good Arguments'. Informal Logic 25.3: 277-287.

[17] Goldman, Alvin I. (1994): 'Argumentation and Social Epistemology.' Journal of Philosophy 91: 27-49.

[18] Goldman, Alvin I.(2003): 'An Epistemological Approach to Argumentation.' Informal Logic 23: 51-63.

[19] Godden, David (2015): 'Argumentation, Rationality, and Psychology of Reasoning'. Informal Logic 25.2: 135-166.

[20] Hitchcock, David (2000): 'The Significance of Informal Logic for Philosophy'. Informal Logic 20.2: 129-138.

[21] Johnson, Ralph H., and J. Anthony Blair (1977): Logical Self-Defense. Toronto: McGraw Hill-Ryerson.

[22] Lumer, Christoph (1991): 'Structure and Function of Argumentations — An Epistemological Approach to Determining Criteria for the Validity and Adequacy of Argumentations.' In F. H. van Eemeren, R. Grootendorst, J. A. Blair and C. Willard (eds.), Proceedings of the Second International Conference on Argumentation, vol. 1A, Amsterdam, SICSAT, pp. 98-107.

[23] Lumer, Christoph (2005a): 'Introduction: The Epistemological Approach to Argumentation — A Map'. Informal Logic 25.3: 189-212.

[24] Lumer, Christoph (2005b): 'The Epistemological Theory of Argument: How and Why'. Informal Logic 25.3: 213-243.

[25] Pinto, Robert C. (2001): Argument, Inference and Dialectic: Collected Papers on Informal Logic. Dordrecht: Kluwer.

[26] Siegel, Harvey (2019): 'Epistemic Rationality: Not (Just) Instrumental'. Metaphilosophy 50.5: 608-630.

[27] Siegel, Harvey and John Biro (1997): 'Epistemic Normativity, Argumentation, and Fallacies'. Argumentation 11.3: 277-292.

[28] Siegel, Harvey and John Biro (2008): 'Rationality, Reasonableness, and Critical Rationalism: Problems with the Pragma-Dialectical View'. Argumentation 22.2: 191-203.

[29] Siegel, Harvey and John Biro (2010): 'The Pragma-Dialectitian's Dilemma: Reply to Garssen and van Laar'. Informal Logic 30.4: 457-480.

[30] Walton, Douglas N. (1989): Informal Logic: A Handbook for Critical Argumentation. New York: Cambridge.

[31] Walton, Douglas N.(1990): 'What Is Reasoning? What Is an Argument?' Journal of Philosophy 87.8: 399-419.

[32] Walton, Douglas N. and Erik C. W. Krabbe (1995): Commitment in Dialogue: Basic Concepts of Interpersonal Reasoning. New York: SUNY Press.

[33] Woods, John (2013): 'Advice on the Logic of Argument'. Revista de Humanidades de Valparaiso 1.1: 7-34.

[34] Woods, John (2016): 'The Fragility of Argument'. In Fabio Paglieri, Laura Bonelli, and Silvia Felletti, (eds.), The Psychology of Argument: Cognitive Approaches to Argumentation and Persuasion, London, College Publications, pp. 99-128.

Aspects of Walton's Theory of Argumentation Schemes

Hans V. Hansen
Department of Philosophy & Centre for Research in Reasoning, Argumentation and Rhetoric, University of Windsor, Windsor, Ontario, Canada N9B 3P4
`hhansen@uwindsor.ca`

Abstract

This paper seeks to further develop the study of argument schemes by focussing on some of the fundamental questions in scheme theory: the relationships between argument types (kinds), argument schemes, and arguments. It is proposed that we can distinguish schemes associated with argument types (generally) and schemes associated with normative types. This distinction has consequences for how we can use schemes for argument evaluation. I attempt to locate Doug Walton's impactful contributions to scheme theory by identifying the particular choices he made.

I knew Doug Walton as a professor when I was a student in Winnipeg in the 1970s. Later I worked with him on an encyclopaedia project through the University of Amsterdam. Subsequently, he gave me the privilege of reading several of his book manuscripts as they were being prepared for publication. We became co-editors of the Critical Reasoning series for Cambridge University Press; and, finally, we became colleagues as fellows in the Centre for Research in Reasoning, Argumentation, and Rhetoric at the University of Windsor. So, for over 40 years, Professor Walton was an influential intellectual presence in my life. He initiated and maintained friendships with a large circle of peers as well as with many younger and emerging scholars. As teacher, mentor and friend, he was unfailingly kind, helpful and encouraging.

I think Doug Walton had a unique and wonderful logical imagination. He found the important building blocks for a new logical theory in the dense thicket of beliefs

---

I wrote this essay during the time of COVID-19 when it was difficult to have access to the many Walton books in my university office. Hence, my references to Walton's key ideas are somewhat eclectic. I am very grateful to Michael Yong-set for his close attention to the first draft. He caught many slips of the pen and made several positive suggestions which I have adopted. I am similarly indebted to Christopher Tindale and Waleed Mebane at CRRAR for their critical observations and helpful suggestions.

and conventions most others took for obstructive and unimportant non-logical noise. In a way, however, Doug's many original insights and innovations were veiled by his plain conversational writing style — as plain as the softness of his voice — which eschewed both literary embellishment and unneeded theoretical language. One is not intimidated by reading any of Walton's many articles or books: he invited us to think along with him along familiar paths. The appreciation and admiration comes later, when we reflect on where he has taken us.

Below, Section 1 gives a brief background to the genesis of argument scheme theory as it appears to informal logicians. Section 2 discusses some of the elements of scheme theory and Section 3 distinguishes three different ways that schemes can be used in argument evaluation. Section 4 delves into aspects of Walton's theory of argument schemes, and Section 5 summarizes our findings about Walton's approach to scheme theory.

1 Background

Walton led the way in the study of argumentation schemes in the community of informal logic scholars. His comprehensive studies of fallacies eventually brought him to see that argumentation schemes underlaid both bad and good argumentation: that the most familiar fallacies could be seen as bad tokens of types of arguments that could also have good tokens.

> ... [It is] possible to solve the identification problem of fallacies by identifying the argumentation schemes that define the types of argumentation corresponding to many of the various individual fallacies. [16, p. 14]

> Many of the most common forms of argument associated with major fallacies, such as argument from expert opinion, *ad hominem* argument, argument from analogy and argument from correlation to cause, have now been analyzed using the device of argumentation schemes. ...in many instances they are reasonable but defeasible arguments. [18, p. 220]

The patterns of argument types that eventually became the major focus of Walton's work in recent years he called 'argumentation schemes'. As a method of evaluating natural language arguments these schemes have great promise and we naturally want to understand them as fully as we can. Hence, we are led to ask questions related to argumentation schemes, and a new field begins to take shape: argumentation scheme theory. J.A. Blair's 2001 *Argumentation* article in which he asks a dozen or more questions about Walton's work on presumptive argumentation schemes may be seen as a founding article in this new area of study. I want to try to add to our understanding of argument schemes by asking additional questions about the field, and locating Doug Walton's work within it.

2 Elements of Scheme Theory

In this section I review the relationships between argument schemes, logical forms, arguments, and argument types.

2.1 Argument schemes and logical forms

One of the first questions that comes up in scheme theory is how schemes are related to the logical forms found in formal logic. The two are sometimes assimilated and sometimes distinguished. Walton was undecided about whether to count logical forms as argument schemes, but in my view he laid the groundwork for a clear distinction which I have tried to developed [5]. Although formal logic and scheme logic share the same ranges of individual and predicate variables, they differ in their syncategorematic elements. Logical forms come from the use of truth-functional connectives and universal and existential quantifiers (and extensions to various modalities). Argument schemes are different in that they include at least one schematic (non-truth-functional) connective (e.g.,'says that') or a schematic (presumptive) quantifier (e.g., 'generally' or 'mostly'), elements not found in Quine's canonical notation. Accordingly, adapting a formulation from Hitchcock [9], we can define argument scheme like this:

> An argument scheme is (i) a pattern of argument, (ii) made of a sequence of sentential forms with variables, of which (iii) at least one of the sentential forms contains a use of a schematic constant [connective] or a use of a schematic quantifier, and (iv) the last sentential form is introduced by a conclusion indicator like 'so' or 'therefore'. [5, p. 444]

It is the third condition that sets argument schemes apart from logical forms. On a higher and more general level, forms and schemes may be seen to belong to the category of argument *pattern*. The differences between formal and scheme logics has consequences for the methods used to evaluate arguments as well as for the expectations that arguers should entertain: formal deductive logic is monotonic, scheme logic non-monotonic.

2.2 Argument schemes and arguments

What is the distinction between argument schemes and arguments? It is the same as that between logical forms and arguments. Schemes, as we defined them above, are linguistic structures ready to have their empty spaces — the variables in their sentential forms — filled by propositions, terms and predicates that will turn them into arguments. But whereas schemes are made of sentential forms, arguments are made of full-fledged sentences or propositions or statements (pick your favourite). The main take-away from the scheme-argument distinction is that the relationship between schemes and arguments is like that between formula and product, and the point to emphasize is that no argument is a scheme and no scheme is an argument. (Why does one "use" a scheme? To produce an argument. Why does one use an argument? To prove or persuade or explore.)

2.3 Argument types and arguments

The passages above from Walton indicate that both fallacious arguments and reasonable arguments can belong to the same argument type. Thus, the same argument type will have tokens some of which are good arguments and some of which are not good arguments, just as the types song and potato will include good and bad tokens of songs and potatoes. Types are abstract objects and the relationship between type and token is like that between platonic forms and object. As Quine said, "Type and token nicely span the abstract and the concrete" [12, p. 218].

2.4 Types and schemes

What keeps us from using 'argument type' and 'argument scheme' interchangeably?

2.4.1 Ontological difference

Schemes are made of sentential forms. An important point about schemes is that they are fully known: they are like poems and unlike arguments in that they can have no unstated components. Schemes can be changed or modified, improved or abandoned, or replaced by other schemes. Arguments come from schemes by substituting for variables. Argument types, by contrast, are abstract objects of thought; their essences are to be discovered. Arguments come from types by imitation, not substitution. Types are to schemes as semantics is to syntax. Schemes are to arguments as syntax is to sentences. Thus, schemes are situated ontologically between argument types and arguments.

2.4.2 Using types and schemes

Schemes have uses including argument generation, argument analysis and evaluation (the subject of Part 3), and searches for argument types [6]. One may consciously use a scheme to produce an argument by making substitutions for all the variables; Practically, argument types are less immediately useful. I think most of us use types inadvertently when we make new arguments by imitating or modifying argument tokens with which we are familiar. So, in argument generation, schemes and types are used in very different ways: in the case of schemes they are used directly, in the case of types they are used indirectly, if not unwittingly.

2.4.3 Schemes and argument production

Having acknowledged that schemes can be used in argument generation I want immediately to introduce a qualification. In some cases, in class-room situations where we make exercises so students can learn what the schemes are, Yes, schemes are directly used that way. But, No, not generally. In my view, the great plurality of people do not use argument schemes to generate arguments. When people make an argument which fits scheme X it is because they have learned to use type X arguments, copying others whom they have observed to use tokens of type X arguments with apparent success, or perhaps they figured it out themselves that type X arguments are worth something in argumentation. In this way, arguments may accidentally be instances of schemes but most arguers have never considered a scheme and could not state one with anything near sufficient detail such that their arguments could be said to be intentional instances of, and guided by, schemes. The analogy here is with rules of grammar. I don't know the rules of grammar of English — they don't teach that stuff anymore — but I make grammatically correct English sentences for the most part (I hope). I learned to make grammatically correct sentences by imitating other people whom I took to be making correct sentences, and being corrected when, in their view, I made a mistake. So, the familiarity is with tokens of a sentence type of which I have an intuitive understanding, not with rules of grammar; similarly, (most) people's familiarity is with tokens of argument types, it is not with argument schemes. Here I am taking my lead from Peter Strawson who gave the following summary of the philosopher's work:

> just as the grammarian ... labours to produce a systematic account of the structure of rules which we effortlessly observe in speaking grammatically, so the philosopher labours to produce a systematic account of the general conceptual structure of which our daily practice shows us to have a tacit and unconscious mastery. [14, p. 7]

The informal logicians working on formulating the schemes for the various types of arguments are like the grammarians working on formulating the rules of grammar. They somehow understand what is in question, and seek to give expression to their findings in schemes. I call this the optimization project — the community of informal logic scholars collectively improve our schemes for argument types by reflection on argument types and exchanging information. So, types are basic and prior to schemes in the sense that schemes depend on types: schemes are as dependent on types as a portrait is on its subject. Schemes, I think, will always be of more interest to logicians and theorists who want to understand and improve argumentation than they will ever be to arguers who just want to argue.

2.4.4 A complication

Consider this case. When person H supports a proposition by appealing to A, an expert or an authority, H uses an appeal to authority argument (AE-argument). *H alleges that A is an authority and that A said that p is the case, and that this is a reason for others to accept p.* Now, it may be that none of the conditions we associate with good AE-arguments are present in the argument H is advancing. Maybe p is not what A really said; maybe p does not belong to the field, F, in which A has expertise; maybe A, although having university degrees in F, is prone to making unwarranted judgments in F; maybe none of the other experts in F agree with A; maybe A has no evidence for p, even though p does belong to F, A's area of expertise. Any or all of these underminers could be true but the argument made by H would still be of the type AE-argument because that was the kind of evidence that H brought forward for the conclusion. That it does not satisfy the good-making conditions associated with AE-arguments is another matter.

2.4.5 Two kinds of argument types

So, we can distinguish being a token of a type and being a good token of the type. We are led to the realization that 'argument type' is ambiguous: it has both a descriptive, or neutral extension, and a normative extension. By 'argument type X' we may mean any argument of type X, good, bad or indifferent, or we may mean only *any good* argument of type X. Thus we will have a set of necessary and sufficient conditions that neutrally characterizes an argument type, and another set of necessary and sufficient conditions that characterize good arguments of that type. The hypothetical situation in the preceding paragraph is meant to show that H's argument may not satisfy the normative conditions for the type good-AE-argument although it did satisfy the neutral conditions for being that type of argument.

2.4.6 Two kinds of schemes

Schemes, we noted, can be used in the analysis of arguments. They can also be used to represent types of arguments ([18, p. 94], or give an analysis of them in the sense of a definition of argument types [16, p. 14]. Since schemes are made of sentence forms, not sentences, and definitions are made of sentences, this needs some explanation. Let us say that for any argument type, T, and any scheme, S, S is a definition of T if, and only if, any correct substitution instance of S is an argument of type T. But now we ask, are schemes to be expressions of the conditions for the neutral, descriptive argument types or the normative argument types? They can be descriptive schemes — D-schemes — neutral characterizations / definitions of the argument type, or normative schemes — N-schemes — schemes that include at least some of the sufficient conditions for good instances of the scheme such that any instance will at least give some initial (*pro tanto*) support for its conclusions. With reference to the story in 2.4.4, we can put it this way: H intentionally imitated tokens of type X which, unbeknownst to H, accidentally coincided with the D-scheme for that type but did not coincide with the N-scheme for that type. (If using schemes for argument mining, D-schemes will cast a wider net than N-schemes.)

3 Schemes and Argument Evaluation

Given that one of our purposes is to develop methods that will assist us in evaluating arguments in natural language discourse, we must consider how the recognition of good and bad tokens of argument types can be included in evaluation procedures using argument schemes. Above, I expressed reservations about the possibility of people who had no familiarity with schemes — and even those who do have the familiarity — using them in argument production. A method of argument analysis and evaluation using schemes, however, will obviously need not only familiarity with schemes but also a good understanding of them. The use of schemes in production and evaluation is therefore, in reality, not symmetrical.

We can envision three alternative programs within argument scheme theory. I will outline two based on D-schemes and one based on N-schemes before turning to a discussion of Walton's theory. First, let us consider one of the D-scheme approaches. We will work with what was also one of Walton's favourite argument types, the AE-argument type, but not his scheme. Instead, let this be the D-scheme for that argument type:

(Sc-1) X is taken to be an expert;
         X is taken as having said that p;
         So, allegedly, p.

3.1 The DS-Approach

Instances of Sc-1 satisfy the descriptive conditions for being AE-arguments. Arguments in which X is not an expert, X never said that p, or p would not follow even if the premises were true, could still be instances of Sc-1. The scheme imports no normative weight to its instances so the goodness of arguments that come from it must be determined by further measures. One way to do it is to state the standards for good AE-arguments along with the scheme. Generally, each D-scheme, defining a different type of argument, will have a unique set of accompanying standards. Merrilee Salmon [13, p. 78], among others, has taken this approach in her discussion of AE-arguments, saying that if these standards are met they "make it reasonable to take the authority's word on the matter":

(S1) The authority invoked is an expert in the area of knowledge under consideration;

(S2) There is agreement among experts in the area of knowledge under consideration;

(S3) The statement made by the authority concerns his or her area of expertise.

We might add: (S4) The authority did say that p; and possibly other requirements as well. This is the *Standards Approach with D-schemes* (DS-approach).[1] The advantage of this version of the DS-approach is that the standards are explicitly stated. They are public, they can be the subject of discussion, they can be refined and improved, even rejected. An argument evaluator will compare instantiations of Sc-1 — intentional or accidental — with these standards and make an evaluation of the argument in light of them. It is also possible to operate the DS-approach without explicitly stated standards. Our acquaintance with argument standards, unless we are argumentation theorists, will likely be intuitive and unstated; hence, they may not be complete or correct.

We should notice, by the way, that Salmon treats AE-arguments, along with appeal to popular opinion and *ad hominem* arguments as inductive arguments; that means that in her use of schemes the illative adverb will be 'probably' or a synonym thereof. More recently, David Zarefsky [23, Ch. 5], who discusses several of the best known argument schemes, also treats them as inductive kinds of arguments.

---

[1] The AE-scheme Salmon actually uses is what I will, in Section 4, call a PN-scheme. The point being made here is that one can evaluate arguments by associating standards with schemes.

3.2 The N-Approach

A second way of connecting argument-type standards to schemes is by incorporating the standards into the schemes. This results in N-schemes which do not neutrally identify arguments but match only *good argument* of the type. We may call this the *Normative Approach to Argument Schemes* (N-approach). Consider this scheme for AE-arguments:

(Sc-2) P1. X is an authority with credentials c, who believes and states p.

P2. Credentials c are positively relevant to p.

P3. X is not biased with regard to X's advocacy of p.

P4. There is wide agreement among the relevant experts that p is acceptable.

P5. p belongs to an appropriate field in which consensus is possible.

C. p should be accepted. (Adapted from Groarke and Tindale, [3, p. 317].[2])

Any argument that is an instance of this scheme, if it has acceptable premises, will give strong support for its conclusion. All the standards mentioned by Salmon and more are included. In building into the scheme the complete set of normative standards for the argument type, we approximate a deductivist approach to the evaluation of natural language arguments: take any instantiation of Sc-2 and the conclusion of the argument follows with great surety, if not with deductive necessity.

3.2.1 Deductivism?

A possible reason to take this approach is that we should consider argument schemes as analogous to valid logical forms, like *modus ponens*, and then we can think of scheme logic as analogous to formal logic because they both provide us with logically good patterns of reasoning. A slightly different consideration is due to natural-language deductivism, one of the approaches to evaluating natural language arguments. It relies not on valid forms but on the semantic definition of deductive validity; arguments that are not valid on first inspection will receive supplements to make them so. Hence, argument schemes should be designed to incorporate this validity. Hence, they will be normative schemes.

When arguments are valid by design, as the method of natural-language deductivism prescribes, questions about their strength will not be about the following-from relation but will be solely focussed on premise acceptability. This seems to some to make argument evaluation easier whereas others see it as distorting the argument

---

[2]Modifications have been made to P2, P3 and P5.

being considered and having the potential of leading to mistaken evaluations. However this disagreement is resolved, we should observe that deductivism changes the locus of defeasibility, the definitional hallmark distinguishing non-monotonic and monotonic (deductively valid) arguments. In its predominant sense an argument is defeasible if it is such that additional information consistent with the given premises can alter the degree of support for the argument's conclusion ([22, p. 34], [19, p. 223]). On deductivists' models, since the following-from question has been decided in favour of deductive validity, defeasibility can only be understood as the possibility that arguments are vulnerable because something may be amiss with one or more of their premises. The locus of defeasibility is thus shifted from the following-from dimension of argument evaluation to the premise-acceptability dimension. But if valid arguments can be considered to be defeasible, then the distinction between monotonic and non-monotonic (defeasible) reasoning is no longer of any value. That a familiar distinction should be retired is not a bad thing in itself, but many who want evaluation of arguments to be closely connected to the arguments actually given rather than reconstructions of them, consider it important to preserve the distinction.

3.2.2 The N-scheme problem

One difficulty that will face all N-approaches is what we are to say when one or more of the necessary conditions (one of the premises) of an instance of an N-scheme is not satisfied; for example, in AE-arguments, if it is false that p belongs to an appropriate field in which consensus is possible. In such a case, it will be unclear whether the argument under inspection is a not-good argument of the AE-type or not an argument of that type at all. N-schemes invariably face the possibility that negative judgments about premiss acceptability will be ambiguous between quality of the instance of the type and membership in the type. This has consequences for evaluators who will wonder what their next step should be. A method that uses D-schemes rather than N-schemes will not land in the same difficulty because the type is identified by the D-scheme apart from the standards to be applied.

3.3 The DQ-Approach

A third way of connecting standards with argument schemes is to connect D-schemes with questions that probe the strength of any of their instances. Call this the *Questions Approach* to argument evaluation with schemes (DQ-approach). The DQ-approach considers arguments to be situated in the context of dialogues that include an essential role for questioning. Such frameworks have been developed by Aristotle, they form part of past and present legal procedures, and Hintikka has even adapted the idea to scientific investigation with scientists putting questions to nature who answers when prompted with experiments [7]. The basic model envisions a proponent making arguments to an interlocutor for a point of view. The interlocutor asks the proponent questions about their argument which they are to answer. There are different kinds of dialogues each with their own starting points and goals [18, p. 9], but this is the core framework around which we can build a method of argument evaluation. The DQ-approach can be clothed in different ways but here we will take it as definitive that it uses D-schemes to identify argument types and then permits questions relevant to evaluation of arguments that fit the scheme. Our focus in this section is on noticing variations in the kinds of questions that can be asked.

3.3.1 Direct and indirect dialectical settings

There will be a distinction between direct (or primary) dialectical settings as just described and indirect (or secondary) dialectical engagements. In direct settings there is an actual dialogue with an argument-giver and an interlocutor, and no one else speaks. The interlocutor asks questions of the argument-giver who answers and the interlocutor makes an argument evaluation subsequent to the answers received. In such settings dialogue rules are very important as they regulate turn taking between the dialoguers and facilitate an orderly generation of commitments [22, pp. 381–82]. In direct dialogical settings the questions will be in the second-person, as in, "What evidence do you have that this is a popular opinion?"

Indirect dialectical settings are those in which one or more of the demands of primary dialectical settings is relaxed. (a) The argument giver, G, does not speak but someone else speaks for them (think of Trotsky's defence of Marxism); (b) G may have directed their argument not to a single interlocutor, H, but to an audience with more than one questioner (think of a news conference with several reporters); (c) H may ask questions of others agents than G (think of checking on G's credentials); (d) G's original argument may not have been directed to H (think of an appeal process to a higher court); or (e) the dialogical protocols may not be strictly followed (think of some recent political debates). And there are other variations possible, for example, symmetrical question and answer roles between dialoguers. The reason to distinguish direct and indirect dialectical settings is that direct dialectical setting are very restrictive. If we want a DQ-approach to argument evaluation that has wide applicability, then we need to loosen the framework of direct dialectical settings in order to make it workable in different contexts. This will affect the kinds of questions that can be asked, of whom they can be asked, who asks them, also how they will be voiced. For example, in indirect settings the questions will be in the third-person, e.g., "What evidence is there — does anybody have — that this is a popular opinion?" And the dialogue protocols will need adjusting as well.

3.3.2 Questions and presuppositions

Questions may be classified in different ways but especially important for dialogical argumentation is the awareness that questions have presuppositions. "Should there be a first question?" is answered by a Yes or No, and presupposes the tautology: either there should be a first question or there should not. "What should the first question be?" is a Wh-question and presupposes the non-tautological claim, "There should be a first question" [8]. Questions do not profitably contribute to argumentation if their presuppositions are not acceptable. One DQ-approach (to argument evaluation with D-schemes) might specify that only Y/N questions can be asked, another approach may have no restrictions, and another might specify a certain order in which questions must be asked.

3.3.3 Interchangeability of DQ- with DS-approach

"Was the authority in a position to see the experiment she reports?" answered with a "Yes" gives the same information as the statement, "The authority was in a position to see the experiment she reports." Any statement can be recast as a question and any question can be recast as a statement. (Witness the television programme "Jeopardy".) So, the DQ-approach can be just another way of bringing standards of the argument type to bear on arguments. But the standards will be basic to the questions: we ask these questions because these are the relevant standards. So, why bother with the DQ-approach? One reason is that evaluation requires information gathering and that is facilitated by the use of questions. Another reason is that the standards for a type of argument may not be fully known to an interlocutor and questioning is a way of probing an argument, feeling one's way forward, hoping to find something relevant. Finally, actively questioning arguments and proponents is closer to real argumentation practice than the direct application of argument standards.

3.3.4 Specified and unspecified questions

When D-schemes are used there will be no evaluation of arguments without questions, only recognition of an argument type. Robert Pinto, defending the DQ-approach to evaluation with argument schemes, wrote as follows:

> ...the point of identifying argument schemes lies in the critical questions associated with them. It isn't the schemes that do the evaluative work,

it's we who do the evaluative work by judiciously challenging premises, identifying the risks of basing conclusions (if only presumptive conclusions) on those premises, and enriching our grasp of potentially relevant fact that might alter the significance of the facts brought to lie in the premises. Evaluating an argument ... is a matter of probing its strengths and weaknesses, and we do well if the initial impetus for our probing comes from the critical questions that have, in the past, proved fruitful. [11, p. 104]

In this passage I take Pinto to have touched on two issues. He shows a preference for D-schemes over N-schemes when he says that it is we, by asking questions, who do the evaluative work and not the schemes. He also recommends the DQ-approach, indicating that the questions we should use to evaluate arguments could come from past experience. This last proposal can be taken two ways. It may be that an individual has their own past experience with arguments of type X and profitably draws on that experience in subsequent questioning. It could also mean that the questions to accompany D-schemes have been drawn up in a list by informal logicians based on their collective experience with type X-arguments, and an interlocutor uses this specified list in their evaluation of an argument. The questions used may be either specified or unspecified. Questions are specified when the argument evaluator has or is given a list of questions to ask about the argument and the evaluator is not expected to stray beyond them. Questions are partially specified when the list does not purport to be complete and interlocutors can ask additional questions on their own initiative as they think will be helpful. Questions are unspecified (completely) when an interlocutor has nothing other than their own experience and intelligence as a basis for addressing the argument.

Those arguers not trained in scheme theory will not have in mind specified questions for an argument type any more than they will be familiar with the standards for the argument type. Hence, they will not have a specified or partially specified list of questions to ask when they turn to argument evaluation. But this does not mean that they have no relevant resources. Their past experience in argumentation both as an observer and as a participant, and their intuitive sense of reasonableness, will provide them with enough critical questions for at least one go-around. Hence, the DQ-approach can be practised without any specified questions. The interlocutor can probe the argument on the table, either in a direct or indirect dialectical setting, with questions that come to mind. Some argumentation participants will be very good at this; others less so, but they are the ones who can be helped by consulting specified sets of questions recommended by logicians for the different D-schemes. My point here, is that the DQ-approach — the Question and Answer Approach with D-schemes — can be, and *is*, practised without given sets of specified relevant questions. Plato's Socratic dialogues may be an example of that, as are our own experiences.

3.3.5 Transference problem

An issue concerning the following-from relation is left over from our discussion of the N-approach. When premises do not entail their conclusions how are we to determine how well they support a conclusions? Since D-schemes are normatively neutral and all evaluation on the DQ-approach must stem from the use of questions, must there not be a question asking how well the premises of an argument supports its conclusion? Johnson and Blair [10, Ch. 2], for example, included both a relevance and a sufficiency condition in their model of a good argument. This suggests that on the DQ-approach we should ask, "Are these premises positively relevant to the conclusion?" and "Are they sufficient for the conclusion?" Hence, when the DQ-approach — in its pure form — is in use it should include an illative question about whether the argument transfers the acceptability of the premises to the conclusion.

In this section we have considered three different ways in which argumentation schemes can be instrumental in natural language argument evaluation. With D-schemes there must be external attachments. These can be either argument standards (the DS-approach) or questions used to probe an argument's strengths and weaknesses (the DQ-approach). Alternatively, instances of N-schemes have at least some of the standards of the N-type of argument built-in to their premises (the N-approach) and do not have any external aids.

4 Walton's Choices

In view of the different approaches to the use of schemes we have considered, let us now explore how Walton's scheme theory fits into the distinctions we have made. He made choices in developing his own theory, and we should try to understand them.

4.1 Forms and schemes

Walton was undecided about whether to consider the forms of formal logic as schemes. Sometimes he excluded them [17, p. 84] and sometimes he included them ([22, p. 1], [18, p. 110]). Since his primary interest was in types of defeasible arguments that he correlated with argument schemes, and arguments fitting valid logical forms are not defeasible, that is a reason to exclude them from his theory of argument schemes. However, we may presume that he wanted as comprehensive a theory of argument patterns as possible, and since valid logical forms are thought to be identifiable patterns of arguments alongside argument schemes, they should be included. This makes sense especially since Walton wanted to work cooperatively with colleagues in AI and automated reasoning projects.

4.2 Schemes as definitions of types

Walton thought of schemes as defining types of argumentation [16, p. 14] and also as representing "the basic structure of each type of argument" [18, p. 96]. I have emphasized the difference between neutral types of arguments and normative types of arguments. Walton, however, distinguishes the valid deductive type, the inductively strong type and the plausible type [20, p. 360]. Any argument pattern that will define any of these three types of arguments will be a normative pattern. Hence, we expect the patterns for plausible (also, sometimes, presumptive) argument types to be a kind of N-scheme.

4.3 Defeasible arguments

Walton distinguished presumptive arguments from deductive and inductive arguments [17, p. 84]. The most basic difference concerns defeasibility. Defeasible arguments are tentative, subject to changes in logical strength and they are candidates for retraction. Deductive arguments (at least the valid ones), in which conclusions follow of necessity, are not defeasible. It is inductive and plausible (or presumptive) arguments that are defeasible. Inductive arguments, as Walton considered them are associated with both physical and social science. They depend on objective gathering of data and their validation procedures involve the use of mathematical calculations and probability theorems. Their conclusions are more or less probable relative to their premises and there is nothing especially dialectical about them. Plausible arguments, by contrast, have their home in ordinary non-scientific discourse and they are validated through an examination guided by critical questions that have little or no mathematical content. It is this last class of arguments that was the main focus of Walton's interest. In such arguments conclusions are plausible (not probable or necessary) relative to their premises. Walton did insert the illative adverb, 'plausibly' in some of his schemes but not in all of them, yet that is what he intended. Although his research was circumscribed by the inductive-plausible distinction there is little doubt that he would agree that some defeasible arguments occupy a borderland and might be treated either way.

Sometimes Walton talked of defeasible arguments as plausible arguments and sometimes as presumptive arguments. A discussion of the exact difference between the two concepts is beyond the scope of this overview essay. But when defeasibility is considered in relation to plausible arguments we are in the neighbourhood of inference to the best explanation, and when it is associated with presumptive arguments the connection is with the concept of burden of proof [19]. An argument fitting one of Walton's schemes creates a presumption for its conclusion and the interlocutor must accept it or take up a burden of proof to show that the argument did not establish its conclusion presumptively. The questions that go with each of the schemes can guide the interlocutor in her efforts.

4.4 Partially normative schemes

Walton's argument schemes are N-schemes rather than D-schemes. The distinction may not have occurred to him since the genesis of his interest in schemes stems from his discovery that some fallacies were also reasonable arguments. Hence, he may have looked for the patterns that made them reasonable and these would have some normative content. This is the AE-scheme Walton favoured as an illustration of his approach.[3]

(Sc-3) Source X is an expert in subject domain D containing proposition p;
X asserts that proposition p in domain D is true;
So, [plausibly], p. ([17, p. 87], [18, p. 67]; [22, p. 310])

This is an N-scheme because some of the standards for the AE-argument type are built-in. But it does not approximate deductive validity as did Sc-2 above; hence, any instances of the scheme will be a defeasible argument. I will say that this scheme is *partially normative* since the transference of acceptability to the conclusion is partially accomplished by the scheme but is then furthered through questioning. Instances of partially normative schemes (PN-schemes), even though they are defeasible, still, *pro tanto*, conditionally transfer some acceptability to their conclusions. (See also [1, p. 374].)

4.4.1 The N- (and PN-) scheme problem

If the premises of Sc-3 are satisfied — if they are true or acceptable — then (i) X *is* an expert in domain D, (ii) X *did* assert that p, and (iii) p *does* belong to D. If any of these conditions are not satisfied one may be inclined to say that the argument is a weak token of the PN-scheme, Sc-3. But that is not right. If the premises aren't satisfied, it is not an argument of the type. This is a problem for any kind of N-schemes whether they be fully or partially normative. The complaint cannot be

---
[3] I have re-lettered it.

dismissed by saying that instances of the scheme are defeasible. Defeasibility does not help here: it can only affect the level of support for a conclusion given that the premises are acceptable, it does not include the question of premise acceptable itself. Any type of argument can have a false premise, including deductively valid arguments and they are not defeasible. Having a false premise makes an argument unsound or uncogent. It also disqualifies the argument from belonging to the N-type to which it was thought to belong. Walton may have adopted a charitable attitude towards the problem of false premises. He could say that the arguer thought the premises were true and then judge it to be an argument of that type with a weakness. But that is just to invoke the spectre of D-schemes in which premises are only alleged to be true and conclusions are only alleged to be supported.

4.4.2 A terminological difference

One more question needs to be addressed under the heading of schemes. Walton preferred the term 'argumentation scheme' to 'argument scheme'. The distinction is related to that between arguments and argumentation. Arguments are sets of propositions; argumentation is an interactive process between people. Still, 'argumentation scheme' is ambiguous between (i) a narrow sense in which it simply means the definition or the structure of a type of argument "used" by a proponent in the course of the activity of arguing, and (ii) a wide sense in which it is a pattern of moves — assertions and questions — in a segment of argumentation. This latter notion Walton sometimes referred to as an argumentation theme [16, pp. 201–202], a generalized profile of a fragment of dialogue [18, p. 235]. But these elaborations and extensions do not fit what Walton has shown us of schemes. His schemes are schemes in the narrow sense. The use of the term, 'argumentation scheme' is perhaps best explained by the hypothesis that he viewed arguments from the perspective of their role in dialogues where they regularly incur resistance. So, Walton's schemes are argument schemes in the narrow sense; they occur primarily in argumentation and so, in his eyes, they were argumentation schemes.

4.5 The Need for Questions

Walton chose external attachments to partially normative schemes as a means of evaluating arguments, and he further chose an approach involving questions rather than one with stated standards [20, p. 358]. He was hugely influenced by Charles Hamblin [4, Ch. 10] who took this to be the best way to make sense of fallacies, and also by van Eemeren and Grootendorst [2] who, in their influential pioneering work in argumentation theory, integrated questioning with their model of a critical discussion. Walton extended Hamblin's idea that fallacies could best be analysed in dialogical contexts to the idea that arguments, generally, could best be studied in dialogical contexts. He further deepened our understanding of argumentation by proposing that there were different types of dialogues, each having their own starting points, goals [18, p. 9] and criteria for relevant questions. Questions, like arguments, are an essential element of argumentation dialogues so Walton partly modelled his theory on what he took to be actual practice, as did the main writers who influenced him.

4.5.1 The W-approach

Walton's approach to argument evaluation is not an N-approach with all the argument standards included in the premises. Nor is it a DS-approach with external scheme-associated standards or a pure DQ-approach with all evaluation dependent on scheme-associated questions. Walton's programme rests on two sources: partly normative (PN-) schemes and supplementary scheme-associated questions. None of the 60 schemes catalogued in the Compendium [22, Ch. 9] are D-schemes and none of them are deductively valid, non-defeasible N-schemes. They are PN-schemes for defeasible arguments with associated questions. (Sometimes we say 'defeasible schemes' but that makes sense only if it means 'schemes for defeasible arguments'.) So, Walton's approach (the W-approach) may be described as a mixed approach. It takes something from the DQ-approach because it gives a central role to questioning and it takes something from the N-approach because the schemes, when instantiated, give some *pro tanto* support to their conclusions.

4.5.2 Walton's Questions

These are the questions Walton associated with Sc-3, his scheme for AE-arguments.

(Q1)	Expertise Question:	How credible is X as an expert source?
(Q2)	Field Question:	Is X an expert in the field that p is in?
(Q3)	Opinion Question:	What did X assert that implies p?
(Q4)	Trustworthiness Question:	Is X personally reliable as a source?
(Q5)	Consistency Question:	Is p consistent with what other experts assert?
(Q6)	Backup Evidence Question:	Is X's assertion based on evidence?

More precisely, these are sentential forms for questions. When 'X' is replaced by a term designating and individual or possibly a group, they become questions. These sentence forms are designed for an indirect dialectical setting since they are in the third-person and are not addressed directly to the argument giver. However, one sees that they are, for the most part, easily adaptable to a direct dialectical setting. The advantage of putting questions in the third-person, as Walton did with all his schemes, is that it makes them general and askable whenever an argument fitting the associated scheme is far or near, and it allows consultation with an indefinite range of sources since the argument artificer need not be the only one to whom questions are directed. That the questions are general and impersonal also makes it possible to direct them to oneself, so we can monitor our own arguments as well as be the judges of other people's arguments. Notice, however, that the further we get from direct dialectical settings the less urgency there is to follow the protocols of dialogue rules. That gives questioning evaluators more leeway.

4.5.3 Kinds of Questions

Q1 and Q3 are (forms for) Wh-questions and the other four are (forms for) Y/N questions. Walton throughout his schemes admitted a mix of both kinds of questions. In the 60 schemes included in the Compendium ([22], not considering the variations) about two-thirds have associated questions, 14 have only Y/N questions, six have only Wh-questions, and about a third have a mixture of Wh- and Y/N questions. It should be considered a strength of Walton's approach that it allows room for both kinds of questions as this will lead to a deeper exploration of the argument's quality. It would be interesting to know for each scheme what led Walton to think that these were the appropriate questions to associate with it. (With reference to the six questions above, notice that Q2 and Q3 are better suited for a D-scheme than an N-scheme since for any argument that is an instance of Sc-3, these question have already been answered.)

4.5.4 Specified Questions

Most of the schemes in the Compendium [22] have associated specified critical questions. A set of questions is specified if they are recommended as the questions to be asked. There is, however, a difference between no specified questions, a few helpful questions to get you started, and a prescribed set of specified questions. The W-approach works with specified questions. There is an advantage to this. A method that leaves less to individual initiative and talent — characteristics that exist in varying quantities — can only help to make bouts of argumentation fairer and more balanced undertakings. But there is also an attendant problem here of which Walton was well aware. When is a set of questions complete, such that if they are all answered satisfactorily, the argument must be accepted as having plausibly (or presumably) established its conclusion? He called this the completeness problem for argumentation schemes [21]. (It is not the same as the problem of whether an inventory of schemes is sufficiently complete to deal with all arguments that can be encountered in a domain of discourse. Walton was also well aware of that problem.)

4.5.5 The Transference Question

When we identified Walton's schemes as partially normative it was because they gave some initial plausibility to the conclusions of any substitution instances with acceptable premises. (D-schemes don't do that.) Still, on Walton's scheme theory a complete argument evaluation has to await a review via specified critical questions. Since his schemes are partially normative it is redundant to ask in connection with the instantiations of any of them whether the premises are relevant to the conclusions. But this still leaves room for a further illative question to be associated with each of Walton's schemes, namely, whether the premises are sufficient for their conclusions. Walton did not regularly include such illative questions among his scheme-associated critical questions. Instead, in some of his schemes, he introduces a defeasible conditional that gives a defeasible *modus ponens* link from premises to conclusion. The inclusion of the conditionals in the schemes vitiates the need for an illative question concerning sufficiency. But only 13 of the 60 schemes in the Compendium include such a defeasible conditional despite the observation that in defeasible arguments a conclusion is accepted provisionally and "the argument used to derive the conclusion has at least one premise that is a generalization that is subject to exceptions" [19, p. 223]. It is puzzling why a scheme, especially a defeasible but partially normative one, would have a linking conditional given that the evaluation is to be completed by a review via questioning.

Here is what I would consider to be a partially normative scheme for appeal to popular opinion:

(Sc-4) p is generally accepted as true;
So, presumably there is a reason in favour of p.

Like Walton, I would consider these two questions to be appropriate:

CQ1: What evidence like a poll or an appeal to common knowledge, supports the claim that p is generally accepted as true?

CQ2: Even if p is generally accepted as true, are there any good reason for doubting that it is true? (from [22, p. 311])

But Walton didn't offer Sc-4. Instead he proposed:

(Sc-5) p is generally accepted as true;
If p is generally accepted as true, that gives a reason in favour of p;
So, presumably there is a reason in favour of p. (Ibid.)

Walton's Sc-5 adds a conditional sentential form to my Sc-4. A conditional would make a great difference if it was a logical conditional or a truth-functional conditional because that would defuse the defeasibility of Sc-4. But the conditional in Sc-5 is a defeasible conditional: notice the quantifier, 'Generally'. It doesn't introduce defeasibility, it duplicates what is already there in Sc-4 and what CQ2 reminds us to consider. Adding the defeasible conditional to Sc-4 is like adding a corresponding truth-functional conditional to an argument that is already deductively valid: it serves no purpose other than to encourage Lewis Carrol and his followers.[4] Maybe Sc-5 is more ready for service in AI than is Sc-4, and that's OK; but from the point of view of argument evaluators, and also that of the theory of defeasible arguments, it only adds clutter.

There is another possibility. Some people think of (normative) schemes as warrants, and warrants can be formulated as conditionals. Recall Toulmin's [15] Bermudian and Swede. I am not convinced of the assimilation of schemes and warrants. Warrants seem more subject matter dependent than schemes, but leave that for another day. With regard to finding the right scheme for arguments from popular opinion, Sc-5 is more than a warrant since it also contains a sentential form for a particular proposition, and if the conditional sentential form alone be the warrant, the warrant is less than the scheme. So, to identify Sc-5 with a warrant will require some realignment of our concepts. Prospects are better if Sc-4 is considered: its corresponding conditional will be a defeasible one and could be a warrant, albeit a very general one.

Walton did not, for the most part, include an illative question with other questions associated with his schemes and in only some of the schemes does he include a defeasible conditional. It seems rather to be his view that the premises of presumptive arguments (arguments instancing one or another of his schemes) give sufficient support for their conclusions if they pass the battery of associated (non-illative) critical questions. In this way, sufficiency supervenes on a thorough specified dialectical examination: the PN-scheme ensures relevance, good answers to questions leads to sufficiency. This is my suggestion of how Walton's scheme theory deals with the transference question (i.e., whether premise acceptability has been transferred to conclusions). I think it is implicit in many of Walton's writings, and that it is the best way for us to make sense of the sufficiency-transference problem on the W-approach.

---

[4]I am thinking of his often reprinted, "What the tortoise said to Achilles".

5 Summary

Scheme theory can be developed in different ways and here I have tried to show that the version developed by one of the leading architects of the field resulted from consequential choices based on reasonable grounds. To summarize:

(a) A very coarse positioning of Walton's approach within general scheme theory can be made by considering two basic distinctions I have made. The one is between D- and N-schemes, the other is whether arguments fitting the schemes are to be evaluated with scheme-associated standards or scheme-associated questions. Using these distinctions, the following quadrants show how we can position various theorists' approaches to argument evaluation with schemes.

	Standards	Questions
D-schemes (descriptive)	Standards for argument types are given in addition to the scheme	Questions for argument types are given in addition to the scheme
N-schemes (normative)	Groarke & Tindale Standards for argument type are built into the scheme	Walton: Schemes are partially normative with additional specified questions

If the emphasis is on standards these can either be eternal to schemes (D-schemes) or internal and built into the schemes (N-schemes). If we think the role of questions should be foremost then we can either join them to D-schemes or attach them as supplementary aids to PN-schemes. Walton chose this last option. (b) Walton was ambivalent about whether logical forms should be included in the category of argument schemes. He chose to make his contribution to logic in the area of schemes but he seems to have thought that both forms and schemes would be needed for a universal theory of argument evaluation. (c) Walton recognized the difference between argument types and argument schemes and considered schemes to be representatives or definitions of argument types. (d) The argument types that Walton studied were defeasible arguments — presumptive or plausible — a proper subclass of normative types of arguments. These types of arguments, Walton thought, belonged to everyday non-technical natural-language argumentation. (e) Accordingly, Walton worked with partially normative schemes (PN-schemes) rather than full-fledged N-schemes or D-schemes. (f) Walton preferred the term 'argumentation scheme to 'argument scheme' not for structural reasons but because he took argumentation dialogues to be the natural home of arguments. He was against context-free argument analysis and evaluation. (g) Walton defended the question and answer method over the method of applying standards. He thought the question and answer method was not only the one that fits natural discourse argumentation the best, it is also suitable for evaluators with different levels of competence. (h) Walton's method formulates questions for indirect dialectical settings and they contain a mix of Y/N questions and Wh-questions. This gives his approach great flexibility and utility. (i) Walton's method puts specified questions at the fore. This is the contribution of logicians to others: to give guidance in argument evaluation. (j) Illative success, for Walton, supervenes on having satisfying answers to other appropriate non-illative questions.

A lot more about Douglas Walton's scheme theory has been said by others and we can be sure there is much more to come. What deserves remembering and celebrating on this occasion, however, is the depth and breadth and originality of his vision. The rest of us will be followers who — like I have done in this essay — will tinker with the details.

References

[1] Blair, J.A. (2001). Walton's argumentation schemes for presumptive reasoning: A critique and development. *Argumentation* 15: 365-79.

[2] Eemeren, F.H. van, R. Grootendorst. (1992). *Argumentation, Communication and Fallacies*. Hillsdale, NJ: Lawrence Erlbaum.

[3] Groarke, L., and C. Tindale. (2013). *Good Reasoning Matters!* 5th ed. Don Mills: Oxford University Press.

[4] Hamblin. C.L. (1970). *Fallacies*. London: Methuen.

[5] Hansen, H.V. (2020). Argument scheme theory, in *Reason to Dissent: Proceedings of the 3rd European Conference on Argumentation*, vol 2. C. Dutilh Novaes, H. Jansen and J.A van Laar, B. Verheij (eds.). London: College Publications. 341-55.

[6] Hansen, H.V., and D.N. Walton. (2013). Argument kinds and argument roles in the Ontario provincial election. *J. of Argumentation in Context* 2: 226 -58.

[7] Hintikka, J. (1992) The concept of induction in the light of the interrogative approach to inquiry. In *Inference, Explanation and Other Frustrations*, J. Earman (Ed.). Berkeley: University of California Press. 23-43.

[8] Hintikka, J., and J. Bachman. (1991). *What if ...? Toward excellence in reasoning*. Mountain View: Mayfield.

[9] Hitchcock, D. (2010). The generation of argument schemes. In C. Reed and C.W. Tindale (eds.), *Dialectics, Dialogue, and Argumentation*. London: College Publications. 157-166.

[10] Johnson, R.H., and J.A. Blair. (1993). *Logical Self-Defence*. New York: McGraw-Hill Ryerson.

[11] Pinto, R.C. (2001). *Argument, Inference and Dialectic*. Dordrecht: Kluwer.

[12] Quine, W.V. (1987). *Quiddities: Am Intermittently Philosophical Dictionary*. Cambridge: Harvard University Press.

[13] Salmon, M. (1984). *Introduction to Logic and Critical Thinking.* New York: Harcourt Brace Jovanovich.

[14] Strawson, P. F. (1992). *Analysis and Metaphysics.* Oxford: Oxford University Press.

[15] Toulmin, S.E. (1958). *The Uses of Argument.* Cambridge: Cambridge University Press.

[16] Walton, D.N. (1995). *A Pragmatic Theory of Fallacy.* Tuscaloosa: University of Alabama Press.

[17] Walton, D.N. (2006). *Fundamentals of Critical Argumentation.* New York: Cambridge University Press.

[18] Walton, D.N. (2013). *Methods of Argumentation.* Cambridge: Cambridge University Press.

[19] Walton, D.N. (2019). The speech act of presumption by reversal of burden of proof, in H.V. Hansen, F.J. Kauffeld, J.B. Freeman, and L. Bermejo-Luque (Eds.), *Presumptions and Burdens of Proof.* Tuscaloosa, University of Alabama Press. 220-33.

[20] Walton, D.N. (2020). Tools for teaching and learning basic argumentation skills, in J.A. Blair and C.W. Tindale (Eds.), *Rigour and Reason.* Windsor: Windsor Studies in Argumentation. 355-78.

[21] Walton, D.N. (Unpublished). Presumptions, critical questions and argumentation schemes.

[22] Walton, D.N., C. Reed and F. Macagno. (2008). *Argumentation Schemes.* Cambridge: Cambridge University Press.

[23] Zarefsky, D. (2019). *The Practice of Argumentation.* Cambridge: Cambridge University Press.

Less Scheming, More Typing: Musings on the Waltonian Legacy in Argument Technologies

Fabio Paglieri
Istituto di Scienze e Tecnologie della Cognizione, Consiglio Nazionale delle Ricerche, Roma, Italy
`fabio.paglieri@istc.cnr.it`

Abstract

The rich and complex legacy of Walton's theoretical work for argument technologies is discussed. This is framed more as a call to action than as a prophecy: by calling attention to certain areas of Walton's work yet to be fully exploited for the development of argument technologies, I hope to prompt scholars to take interest in them; conversely, by casting some reasonable doubts on the prospects of other lines of research inspired by Walton, I try to redirect resources and efforts towards more productive ambitions. My musings on this topic lead me to articulate two versions of argument scheme theory, discuss current limitations and future improvements of dialogue type theory, and critically assess the state-of-the-art of argument mining, as well as the potential for other types of argument technologies.

1 Introduction

The year 2020 begun in the worst possible way for argumentation scholars: the death of Douglas Walton was an enormous blow to us all, and doubly so for those interested in computational models of arguments. Indeed, Walton was not only one of the leading figures in argumentation theory, but also the single largest contributor to that joint venture of philosophy, computer science, and cognitive studies variously referred to as "computational study of argument", "argument technologies", "computational argumentation", or whatever moniker works best for you. Doug's role in this interdisciplinary enterprise cannot be stressed enough: before almost anyone else in the argumentation theory community, he was the first to pay close attention and serious interest in what was transpiring in AI and computer science, and to appreciate the rich opportunities for cross-fertilization available on both sides. Thanks to his early presence and seminal contribution in this burgeoning new field, his theoretical models were since the beginning, and remain to this day, the North Star of many research agendas on computational argumentation. As a result, of all theoretical frameworks that were not "natively computational",[1] Walton's argument schemes are by far the most frequently used in computational work, with Toulmin's model probably being a distant second [70], whereas other authoritative approaches in argumentation theory, such as pragma-dialectics [22] or the argumentum model of topics [62], have only recently begun to be taken into consideration in computer science (e.g., [46, 71]).

Given the stature of Walton's contribution to the computational study of argument, I propose to honor his memory here by offering some personal considerations on what will be his rich legacy in computer science and AI — whereas others will no doubt dwell on Walton's equally momentous impact in other areas. Even though Doug is no longer with us, it is clear that his voice will keep being heard in the field, thanks to his many works and all that came out of them. But to what facet of his message should we listen most carefully, and how should we understand his lesson?

I will articulate my musings in four segments. First, I will discuss some potential pitfalls inherent to (a certain version of) the widespread success of argument schemes in computational approaches to argumentation (section 2). This will lead me to put forward a slightly heretical proposition, to wit, that it can be very productive to focus mostly on critical questions, without necessarily paying (too) much attention to argument schemes per se (section 3). Then I will move to consider another crucial aspect of Walton's theoretical contribution on argumentation, i.e., the theory of dialogue types: I will show how this has received comparatively little interest in computer science, and I will argue that this lack of interest is a mistake, worth rectifying in future works (section 4). Finally, I will bring to bear these considerations on dialogical interactions in the wild, to provide some suggestions on how Walton's teachings help us making sure argument technologies may actually improve the quality of our everyday discourse ecology, instead of remaining largely irrelevant — as they appear to be now, to be honest (section 5).

Before we begin, a quick note of caution to my readers. Perhaps unusually for a paper submitted to a scientific journal, this contribution is intended mostly as an op-ed: I will present arguments, hopefully sound ones, to defend a specific, highly personal view on Walton's legacy for computational models of argument. While I firmly stand by this view, I do not believe for one minute that it is the only viable

---

[1] As an example of an influential theoretical approach that was natively computational, instead of being imported from philosophical reflection, consider Dung's abstract argumentation frameworks [20].

one, and I welcome and applaud the plurality of opinions that is bound to flourish on the exact scope and nature of Doug's impact on argument technologies: in fact, this polyphony will provide further proof (although none was needed) of his enduring influence in the field.

2 Too Many Schemes Will Kill You: On the Dark Side of Argument Schemes in Computational Argumentation

In order to avoid any misconception, let me be clear since the onset: Walton's argument schemes had a widespread, profound, enduring, and (mostly) positive effect on the computational study of argumentation. In the early days of the love affair between argumentation theory and computer science, argument schemes were instrumental to the development of several pioneering technologies in the field, such as the argument mapping tool Araucaria [60] and the argumentation software Carneades [28], and they remained central to many subsequent refinements of those early attempts. Moreover, argument schemes spawned a wealth of practical applications in a variety of domains: multi-agent communication [61], legal reasoning [55], collaborative decision making [68], democratic deliberation [6], science education [47], reasoning about trust [53], to name but a few. In light of such an impressive track record, there is no denying that Walton's argument schemes were (and still are) a huge hit in computer science; claiming otherwise would simply constitute a pedestrian attempt at rewriting history.

If so, then what is this disturbing "dark side" of argument schemes, which is alluded to in the heading of this section? In a nutshell, I believe Walton's theory occasionally became victim of its own success: this does not cancel out its many crucial achievements, of course, yet it invites caution in how we approach it for future forays on computational models of argument. To substantiate this claim, it is important to appreciate that there are at least two[2] different ways of using Waltonian argument schemes: let us label them as, respectively, the *Generative Approach* and the *Taxonomic Project*. These are not antithetical doctrines, and indeed they can

---

[2]The fact that I focus my attention on these two versions of argument schemes should not be taken to imply that this is the only distinction worth articulating, with respect to argument schemes; on the contrary, the notion is quite complex and multi-faceted, so much so that Yu and Zenker recently noted the absence of a consensus definition for it [82]). Anyone interested to further explore this complexity, well beyond the distinction highlighted here, will find food for thought in Blair [8], Garssen [27], Walton et al. [80], Prakken [58], Lumer [39], Shecaira [63], Wagemans [73], among others.

peacefully co-exist, to some extent, within the same research agenda: yet it seems to me that most applications of Walton's argument schemes in computer science ended up endorsing only, or mostly, one or the other of these two options. It is thus worth appreciating how they differ.

On the Generative Approach, argument schemes are conceived as a framework to model a reasoning pattern in a way that is both prescriptive (it can be used to discriminate between good and bad instances of that pattern) and realistic (it is derived from the observation of empirical occurrences and its validity is presumptive, i.e. subject to several conditions). In particular, the application of argument schemes to a phenomenon requires that such phenomenon can be meaningfully described by a premises-conclusion model, and that a finite set of critical questions can be identified to adjudicate on its (presumptive) validity: this strategy has been often successfully exploited in the computational literature on argument schemes, e.g. in modelling legal reasoning [55], democratic deliberation [6], reasoning about trust [53], among others. This is a "generative" view because the general machinery of argument schemes is used to identify new schemes, based on careful observation of reasoning patterns within a certain domain. Since reasoning practices are not bounded a priori, i.e. new patterns can be invented at any time by reasoners, this approach is not committed to the existence of a finite set of argument schemes — even though, typically, within a certain domain and a specific frame of reference, it is possible to distil a limited set of argument schemes, considered as the most representative for that particular application. Yet new argument schemes can always "pop up", and indeed in the Generative Approach argument schemes are seen as a precious tool for theorizing about reasoning, not as a comprehensive bestiary of argument types.

In contrast, the Taxonomic Project tackles head on the challenge of enumerating and organizing argument schemes as a list of all known reasoning patterns, divided by types and sub-types, to be meticulously described. Although the discovery of new species, i.e., new schemes, is always possible, the value of this approach is measured on the exhaustiveness of the resulting taxonomy: nobody ever claimed to have achieved a complete list, yet it is understood that any omission is culpable, albeit excusable — that is, it is a defect waiting to be redeemed, not a feature. This is also true for all applications that rely on this version of argument schemes, most notably argumentation mining, i.e. the attempt of automatizing argument retrieval. For argument schemes to be significantly useful for mining, we need to possess a list of them that is clear enough to be operationalized, and complete enough to avoid missing too many instances. Due to its commitment to providing a full-blown classification of various reasoning patterns, the Taxonomic Project is also the favored target of those who complain about the unbridled proliferation of argument schemes, as well as the somewhat arbitrary character of some of its articulations, and thus call for a regimentation inspired by some general normative principle or framework (for an interesting proposal on how to use Bayesianism to achieve that goal, see Hahn & Hornikx [30]).

While it is not my intention to adjudicate which version of argument schemes theory has been most successful overall, between the Generative Approach and the Taxonomic Project, I want to stress that the latter is the one facing the hardest problems, when it comes to real-life computational applications. This is especially apparent if we turn back our attention to argument mining: in spite of its theoretical promises [77] and some preliminary success [25, 33], the more we try to use argument schemes as a blueprint to automatically detect their instances in natural language, the more we run into the same wall, over and over — namely, the fact that the proposed taxonomy is too complex to provide efficient guidance for that particular task, especially at scale. Importantly, the problem is not computational complexity per se, since the theory of argument schemes is not formalized enough to even allow to precisely estimate its complexity (but see [21] for an interesting proposal on how to model computationally-friendly argument schemes in multi-agent communication — nothing to do with argument mining, alas!). The difficulty is much more basic: to put it bluntly, not even human coders can be trusted to reliably detect argument schemes in natural language, which in turn prevents us from having a clear insight on what processes they might use to do it (at most, we get some indications on how not to do it, looking at human reasoners), as well as making it extremely time-consuming to develop high-quality, scheme-based, human-annotated corpora, which is what we would need to attempt any brute-force solution to the hurdle of mining for argument schemes.

Notice that this bleak outlook on the prospects of Waltonian schemes mining is not based on the views of naysayers, such as myself [49]: it is actually the well-reasoned conclusion of those who tried to boldly go where no scholar has gone before, and lived to tell the tale of their very instructive failure. As a case in point, consider the work of Musi and collaborators, who used the argumentum model of topics [62] to help coders coping with the complexity of argument schemes annotation, since it provides "a hierarchical and finite taxonomy of argument schemes as well as systematic, linguistically-informed criteria to distinguish various types of argument schemes" [46, p. 82]. In spite of their efforts, minimally trained non-expert annotators still failed to reach a decent level of inter-coder agreement, whereas obtaining acceptable convergence on annotations involved refining the guidelines and using expert annotators, giving them specific training on that particular annotation task. As a result, Musi and colleagues were forced to conclude that "the annotation of argument schemes requires highly trained annotators" [46, p. 82]. To the best of my knowledge, this is the unofficial consensus position on argument scheme annotation in the field: to work decently, it requires expert annotators with a lot of training, hence it is not expected to produce scalable results any time soon [49]. In fact, some scholars think that argument mining should bypass argument schemes entirely, at least for the time being, and focus instead on argument structure at a different level of granularity (e.g., [54]). And even those who keep working to improve argument scheme annotation honestly acknowledge the limitations of the resulting corpora, "which tend to suffer from a combination of limited size, poor validation, and the use of ad hoc restricted typologies" [72, p. 1].

In saying this, I do not want to position myself as the Cassandra of argument scheme mining: I have no special prophetic talent, thus I cannot claim that this approach will never work. Instead, I am simply noting that it has had limited success so far, and that any small result came at a high price, especially in terms of annotation efforts. I suspect even the staunchest supporters of Waltonian argument schemes will be willing to concede that scheme-based argument mining is, to say the least, a very costly enterprise. This in turn invites cost-benefit considerations, which is indeed the main point I am making here: given the manifest costs of using Waltonian argument schemes for the sake of annotation and as a target for argument mining, do the expected benefits justify footing the bill? In fact, what exactly are these benefits?

Keep in mind that here we are talking of the benefits associated with using Waltonian argument schemes for the sake of argument mining, not in general terms. I have no doubt whatsoever that argument schemes have been, and will continue being, a very productive theoretical framework, both along the lines of the Generative Approach and as a tool to theorize and teach about arguments. But when it comes to processing large bodies of naturally occurring dialogues, with the aim of automatically detecting and retrieving argumentative features of that discourse, the fundamental question of Waltonian schematism, i.e. "What type of argument is this?", strikes me as having relatively low priority. Instead, many more pressing inquiries spring to mind, such as: How many parties are represented in the interaction (in real life, more than two is the norm rather than the exception, and this matters for evaluation; see [35])? What are the main standpoints? What are the key assumptions, and how are they justified? Who has the burden of proof, and how is it being shifted as the dialogue unfolds? What is the relevant standard of inference and/or proof? Even more basically, what type of dialogue is this, and what are the implications of its typology for the evaluation of individual positions and arguments (more on this in section 4)? Granted, Walton's theoretical insights have a lot to offer on many of these issues, or even on all of them. But this contribution is largely independent from the theory of argument schemes, which is why I think that particular theoretical tool should be abandoned, or at least temporarily benched, in our quest for a better computational handling of natural language argumentation (communication protocols for multi-agent interaction are a different proposition altogether, of course).

Two main objections to my claim can be anticipated. Firstly, argument schemes remain needed to individuate hidden assumptions of naturally occurring arguments, whenever they are presented elliptically — which happens often enough. Secondly, different argument schemes entail different sets of critical questions, which in turn are essential to assess the validity of the argument: unless we want to throw away the baby of evaluation with the bathwater of Waltonian schematism, we have to keep using argument schemes. The first objection can be answered in a twofold manner. On the one hand, it is fair to say that Waltonian schematism is not the only game in town, when it comes to interpreting enthymemes, hence we need not be committed to it, just because we want to be able to incorporate missing premises (as we do) in our argumentation mining effort: for some alternative and simpler options on how to deal with enthymemes, see [51]. On the other hand, the fact that some structural understanding of arguments is needed to deal with assumptions (and with many other features of argument, by the way) is both obvious and irrelevant: the question is not whether structure matters (of course it does!), but rather what targets we ought to use in our annotation efforts. Consider the following naturally occurring argument: "The opinion voiced by Wilson, Corbett and Tobey on the BMJ proves that airborne transmission of COVID-19 is a distinct possibility". Here the fact that Wilson, Corbett and Tobey are experts in pathogen transmission and that the BMJ is a trustworthy outlet of medical information are key assumptions for the conclusion to have presumptive weight, and these assumptions bear a clear connection to the text being an instance of an argument from expert opinion, on Walton's taxonomy. Nonetheless, naïve interpreters do not need to associate the text to that particular scheme (in fact, they may have no inkling of the existence of argument schemes at all), for them to be able to appreciate what are the relevant assumptions determining its presumptive force: in other words, competent arguers have an intuitive understanding of arguments based on expert opinion prior to, and independently from, the specifics of Waltonian argument schemes. Moreover, what ultimately bears on the evaluation of the argument are the relevant assumptions, not the underlying schemes. Hence it seems both clearer and more relevant to ask annotators to directly individuate those assumptions, instead of tasking them with ascertaining the corresponding argument scheme.

This brings us to the second objection: if we relinquish argument schemes in text annotation and argument mining, are we also giving up on critical questions? My answer to that dilemma is a resounding "no!", for reasons that I will discuss in the next section.

3 Flirting with Heresy: Critical Questions without Argument Schemes

The standard presentation of a Waltonian argument scheme first introduces the building blocks of the scheme, i.e. a set of premises and their conclusion in fairly abstract terms, then it enumerates a list of critical questions (CQs), i.e. reasonable doubts that the argument must be able to positively answer, in order to have (higher or lower) presumptive force. CQs typically target key assumptions of the argument, so that alternative approaches to argumentative schematism treat them as additional premises to the argument (e.g., [29]). The fact that CQs come after the argument scheme per se and that they are specific to it tends to give the impression that CQs are a sort of byproduct, something that would have no sense in the absence of schemes. However, the simple example we just discussed at the end of the previous section casts doubt on the derivative nature of CQs: in fact, it seems empirically false that argument schemes need to be mastered, for appropriate CQs to be asked. Quite the contrary: any competent reasoner is well equipped to challenge a claim from expert authority, or from sign, or from consequence, without necessarily being able to properly associate those claims with the corresponding Waltonian schemes. CQs are not read off our theoretical appreciation of argument schemes; it is the other way around — the challenges that are intuitively relevant to question an argument are a powerful clue to establish the reasoning pattern being instantiated in that argument.

Nothing of this entails a critique of Waltonian schemes as a theory of argument: on the contrary, the fact that our native intuitions align well with them goes to show that those schemes nicely capture recurring patterns of reasoning, as Walton intended them to. But being appropriate description of certain types of argument does not make schemes particularly salient or easy to annotate and detect in natural language texts. By the way, this is not a weird or isolated phenomenon: good taxonomies are typically not meant for mundane description of state of affairs. After all, this is exactly what happens also with the taxonomy of the natural world: anyone with normal vision and in possession of basic cognitive skills is capable of telling apart a *Psittacula krameri* (a small parrot) from an *Alces alces* (a moose), and even listing many of the relevant features that make these two animals different from each other, without having any clue on what their exact species are, and possibly even without having any knowledge of the underlying taxonomy. Similarly, telling apart "p because X says so" and "x because otherwise Y", as well as being able to properly evaluate their respective presumptive force in context, does not require having any notion of concepts such as "argument from expert opinion" and "argument from consequences".

This simple fact has far-reaching implications for argument coding and mining: if we want to successfully leverage people's intuitions in analyzing large corpora of argumentative discourse, we need annotation criteria that match those intuitions. Instead of asking coders to identify argument schemes, we should directly instruct them to (i) individuate premise-claim pair and (ii) articulate what gives presumptive force to that pair, possibly (iii) analyzing what assumptions need to be in place for that force to be justified. The latter aspect is precisely the province of CQs, and it is a space that can be explored without any explicit reference to argument schemes. Again, let us consider our previous example: "The opinion voiced by Wilson, Corbett and Tobey on the BMJ proves that airborne transmission of COVID-19 is a distinct possibility". You do not need to label this as an argument from expert opinion to be able to grasp what are the key options for challenging its claim. Are Wilson, Corbett and Tobey actual expert in this domain? Is BMJ a credible source, i.e. trustworthy and impartial? Does their opinion truly support airborne transmission of COVID-19? And what "distinct possibility" means exactly — are we talking about high, medium, or low risk of contagion? How high, or how low? And what about dissenting opinions on this subject? Are there any, and how authoritative are they?

Thus, the point worth stressing is that our capacity to critically question an argument is independent from, and arguably better developed than, our ability to correctly label it as a certain type of argument scheme (incidentally, this is very much in line with the tenets of the argumentative theory of reasoning, see [43, 44, 45]). We should embrace this fact and let it guide our efforts at argument annotation. Moreover, we should keep in mind that the function of CQs is primarily to test the mettle of an argument, and thus make it as strong as it can be — or, alternatively, reveal its inherent weakness. This has important applications in computer science, well beyond the scope of argument mining: CQs can provide inspirations for argument evaluation support systems, by suggesting ways to cross-examine the alleged validity of target arguments (e.g., in critiquing systems, see [17]), and even offer guidance on how to scaffold better arguments, thus enabling argument production support systems (for early proposals in this regard, see [69, 78]).

Freeing CQs from the embrace of argument schemes theory and moving beyond argument mining in their application to computational tools for better argumentation are just two ways of doing justice to Walton's legacy in computer science. A third fundamental move entails turning our attention to the other cornerstone of Doug's contribution to argumentation theory: namely, his typology of dialogues [79, 74]. Waltonian dialogues types have been extensively used in specific areas of computational models of argument (most notably, multi-agent communication, e.g.

[59, 1, 41, 42, 3, 56, 7]), while remaining relatively understudied in other domains of application, such as argument mining and user support systems. The remainder of this contribution is dedicated to explain why and how dialogue types can (and therefore should) provide valuable insights also in these areas.

4 The Lay of the Land: Dialogue Types as Orienteering Devices

According to Walton, dialogue types can be classified based on three distinctive features: the initial situation from which they originate, the individual goals of the participants engaged in the interaction, and the goal of the dialogue (which would be best characterized as its function, more on that later). Table 1 shows a summary of Waltonian dialogue types, taken from the work of Fabrizio Macagno [40] and only minimally adapted from the original proposal [79, 74]. The resulting list is not exhaustive, nor did Walton ever intended it to be: indeed, additional types of dialogue have since been analyzed, such as examination dialogues [75] and adjudication dialogues [57], while the possibility of mixed dialogues and shifts from one dialogue type to another was already contemplated in the original theory — so much so that the understanding of informal fallacies in this context hinges on discriminating between licit and illicit dialogical shifts (for discussion, see [67]).

Type of Dialogue	Initial situation	Participant's Goal	Goal of Dialogue
Persuasion	Conflict of opinions	Persuade the party	Resolve or clarify issue
Inquiry	Need to have proof	Find and verify evidence	Prove (disprove) hypotheses
Information-Seeking	Need of information	Acquire or give information	Exchange of information
Deliberation	Dilemma or personal choice	Co-ordinate goals and actions	Decide the best available course of action
Eristic	Personal conflict	Verbally hit out at opponent	Reveal deeper basis of conflict

Table 1: Waltonian dialogue types (from Macagno [40])

In spite of its intuitive appeal, Walton's taxonomy of dialogue types has been challenged both on theoretical grounds and in terms of practical applicability: here

I will briefly discuss the former set of criticisms, whereas the latter will be dealt with in the next section. Much theoretical perplexity with Waltonian dialogue types is associated with their aptitude, or lack thereof, in delivering a workable account of fallacies as illicit dialogue shifts: since fallacies are not our current topic of interest, and since I endorse a radically different understanding of them with respect to Walton [81, 9, 48, 50], we can dispense further discussion of this point (but see [67, 23, 24]). Instead, following Lewinski [34], I will focus on three other aspects of theoretical dissatisfaction with Waltonian dialogues types: (i) unclear status of the taxonomy, (ii) confusions on the notion of "goal of the dialogue", and (iii) weak link between some of these conversational types and argumentation. The first concern boils down to asking whether Waltonian dialogue types are to be interpreted as normative or descriptive notions: their treatment in Walton's work provides justification for either interpretation, yet some have insisted that this ambiguity does not do any good to the theory [23, 24, 34]. Moreover, unless we interpret dialogue types as being endowed with normative force, it remains unclear how we can make use of them to adjudicate on purely normative issues, such as, once again, fallacy evaluation. I happen to agree with this line of criticism of the notion, yet it has no traction on my own interest for dialogue types: as it will become clear in a moment, I find the notion useful first and foremost as a descriptive device, so I have no problem giving up any pretention of normativity for it (although I suspect Walton would have had something different to say in that regard).

The second criticism targets the notion of "goal of the dialogue": Lewinski wonders whether these goals are "based on empirical analyses or stipulated on the basis of theoretical considerations? In other words, are these goals familiar to, or at least reflectively recognisable by, the discussants concerned or are they formulated by some theorist, in this case Walton himself?" [76, p. 21]. Even more drastically, McBurney and Parsons notice that, "although Walton and Krabbe talk about the goal of a dialogue and the goal of a dialogue type, only participants can have goals since only they are sentient. (...) Instead of dialogue goals it makes sense only to speak of participant goals and dialogue outcomes" [76, p. 263]. This comment is revealing of a pernicious misconception, one that was invited by the somewhat relaxed, pre-theoretical usage of the notion of "goal" in the hands of Walton and Krabbe: the idea that a goal can only be a mental representation of a desired state of affair, thus something that can be predicated only of sentient beings. However, cognitive scientists working on goal-directed action have clarified that goals, in the cybernetic sense of a set-state capable of orienting action, can also be externalized, and that a prime example of external goals, i.e. goals that need not be represented in any specific mind to work, are functions — either natural (e.g., the function of an organ) or artificial (e.g., the function of a chair, or of an institution). Without going into the details of the theory of functions as external goals (but see [19, 12, 14, 15]), looking at the list of items indicated as "goals of the dialogue" by Walton and others, it is quite clear that what they actually intend to capture is the function of the corresponding dialogue type: the function of an inquiry, for instance, is to prove or disprove certain hypotheses, while the function of a negotiation is to reach a settlement that all parties can accept. Succeeding or failing to achieve these outcomes will determine whether that particular interaction is socially successful or not, and such outcomes are typically obtained by agents that are pursuing their own agenda, i.e. their own individual goals: once we appreciate the difference between the speakers' goals and the dialogue functions, this dialectic is neither problematic nor novel — in fact, it is how social order is normally achieved by autonomous agents in all walks of life [12, 13]. To paraphrase Adam Smith, it is not from the benevolence of the arguers that we expect our conversational achievements, but from their regard to their own self-interest.

The final theoretical worry with Waltonian dialogue types concerns to what extent they are supposed to represent genuine instances of argumentation. Lewinski [34, pp. 21–22] articulates this as a dilemma: either dialogue types are not inherently argumentative in nature and instead simply include an argumentative component, in which case it is unclear how they can offer reliable guidance on normative issues; or they are purely argumentative types, in which case they require us to stretch unreasonably the notion of "argumentativeness", in order to cover also confrontation-free types of discourse, such as inquiry and information-seeking dialogue. Since I do not see dialogue types as a normative tool, I am happy to embrace the first horn of this dilemma: sure, Waltonian dialogue types make sense only as an empirically based, open-ended taxonomy of conversational interactions, in which argumentation may occur (and often does), alongside many other types of discourse. Does this mean dialogue types have no pull in adjudicating on the normativity of arguments that take place within their boundaries? Contra Walton, I agree with Lewinski and others in claiming that dialogue types are ineffective in deliberating on the internal rationality of argument (for instance, on establishing whether something is fallacious or not); but I also hasten to add that dialogue types have much to contribute on the *external validity of arguments* — in other words, on whether arguing is appropriate in the first place in that conversational context, under what conditions and in which forms.

Appreciating this aspect brings us back to how dialogue types may yet have to offer further insight to argument technologies: determining what type of conversational interaction we are engaged in is essential to better gauge how to make an optimal argumentative contribution to it — or, if you want to look at this from an evaluative perspective, knowing the dialogical context is instrumental to assess the appropriateness of the argumentative moves that are made in it. Again, contextual appropriateness is not to be confused with internal validity: a fallacy, such as an ad hominem attack, will remain fallacious regardless of its dialogical context, yet in some domain its use may be conversationally appropriate, while in others it is not — as a case in point, the function of eristic confrontation is often well served by ad hominem attacks, insofar as they uncover personal animosity as the ultimate basis of disagreement, whereas they merely derail the interaction in persuasion or inquiry. Conversely, rationally sound argumentative moves may nevertheless turn out to be contextually inappropriate in certain dialogue types, in spite of their rational pedigree.

Should you be wondering how a rational argumentative strategy may be unwelcome in any conversational context worth its salt, let me introduce you to a relatively newfangled type of trolling: sea-lioning. If you are conversant with online debates, you may be already familiar with the term: on Wikipedia it is characterized as "a type of trolling or harassment that consists of pursuing people with persistent requests for evidence or repeated questions, while maintaining a pretense of civility and sincerity", whereas other sources describe it as "the confrontational practice of leaping into an online discussion with endless demands for answers and evidence" [16], "an intentional, combative performance of cluelessness" [32], or "incessant, bad-faith invitations to engage in debate" [66]. A caricature of this practice can be found in the webcomic strip that originated the term (see Figure 1), in which an actual sea lion suddenly pops up during a discussion between two human characters and start pestering them with overly polite questions on the motivations of their alleged dislike for sea lions.

I am sure the natural inclination of many argumentation scholars will be to sympathize with the sea lion here: after all, the standpoint expressed by the female character in the first panel (her dislike for sea lions) remains unjustified throughout the exchange, thereby vindicating the persistent, yet polite requests for justification by the rightfully offended sea mammal. However, the point that the cartoonist is making here is twofold: firstly, in certain dialogical contexts (here, a private conversation) the manifestation of one's standpoint does not necessarily require any justificatory backing, especially if the one asking for it was not included in the original exchange; secondly, and more importantly, even justificatory requests may become a tool for malicious sabotage of the conversation, when they are reiterated aggressively and ad nauseam.

Sea-lioning is interesting for us because it provides a good example of a rational practice perverted into an instrument of evil by applying it to the wrong, i.e. inappropriate, dialogical context. Instead of relaxing the rationality status of justificatory requests, we have to live with the fact that such requests may be woefully

Figure 1: The origins of sea-lioning (David Malki!, September 19, 2014, `http://wondermark.com/1k62/`)

inappropriate in certain contexts, no matter how rational they are in principle. In fact, the prima face rationality of these argumentative moves is precisely what makes them so helpful to trolls: whenever accused of doing something wrong, malicious sea-lioners typically reply by casting themselves in the role of maligned truth-seekers. For instance, in the context of the controversial Gamergate, i.e. violent episodes of trolling against minorities in the gamer community, participants accused of sea-lioning claimed that their opponents had misrepresented their expressions of sincere disagreement as harassment, ultimately using the accusation of sea-lioning to silence legitimate requests for proof [31]. As Daniel Cohen put it, argumentation theory "needs to catch up to the new reality. Successful strategies for traditional contexts may be counterproductive in new ones; classical argumentative virtues may be liabilities in new situations" [18, p. 179].

All of these point to the importance of dialogical context in assessing the appropriateness of argumentative contributions, both valid and invalid ones. On this interpretation, dialogue types work as rough-and-ready maps of the conversational context we are supposed to navigate: by telling us where we are in the discourse space and by providing us with its basic coordinates (i.e., from where this interaction started, what is at stake for individual participants, and what social function our exchange is supposed to serve), they point our argumentative contributions in the right direction. In order to do that, we need to be able to detect the relevant conversational context, and this is precisely where Waltonian dialogue types can help breaking new ground in argument technologies. This will require including the automatic detection of dialogue types among the priorities of argument mining. Unfortunately, at the moment this aspect does not seem to register often enough on the radar of argument miners: even though some understanding of dialogue dynamics have been indicated as a requirement for successful argument mining [11], recent reviews of the abundant work done in this area reveal very little in the way of dialogue type detection [37, 65]. Thus, not surprisingly, in a recent assessment of Walton's impact in AI and law, Atkinson and collaborators [4] highlight how dialogue types featured prominently in communication protocols for autonomous agents, whereas they still remain conspicuously silent in scaffolding argument-based NLP technologies, and make only cursory appearances in other sub-fields of AI (e.g., the Social Internet of Things, where they are used to capture different forms of engagement between users and intelligent devices; see [38]).

Of course, this notable lack of computational work on automatic dialogue type detection may have an innocent and convincing explanation: namely, that no such detection is possible for AI systems — either because the problem is too complex and ill-defined, or because we currently lack the required technology. Dispelling this reasonable doubt is the business of the next section, in which we will consider some of the practical criticisms that have been levied against Waltonian dialogue types.

5 Argument Typing in the Wild: Theoretical Perfection vs. Practical Efficacy

In his review of off-the-shelf theoretical models to analyze online discussion, Lewinski provides a rather comprehensive argument against the usefulness of Waltonian dialogue types for that purpose, as follows: "Walton's approach is ill-suited as a starting point for carrying out empirical research focused on actual contexts of argumentation, online political forum discussions in particular. This is partly due to the fact that the characteristics of various dialogue types are meant to be normative, and may thus be seen as deliberately not grounded in a methodical research of empirical reality. However, (...) it is hard to envisage how the concept of dialogue types can enrich our knowledge of actual conversational contexts of argument. Notably, since Walton's theoretical concepts and analytical tools are vague, it is very difficult to apply them in a way that would produce empirical results giving new insights into the argumentative qualities of online discussions. Simply put, once confronted with actual online discourse Walton's concepts generate difficulties (if not inconsistencies) that seriously impede efforts at drawing a methodologically strong, coherent picture of conditions for argumentation in online discussions" [76, p. 30]. In other words, when confronted with the nuanced complexities of real-life arguments (as opposed to artificial examples invented by the theoreticians themselves), Waltonian dialogue types fail to offer guidance for a secure interpretation of the relevant conversational context: too many options are available to the analyst (e.g., mixed dialogue, imperfect realization of a single pure type, multiple pure types with frequent shifts from one to another — not to mention whether such shifts are to be considered licit or not), and the theory does not provide any principled reason to adjudicate among them. Hence, according to Lewinski, invoking dialogue types as an empirically useful notion is a non-starter, especially when it comes to analyzing large corpora of online discussion.

These are serious and well-argued allegations, which require proper attention and go a long way in explaining why dialogue types have not received more interest from scholars trying to automatize the detection of relevant argumentative features. Lewinski is right on the money when he notes that Waltonian dialogue types are "theoretical categories, rather than ordinarily recognisable conversational contexts" [76, p 23]. Indeed, if we look at the defining features provided in the Waltonian taxonomy (Table 1), empirical scholars may find reason to despair: one item refers to the antecedents of the observed dialogical exchange, another invokes non-observable internal states of the arguers, and the last one requires divining the social function of the interaction. None of these characteristics are easy to pin down for the average human analyst, let alone for an automated classifier. Yet the appropriate reaction is not to throw away the baby of Waltonian dialogue types with the bathwater of their poor empirical characterization, but rather to acknowledge them for what they truly are: a seminal starting point — nothing less, nothing more. If we want to give dialogue types new life in computational models of argument, the first thing we have to do is to turn them into ordinarily recognisable conversational contexts, as Lewinski put it.

"But can it be done, really?" — now, this is the crucial question. Is there a "conversational footprint" of a specific dialogue type, something empirically analyzable that can be reliably tied to the corresponding theoretical notion, or to some reasonable approximation of it? Needless to say, this is largely an empirical question, yet theory and practice will have to meet halfway, if we want to make some progress on it. What I mean is that we should not enter this venture assuming that Walton's original taxonomy is non-negotiable. As mentioned, Walton himself did not intend it to be set in stone; even more importantly, our quest for an empirically grounded typology of dialogical contexts may lead us astray from the six-fold canon originally envisioned by Walton and Krabbe — and that is fine. Some of these types may turn out to be relatively irrelevant in some contexts of analysis, whereas other new typologies may be so prominent as to deserve their own category (something that we have already seen happening in some domains, e.g. legal reasoning; [55, 5]); in fact, even some of the defining features may have to go, as far as detection of concrete instances is concerned (e.g., private goals of the arguers may be so hard to establish from textual evidence alone as to be useless), whereas other characteristics will become central (e.g., lexical choices, tone, persistency of engagement). We should remain open minded about the vigorous theoretical and methodological tinkering required to turn Waltonian dialogue types into a productive tool for the automatic analysis and detection of real-life conversational patterns: far from showing lack of respect for this particular bit of Doug's many-folded legacy, such efforts have the potential to open new domains of application for it.

Beyond any methodological adjustment, I believe a fundamental shift needs to happen in how we look at Waltonian dialogue types, for them to became more relevant for argument mining. Instead of obsessing over the problem of how to reliably classify concrete instances of dialogue within our own favored typology, we should leverage the insight offered by a good taxonomy to formulate theory-driven predictions on how a conversational interaction is most likely to unfold, given its current features. This *shift from classification to prediction* requires a bit of a paradigm change in how we think about dialogue types, yet it is well worth the effort.

Here is the basic rationale for it: given a certain dialogical context, we can make an educated guess on how a certain conversational move will impact the prosecution of the interaction. Ironically, in online discussion the most paradigmatic argumentative contributions, such as questioning hidden assumptions or pointing out logical inconsistencies, will typically *decrease* the argumentative quality of the ensuing discussion, rather than elevating it. In particular, similar contributions on social media tend to have an inflammatory effect, increasing polarization and promoting a shift from reasonable debate to mere bickering — what Walton would have called, somewhat euphemistically, "eristic confrontation". Competent online arguers are familiar with this rule of thumb: if you really want to engage in argumentation with your contacts (as opposed to simply posturing as the solitary voice of reason in a sea of idiocy), you need to master the art of *covert arguing* — how to nudge all parties, yourself included, towards better standards of reasoning and debate, without being too obvious about it. On social media, manifest opposition closes the window of rationality and opens the door to verbal sparring: this is often true also of face-to-face interactions, but it is much more pronounced online, mostly due to the eminently public nature of that sphere of interaction, in which every dialogue tends to be (and be perceived as) a showdown in front of a vast audience.

However, appreciating the nuances of the dialogical context in which our conversations are embedded allows for much more fine-grained predictions on the likely consequences of new argumentative contributions. For instance, even a moderately dissenting voice can be taken as an aggressive threat, in the context of a group of friends pandering to each other and mutually reinforcing their own convictions (a frequent occurrence, in online "filter bubbles"; see [52, 10, 26, 64]); yet the same moderated dissent may have the opposite effect in a discussion already characterized by bitter opposition between two factions, where that contribution may actually open up a viable compromise and thus help making the conversation less polarized. Similarly, carefully fact-checking an opinion voiced by one of our contacts can either lead to resentment and escalation, if the original post was made in a spirit of partisanship or to articulate the poster's ideology [83], or on the contrary produce an amicable and timely correction of the record, if it is framed within a collaborative attempt of broadening the scope of discussion on an unresolved issue. As for questioning, e.g. asking further reasons, proofs, or references on a controversial point, whether it is interpreted as legitimate cross-examination or malicious sea-lioning will often depend on the relevant dialogical context: with respect to online debates, using private messaging instead of public posting and manifesting respect (rather than contempt) for the position being questioned go a long way in prompting a real answer to our queries.

In other words, paying attention to dialogue types help us understand the manifest destiny of our conversational interactions: this is relevant not only for descriptive purposes, but also (and possibly mostly) to foster prescriptive aims, for good or for bad. Let us first take a short walk on the dark side of large-scale debates, to see how we can use Walton's taxonomy to characterize the simple yet powerful tactics of those who deliberately try to cripple public debate. With some simplification here and there, the following is a pretty accurate Waltonian sketch of the agenda of most information operations currently occurring in online discussion (e.g., those of the infamous, Russia-sponsored Internet Research Agency; see [2]): by parroting critical discussion and persuasion, malicious agents aim to stop inquiry and information search, marginalize negotiation and undermine collective deliberation, in order to ensure that our day-to-day interactions reliably degenerate into eristic confrontations. Regarding this last aspect, it is essential to understand that the endgame for the professional saboteur is not to instigate prolonged and vicious quarrels, but rather to *stop the conversation* altogether among dissenting parties: the quarreling is just an expedient means to that end, to be complemented with other tactics, such as feeding us likeable contents designed to further radicalize our pre-existing convictions, while posturing as like-minded "friends" (on this particular tack of online manipulation, see [36]).

Consequently, the prime directive for all well-meaning arguers must be to keep the conversation going, as well as shaping it as an open-minded discussion among parties with a reasonable variety of standpoints and mutual respect for each other. The alternative, i.e. closeting ourselves in our own favorite online bubble and labeling everyone outside of it as a dangerous moron, is the nightmare scenario of collective rationality, as well as the target state of professional online saboteurs. The relevant counter-tactics of "argument patriots" are also amenable of being described using Walton's categories for dialogue types: responsible online arguers resist the temptation of eristic confrontation and tread carefully with critical discussion and persuasion, trying instead to frame all interactions as collaborative inquiry and information search, in order to enable large-scale deliberation and negotiation among different segments of the public. Of course, all of this is necessarily premised on our ability to detect the relevant differences among the various dialogical types, which brings us back to the enduring relevance of Walton's insight for argument analysis and evaluation.

However, Lewinski's original concern still remains to be addressed, and in fact it is made all the more urgent by our renewed interest in dialogue types as a blueprint for argument technologies: how can we translate the idealized Waltonian categories into something that human and artificial coders may be able to reliably detect? As it stands now, the characterization of dialogue types offered by Walton is woefully inadequate to the task: given a discussion corpus, it would offer very little guidance on how to parse and analyze it in terms of different dialogical segments. To me, this sounds like a call to arms, not a declaration of defeat: the combination of our need for dialogue type characterization with the current lack of principled methods to obtain it is a powerful motivation to launch a new line of research, at the interface between argument theory and computer science — one devoted to turn abstract theories of dialogue types into actionable concepts and reliable tools.

This may entail providing a mapping between Waltonian types and ordinarily recognisable conversational contexts, as Lewinski envisioned (classification); or it may lead us to abandon completely the taxonomic project, developing instead a *feature-based approach*, in which dialogical segments are not labeled as belonging to any specific type, but rather as exhibiting various key properties, possibly with certain degrees (characterization). Relevant features that we may want to operationalize will include traditional aspects of argumentative exchanges, such as open-mindedness, tolerance for dissent, argumentativeness, polarization, and mutual respect, as well as more practical characteristics, such as level of publicity and time constraints. Whatever factors we deem relevant, we would have to ensure that they are detectable with the required ease and accuracy, so that exploiting them for automatic dialogue type detection does not remain a pipe dream. And we should also make sure that they provide valuable insights to the arguers themselves, both in terms of analyzing and predicting their dialogical interactions: this in turn will inspire tools that can help users scaffolding their conversations and provide them with valuable feedback on the kind of dialogical context they are engaged with.

At present, this offers just a sketchy roadmap for integrating dialogue types within the broader agenda of argument technologies. Yet it has the potential of opening up broad theoretical perspectives and paving the way to innovative tools for argument support: an alluring panorama of options, one that we first glimpsed from the vantage point of Douglas Walton's seminal taxonomy of dialogue types. Abandoning the security of that starting position and moving on to explore the landscape it revealed us should figure prominently in the future agenda of argument scholars, knowing well that the farther away we move from Doug's initial intuitions, the more we honor his legacy.

6 Conclusions: What Waltonian Legacy?

If nothing else, this contribution should have confirmed what many readers already suspected since the onset, or knew for a fact: Doug Walton's contribution to the computational study of arguments was both vast and momentous. When it comes to cherry-picking what aspects of it will prove most fruitful in decades to come, I am sure the previous sections gave away my personal preferences. While I am a big fan of the Generative Approach to argument schemes as a way of scaffolding useful intuitions on recurrent patterns of reasoning, I am highly skeptical on the prospects of the Taxonomic Project, especially when it comes to leveraging argument schemes for the sake of argument mining. Conversely, while I applaud the extensive use of dialogue types in developing multi-agent communication protocols, I am dismayed by the relative lack of interest for the annotation of dialogical context: granted, there are solid reasons for that, the chief being that Walton's original taxonomy of dialogue types does not easily lend itself to reliable annotation of real-life debates. However, my reaction to that is to urge renewed efforts in that direction, possibly with greater deviation from Doug's original proposal, instead of simply dismissing the challenge.

In saying this, I am under no prophetic delusion: on the contrary, I am sure

Doug's ideas will inspire argument technologies in ways that will surprise and delight me, and I look forward to being proven wrong on several counts. Yet I believe that having an in-depth conversation on what are the most promising or less developed insights in the rich Waltonian canon will ultimately benefit the computational study of arguments, no matter what the initial convictions of the discussants happen to be. Thus, it is in that spirit that I offer these notes to the community, while at the same time urging all of us to be creative in how we take advantage of the many scholarly gifts that Doug generously bestowed upon our field, instead of lazily reaching for the low-hanging fruits. Because there lies the danger, the risk of making Walton's theory a victim of its own success: precisely because many of its contributions have become mainstream in argument technologies (once again, argument schemes come to mind), we may be tempted to be too carefree in how we use them, adopting them out of habit and applying them by rote, without proper critical reflection and with little interest in improving on them.

That would do Doug's memory a bitter disservice: his work had often a programmatic character, as if it was designed to act more as a springboard for future developments than as an end result, and his writings are ripe with intuitions yet to be fully explored, both theoretically and practically. In fact, when it comes to argument technologies, I do believe that one of the more underdeveloped and less applicative aspects of Doug's work, the theory of dialogue types, may have the most to offer in terms of potential for future research. Hence my titular invitation to stop scheming and start typing, in the same spirit of bold intellectual exploration that was the hallmark of our dearly departed friend, Douglas Neil Walton.

References

[1] Amgoud, L., Parsons, S., & Maudet, N. (2000). Arguments, dialogue, and negotiation. In W. Horn (Ed.), *Proceedings of the Fourteenth European Conference on Artificial Intelligence (ECAI 2000)* (pp. 338 — 342). Amsterdam: IOS Press.

[2] Arif, A., Stewart, L., & Starbird, K. (2018). Acting the part: examining information operations within ♯BlackLivesMatter discourse. In *Proceedings of the ACM on Human-Computer Interaction*, Vol. 2, CSCW, Article 20. New York: ACM, 1 — 26.

[3] Atkinson, K., Bench-Capon, T., & McBurney, P. (2005). A dialogue game protocol for multiagent argument for proposals over action. *Autonomous Agents and Multi-Agent Systems*, 11(2), 153 — 171.

[4] Atkinson, K., Bench-Capon, T., Bex, F., Gordon, T., Prakken, H., Sartor, G., & Verheij, B. (2020). In memoriam Douglas N. Walton: the influence of Doug Walton on AI and law. *Artificial Intelligence and Law*, 28(3), 281 — 326.

[5] Bench-Capon, T., & Prakken, H. (2010). Using argument schemes for hypothetical reasoning in law. *Artificial Intelligence and Law*, 18(2), 153 — 174.

[6] Bench-Capon, T., Prakken, H., & Visser, W. (2011). Argument schemes for two-phase democratic deliberation. In K. Ashley & T. van Engers (Eds.), *Proceedings of the 13th International Conference on Artificial Intelligence and Law (ICAIL'11)* (pp. 21 — 30). Menlo Park: AAAI Press.

[7] Black, E., & Hunter, A. (2007). A generative inquiry dialogue system. In M. Huhns, O. Shehory, E. H. Durfee, & M. Yokoo (Eds.), *Proceedings of the Sixth International Joint Conference on Autonomous Agents and Multi-Agent Systems (AAMAS 2007)*. Honolulu: IFAAMAS, ACM Press.

[8] Blair, A. (2001). Walton's argumentation schemes for presumptive reasoning: a critique and development. *Argumentation,* 15(4), 365 — 379.

[9] Boudry, M., Paglieri, F., & Pigliucci, M. (2015). The fake, the flimsy, and the fallacious: demarcating arguments in real life. *Argumentation,* 29(4), 431 — 456.

[10] Bozdag, E., & van den Hoven, J. (2015). Breaking the filter bubble: democracy and design. *Ethics and Information Technology,* 17(4), 249 — 265.

[11] Budzynska, K., Janier, M., Kang, J., Reed, C., Saint-Dizier, P., Stede, M., & Yaskorska, O. (2014). Towards argument mining from dialogue. In S. Parsons, N. Oren, C. Reed & F. Cerutti (Eds.), *Proceedings of the Fifth International Conference on Computational Models of Argument (COMMA 2014)* (pp. 185 — 196). Amsterdam: IOS Press.

[12] Castelfranchi, C. (2000). Through the agents' minds: cognitive mediators of social action. *Mind & Society,* 1(1), 109 — 140.

[13] Castelfranchi, C. (2001). The theory of social functions: challenges for computational social science and multi-agent learning. *Cognitive Systems Research,* 2(1), 5 — 38.

[14] Castelfranchi, C. (2012). Goals, the true center of cognition. In F. Paglieri, L. Tummolini, R. Falcone, & M. Miceli (Eds.), *The goals of cognition. Essays in honour of Cristiano Castelfranchi* (pp. 825 — 870). London: College Publications.

[15] Castelfranchi, C., & Paglieri, F. (2007). The role of beliefs in goal dynamics: prolegomena to a constructive theory of intentions. *Synthese,* 155(2), 237 — 263.

[16] Chandler, D., & Munday, R. (2016). *A dictionary of social media.* Oxford: Oxford University Press.

[17] Ches| nevar, C., Maguitman, A., & Simari, G. (2006). Argument-based critics and recommenders: a qualitative perspective on user support systems. *Data & Knowledge Engineering,* 59(2), 293 — 319.

[18] Cohen, D. (2017). The virtuous troll: Argumentative virtues in the age of (technologically enhanced) argumentative pluralism. *Philosophy & Technology,* 30(2), 179 — 189.

[19] Conte, R., & Castelfranchi, C. (1995). *Cognitive and social action.* London: Psychology Press.

[20] Dung, P. M. (1995). On the acceptability of arguments and its fundamental role in nonmonotonic reasoning, logic programming and n-person games. *Artificial intelligence,* 77(2), 321Ð357.

[21] Dunin-Kęplicz, B., & Strachocka, A. (2014). Computationally-friendly argumentation

schemes. In D. ?lęzak, B. Dunin-Kęplicz, M. Lewis, & T. Terano (Eds.), *Proceedings of the 2014 IEEE/WIC/ACM International Joint Conferences on Web Intelligence (WI) and Intelligent Agent Technologies (IAT)* (pp. 167 — 174). Washington: IEEE Press.

[22] Eemeren, F.H. van, & Grootendorst, R. (2004). *A systematic theory of argumentation: The pragma-dialectical approach.* Cambridge: Cambridge University Press.

[23] Eemeren, F.H. van, & Houtlosser, P. (2007). The contextuality of fallacies. *Informal Logic*, 27(1), 59 — 67.

[24] Eemeren, F.H. van, Houtlosser, P., Ihnen, C., & Lewiński, M. (2010). Contextual considerations in the evaluation of argumentation. In C. Reed & C. Tindale (Eds.), *Dialectics, dialogue and argumentation. An examination of Douglas Walton's theories of reasoning* (pp. 115 — 132). London: College Publications.

[25] Feng, V. W., & Hirst, G. (2011). Classifying arguments by scheme. In D. Lin (Ed.), *Proceedings of the 49th Annual Meeting of the Association for Computational Linguistics: Human Language Technologies* (pp. 987 — 996). Stroudsburg, PA: ACL.

[26] Flaxman, S., Goel, S., & Rao, J. M. (2016). Filter bubbles, echo chambers, and online news consumption. *Public opinion quarterly*, 80(S1), 298 — 320.

[27] Garssen, B. (2001). Argument schemes. In F. van Eemeren (Ed.), *Crucial Concepts in Argumentation Theory* (pp. 81 — 99). Amsterdam: Amsterdam University Press.

[28] Gordon, T., & Walton, D. (2006). The Carneades argumentation framework Ñ using presumptions and exceptions to model critical questions. In P. E. Dunne & T. Bench-Capon (Eds.), *Proceedings of the First International Conference on Computational Models of Argument (COMMA 2006)* (pp. 195 — 207). Amsterdam: IOS Press.

[29] Groarke, L., & Tindale, C. (2012). *Good reasoning matters! A constructive approach to critical thinking* (5th ed.). Oxford: Oxford University Press.

[30] Hahn, U., & Hornikx, J. (2016). A normative framework for argument quality: argumentation schemes with a Bayesian foundation. *Synthese*, 193(6), 1833 — 1873.

[31] Jhaver, S., Chan, L., & Bruckman, A. (2018). The view from the other side: the border between controversial speech and harassment on Kotaku in Action. *First Monday*, 23(2). doi:10.5210/fm.v23i2.8232

[32] Johnson, A. (2017). The multiple harms of sea lions. In N.J. Reventlow et al. (Eds.), *Perspectives on harmful speech online* (pp. 13 — 15). Cambridge: Berkman Klein Center for Internet & Society Research Publication.

[33] Lawrence, J., & Reed, C. (2016). Argument mining using argumentation scheme structures. In P. Baroni, T. Gordon, T. Scheffler, & M. Stede (Eds.), *Computational Models of Argument: Proceedings from the Sixth International Conference on Computational Models of Argument (COMMA)* (pp. 379 — 390). Amsterdam: IOS Press.

[34] Lewiński, M. (2010). *Internet political discussion forums as an argumentative activity type: a pragma-dialectical analysis of online forms of strategic manoeuvring in reacting critically.* Amsterdam: Rozenberg Publishers.

[35] Lewiński, M., & Aakhus, M. (2014). Argumentative polylogues in a dialectical framework: A methodological inquiry. *Argumentation*, 28(2), 161 — 185.

[36] Linvill, D. L., & Warren, P. L. (2020). Troll factories: manufacturing specialized disinformation on Twitter. *Political Communication*, 3(4), 447 — 467.

[37] Lippi, M., & Torroni, P. (2016). Argumentation mining: state of the art and emerging trends. *ACM Transactions on Internet Technology (TOIT)*, 16(2), 1 — 25.

[38] Lippi, M., Mamei, M., Mariani, S., & Zambonelli, F. (2018). An argumentation-based perspective over the social IoT. *IEEE Internet of Things Journal*, 5(4), 2537 — 2547.

[39] Lumer, C. (2011). Argument schemesÑan epistemological approach. In F. Zenker (Ed.), *Proceedings of the 9th OSSA Conference, Argumentation: Cognition & Community*. Windsor: OSSA.

[40] Macagno, F. (2008). Dialectical relevance and dialogical context in Walton's pragmatic theory. *Informal logic*, 28(2), 102 — 128.

[41] McBurney, P., & Parsons, S. (2002). Games that agents play: a formal framework for dialogues between autonomous agents. *Journal of Logic, Language and Information*, 11(3), 315 — 334.

[42] McBurney, P., & Parsons, S. (2009). Dialogue games for agent argumentation. In I. Rahwan, & G. Simari (Eds.), *Argumentation in Artificial Intelligence* (pp. 261 — 280). Berlin: Springer.

[43] Mercier, H., & Sperber, D. (2011). Why do humans reason? Arguments for an argumentative theory. *Behavioral and Brain Sciences*, 34(2), 57 — 74.

[44] Mercier, H., & Sperber, D. (2017). *The enigma of reason*. Cambridge: Harvard University Press.

[45] Mercier, H., Boudry, M., Paglieri, F., & Trouche, E. (2017). Natural-born arguers: teaching how to make the best of our reasoning abilities. *Educational Psychologist*, 52(1), 1 — 16.

[46] Musi, E., Ghosh, D., & Muresan, S. (2016). Towards feasible guidelines for the annotation of argument schemes. In C. Reed (Ed.), *Proceedings of the 3rd Workshop on Argument Mining* (pp. 82 — 93). Stroudsburg, PA: ACL.

[47] Nussbaum E.M., Sinatra G.M., & Owens M.C. (2012). The two faces of scientific argumentation: applications to global climate change. In M. Khine M. (Ed.), *Perspectives on Scientific Argumentation* (pp. 17-37). Springer: Dordrecht.

[48] Paglieri, F. (2016). Don't worry, be gappy! On the unproblematic gappiness of alleged fallacies. In F. Paglieri, L. Bonelli, & S. Felletti (Eds.), *The psychology of argument: Cognitive approaches to argumentation and persuasion* (pp. 153 — 172). London: College Publications.

[49] Paglieri, F. (2017). A plea for ecological argument technologies. *Philosophy & Technology*, 30(2), 209 — 238.

[50] Paglieri, F. (2019). The scaremongering fallacy of fallacy theory: how to improve reasoning without fear of error. In D. Gabbay, L. Magnani, W. Park, & A.V. Pietarinen (Eds.), *Natural arguments: A tribute to John Woods* (pp. 79 — 101). London: College Publications.

[51] Paglieri, F., & Woods, J. (2011). Enthymematic parsimony. *Synthese*, 178(3), 461 —

501.

[52] Pariser, E. (2011). *The filter bubble: What the Internet is hiding from you.* New York: Penguin Press.

[53] Parsons, S., Atkinson, K., Li, Z., McBurney, P., Sklar, E., Singh, M., Haigh, K., & Rowe, J. (2014). Argument schemes for reasoning about trust. *Argument & Computation*, 5(2-3), 160 — 190.

[54] Peldszus, A., & Stede, M. (2013). From argument diagrams to argumentation mining in texts: a survey. *International Journal of Cognitive Informatics and Natural Intelligence (IJCINI)*, 7(1), 1 — 31.

[55] Prakken, H. (2005). AI & Law, logic and argument schemes. *Argumentation*, 19(3), 303 — 320.

[56] Prakken, H. (2006). Formal systems for persuasion dialogue. *The Knowledge Engineering Review*, 21(2),163 — 188.

[57] Prakken, H. (2008). A formal model of adjudication dialogues. *Artificial Intelligence and Law*, 16(3), 305 — 328.

[58] Prakken, H. (2010). On the nature of argument schemes. In C. Reed & C. Tindale (Eds.), *Dialectics, dialogue and argumentation. An examination of Douglas Walton's theories of reasoning and argument* (pp. 167 — 185). London: College Publications.

[59] Reed, C. (1998). Dialogue frames in agent communications. In Y. Demazeau (Ed.), *Proceedings of the Third International Conference on Multi-Agent Systems (ICMAS-98)* (pp. 246 — 253). Washington: IEEE Press.

[60] Reed, C., & Rowe, G. (2001). Araucaria: software for puzzles in argument diagramming and XML. Technical report, Department of Applied Computing, University of Dundee.

[61] Reed, C., & Walton, D. (2005). Towards a formal and implemented model of argumentation schemes in agent communication. *Autonomous Agents and Multi-Agent Systems*, 11(2), 173 — 188.

[62] Rigotti, E., & Greco, S. (2010). Comparing the argumentum model of topics to other contemporary approaches to argument schemes: The procedural and material components. *Argumentation*, 24(4), 489 — 512.

[63] Shecaira, F. (2016). How to disagree about argument schemes. *Informal Logic*, 36(4), 500 — 522.

[64] Spohr, D. (2017). Fake news and ideological polarization: filter bubbles and selective exposure on social media. *Business Information Review*, 34(3), 150 — 160.

[65] Stede, M., & Schneider, J. (2018). *Argumentation mining.* Williston, VT: Morgan & Claypool.

[66] Sullivan, E., Sondag, M., Rutter, I., Meulemans, W., Cunningham, S., Speckmann, B., & Alfano, M. (2020). Can real social epistemic networks deliver the wisdom of crowds? In T. Lombrozo, J. Knobe, & S. Nichols (Eds.), *Oxford Studies in Experimental Philosophy*, Volume 3. Oxford: Oxford University Press.

[67] Tindale, C. (1997). Fallacies, blunders, and dialogue shifts: Walton's contributions to the fallacy debate. *Argumentation*, 11(3), 341 — 354.

[68] Tolchinsky, P., Modgil, S., Cortés, U., & Sànchez-Marrè, M. (2006). CBR and argument schemes for collaborative decision making. In P. E. Dunne & T. Bench-Capon (Eds.), *Proceedings of the First International Conference on Computational Models of Argument (COMMA 2006)* (pp. 71 — 82). Amsterdam: IOS Press.

[69] Verheij, B. (2003). Artificial argument assistants for defeasible argumentation. *Artificial intelligence*, 150(1-2), 291 — 324.

[70] Verheij, B. (2009). The toulmin argument model in artificial intelligence. In I. Rahwan, & G. Simari (Eds.), *Argumentation in Artificial Intelligence* (pp. 219 — 238). Berlin: Springer.

[71] Visser, J. (2017). Speech acts in a dialogue game formalisation of critical discussion. *Argumentation*, 31(2), 245 — 266.

[72] Visser, J., Lawrence, J., Reed, C., Wagemans, J., & Walton, D. (2020). Annotating argument schemes. *Argumentation*, in press. https://doi.org/10.1007/s10503-020-09519-x

[73] Wagemans, J. H. (2019). Four basic argument forms. *Research in Language*, 17(1), 57 — 69.

[74] Walton, D. (1998). *The new dialectic: Conversational contexts of argument*. Toronto: University of Toronto Press.

[75] Walton, D. (2006). Examination dialogue: an argumentation framework for critically questioning an expert opinion. *Journal of Pragmatics*, 38(5), 745 — 777.

[76] Walton, D. (2010). *Types of Dialogue and Burden of Proof*. Paper presented at the Computational Models of Argument Conference, Desenzano del Garda, Italy.

[77] Walton, D. (2011). Argument mining by applying argumentation schemes. *Studies in Logic*, 4(1), 38 — 64.

[78] Walton, D., & Gordon, T. (2005). Critical questions in computational models of legal argument. In *Argumentation in artificial intelligence and law* (pp. 103 — 111). Windsor: CRRAR Publications.

[79] Walton, D., & Krabbe, E. (1995). *Commitment in dialogue: basic concepts of interpersonal reasoning*. Albany, NY: SUNY press.

[80] Walton, D., Reed, C., & Macagno, F. (2008). *Argumentation schemes*. Cambridge: Cambridge University Press.

[81] Woods, J. (2013). *Errors of reasoning. Naturalizing the logic of inference*. London: College Publications.

[82] Yu, S., & Zenker, F. (2020). Schemes, critical questions, and complete argument evaluation. *Argumentation*, 34(4), 469 — 498.

[83] Zollo, F., Bessi, A., Del Vicario, M., Scala, A., Caldarelli, G., Shekhtman, L., Havlin, S., & Quattrociocchi, W. (2017). Debunking in a world of tribes. *PLOS ONE*, 12(7). https://doi.org/10.1371/journal.pone.0181821

The Waltonian Foundations of Argument Technology

Chris Reed
Dundee University, UK.
c.a.reed@dundee.ac.uk

1 Introduction

Argument Technology lies at the intersection of research into argumentation and artificial intelligence. It draws on theories from philosophy, linguistics, cognitive science, sociology, mathematics and law and integrates them with recent advances in computational models of argument. It lies at the applied end of the spectrum of AI research, and has a strong engineering orientation, aiming to develop and deploy technology that has direct impact on people's lives. Argument technology systems are used in domains as diverse as intelligence analysis and couples mediation, but perhaps one of the most significant areas of application is education.

The aim of this paper is to explore the ways in which two threads of the work of Douglas Walton — argumentation schemes and formal dialectics — have helped to shape the foundations of argument technology, and how they continue to contribute to cutting edge results across the field. With a particular focus on applications of argument technology to pedagogy, the paper reflects on the ways in which Walton's insights have not only provided theoretical backdrop and practical underpinning to research in the area, but how his approach to the study of argumentation has shaped the field and formed many of its central tenets.

2 Argumentation Schemes and their Annotation: Pedagogy and Research

From Walton's original manuscript [36] onwards, argumentation schemes have been closely connected to the teaching of critical thinking. Schemes provide a scaffolding first for recognising the structure of reasoning and subsequently for evaluating and critiquing that structure. Students find the analysis of argument structure extremely demanding. There are at least four distinct tasks each of which presents significant challenge. First, there is working out which passages are argumentative at all. In a textbook, of course, this is often simple because the task has usually already been carried out by the textbook author. But in naturally occurring discourse, deciding whether or not a passage includes an argument is surprisingly tricky: authors and speakers often use a veil of apparent argumentative structure to disguise all manner of rhetorical and procedural devices — from slander and mere opinion-giving, through humour and wit to rule-following and narrative. Letters to the editor of a newspaper, for example, constitute a domain in which one might expect to find plenty of argumentation, yet humour, irony, complaint and attention-seeking are far more common than anything that might be familiar to scholars of argumentation. Parliamentary records such as Hansard might similarly be expected to yield a rich source of argumentation, yet once again, character attacks and other ethotic devices are the mainstay. Legal proceedings (with the possible exception of Hollywood dramatisations) are again bereft of much true argumentation, characterised instead by execution of legal procedure and reference to cases and statute.

Even once argumentative passages are identified, the second challenge is to separate out the individual components. Such individuation is again surprisingly difficult: natural discourse does not naturally fall into clearly punctuated propositions, but is instead contaminated with complex syntactic transformations, implicated meanings, and half-expressed ideas. Transcripts of spoken discourse are even more challenging, with corrections, interruptions and disfluencies of many kinds.

Then with individuation accomplished, the third challenge is to distinguish premises from conclusions; to identify both the connections between components and the directionality of those connections. It is a surprise to many educators that this task too is surprisingly challenging for the novice. Once again, though textbook examples are often clear cut, one need only reflect for a moment on, for example, the links between abduction and causality to see that conclusion-hood and premisehood are far from intrinsic properties of spans of discourse, but depend on slippery, subjective interpretations of speaker intentions and audience expectations. Connections between argument components are also not simple binary relations: informal logic has long recognised that there are complex configurations, not least of which is the distinction between linked and convergent argumentation. Here even textbooks struggle to exemplify the distinction crisply, and scholars (most notably J. B. Freeman [9])) have written extensively on the challenges that face any analyst in trying to distinguish between these two structures. So what hope for the beleaguered student of critical thinking?

And yet still the task is not complete: there remains the challenge of assembling fragments of argumentation into larger scale structures: whether serial or convergent or divergent, and then, even more elusive, the reconstruction of enthymemes, again a problem upon which scholarly study (see, e.g. [13]) has unpacked myriad problems facing the would-be analyst. And I leave to one side entirely the further challenges of trying to perform these tasks in the context of discourse that is conducted dialogically.

Some educators have had success using the Toulmin scheme as a scaffold by which to introduce students to argumentation concepts, and then to help them in the tasks of argument analysis, reconstruction and evaluation. The didactic approach is usually founded upon a single page of [30] to introduce role distinctions for the components of arguments (data, claim, warrant, backing, rebuttal), yet we should wonder at the success that these approaches have when even the fundamental data-warrant distinction is fraught with difficulties (Freeman, *ibid*). Toulmin's pattern is also, without extension, poorly fitted to the task of dealing with real-world argumentation, focusing as it does on the small scale and the need for inferential steps to be rooted in appropriate field-specific backing.

Walton's insights offer the educator a breath of fresh air. For each of the four steps, argumentation schemes offer the student a supportive hand. In recognising that argumentation is happening at all, the breadth of his account of schemes means that a student has a lot to go on: is there testimony being adduced? Is there an expert to whom reference is being made? Are consequences discussed? Is classification performed? Is an analogy laid out? Once a student gets a whiff of one of these flavours, they can be off, recognising not only that there might be argumentation in play, but then where the bounds might lay upon the components of that argumentation — the testimony, the expert evidence, the consequences, the classification, the analogy; and then immediately, in addition, the connection and directionality: which are premises and which conclusions. Argumentation schemes also provide a no-nonsense account of the linked-convergent distinction, providing a template of linked premises that together support a conclusion, the challenge remaining to merely distinguish those premises that are mentioned from those that are not — and thereby offering the student a route to identifying (at least some) enthymemes too.

One of the signal strengths of Walton's approach lies in its extensibility. Quite apart from the evolution of his thinking from [36] to [38], he was ever the pragmatist, seeing the fecund extension of his core set of argumentation schemes by scholars and educators alike as nothing but a good thing. Those working with argumentation and critical thinking in, say, law, could extend and refine the compendium of schemes so as to offer a more nuanced account of legal reasoning; those in health sciences could do the same for their field; and likewise those in decision making (Atkinson and Bench-Capon's [1] work on refining schemes for practical reasoning can serve as a good example of particularly elaborate extension in this style). This unconstrained potential for extension of the set of schemes allows educators to develop a model that offers hand-holding for the student tailored precisely to their concerns and context.

Unfortunately, this strength of the approach is also one of its greatest weaknesses. With several dozen argumentation schemes, one might (perhaps, just) get away with a merely compendious approach. When the number approaches one hundred — or, across all of these domains and extensions and applications, many hundreds — more theoretical work is required. Specifically, it becomes necessary to taxonomise; to arrange the schemes in some kind of principled order by which new subspecies of schemes can be accounted for and integrated. Though Walton and colleagues worked hard on such cladistic thinking, it has been argued that argumentation schemes by their very nature do not yield well to such taxonomic efforts. Or rather, and more problematically, they yield simultaneously to many different taxonomic approaches, with no clear way to resolve which taxonomy is better or 'right' [15]. For the student, the challenge is not how to taxonomise, but how to classify: with which of the hundred schemes am I faced here?

The challenge of how best to arrange argumentation schemes (and relatedly, the questions of exhaustivity, comprehensiveness and distinctness) remain open research questions. For practical purposes, however, the very idea of a taxonomy of schemes hints at a way forward. The biological analogy upon which it rests can be further exploited not just in arranging the schemes but also in aiding in their identification. Whilst the Linnean approach to systematics lays down how species can be arranged consistently, for quotidian identification; the student naturalist is taught to use a dichotomous key: a series of binary questions which can lead to a unique species. If one is faced with a fallen leaf, it might be difficult to identify the species of tree from which it has fallen. But by answering a series of questions (is it a needle or a broad leaf?; if the latter, is the leaf simple or compound? if the former, is it lobed or whole? if the former, does it have three lobes or five?) it becomes straightforward to realise one is looking at the leaf of a sugar maple tree. The same approach can serve for argumentation schemes. Recent work in large-scale annotation of argumentation has demanded a consistent, reliable and easy-to-use mechanism for the identification of schemes, and the dichotomous key has been demonstrated to be a powerful technique. Visser et al. [33] lay out a practical key involving some sixty branching questions, which lead an analyst to determine which of the schemes in [38] is being used in a given example. Early results suggest that this approach aids learners in quickly and consistently identifying scheme usage in unconstrained naturally occurring discourse.

Argumentation schemes similarly have an important role to play not just in the identification of argument, but also in the other side of the same coin: supporting the construction of argument. Though argument technology for assistance in the process of crafting arguments remains a rarity (despite valiant battle cries such as that of Moor and Aakhus [19]), the potential for harnessing scheme structure in order to provide scaffolding and feedback for the novice arguer is clear and represents an open direction for future research.

3 Harnessing Argumentation Schemes for Automation

It is not just students and researchers that need to be able to pick apart the structure of argumentation — and struggle to do so. Increasingly, it is also a challenge faced by Artificial Intelligence systems. For more than a decade, a subfield of AI, natural language processing, has been growing in sophistication and applicability, and the products of its success are increasingly visible beyond academia — Alexa and Siri and their kith are all demonstrations of the newfound capabilities of such AI research. One keystone functionality is the ability to detect sentiment: automatically determining whether a sentence is expressing positivity or negativity, broadly construed. This technique of 'opinion mining' has found application in automatic processing of reviews of hotels, restaurants and shopping, and is used at scale in automatically processing vast datasets to provide insight for marketing and public relations. Given the enormous academic and commercial success in such techniques for understanding what opinions people hold, it is only natural that the question has arisen of whether it might be possible similarly to automatically identify why people hold the opinions they do. That is to say, can we build software that automatically identifies opinions that people hold, along with the reasons that they give for those opinions? This is the new specialism of 'argument mining' which sets out to build AI systems that automatically analyse the structure of argument — that is to say, that mimic the task that students perform when they reconstruct argument. It will surprise no-one in argumentation theory to learn that this is an extraordinarily challenging task. Argument mining is one of the most demanding open challenges in NLP today because it requires an integration of language processing with background knowledge. As a result, though the specialism now represents a significant academic industry with many dozens of research labs working on the problem, performance is currently, in the main, rather poor (at least by comparison to other tasks in NLP and to opinion mining in particular). There are currently very few techniques that even try to tackle the problem in the general case in which domain and genre are not known beforehand, and even those approaches that are constrained a priori struggle to reach performance anywhere near human level (for a review of recent techniques and trends and comparison of performance, see [17]).

One challenge facing argument mining is the need for data. Almost all NLP successes are founded upon examples — lots of examples. The strongest language models in NLP as a whole are based on corpora of many billions of words. For argument mining, at least one part of the strategy mirrors that of NLP as a whole, viz., to use large datasets of argumentation. The intuition is that with many examples of how arguments are put together by humans, it should be possible to train AI systems to recognise the regularities. The problem is that we don't have billion-word corpora of argumentation. In fact, until very recently we have barely had thousand-word corpora of argumentation. And the billion-word corpora that have delivered such transformative change in NLP are designed to capture regularities that are far simpler than those latent in argumentation. So ceteris paribus, we would need even larger corpora to model argumentation.

There is certainly a trend, and one that is accelerating, to build machine readable corpora of argument. Some come from commercial and social enterprise initiatives such as Kialo (www.kialo.com), a sophisticated platform for conducting semi-structured debate online. From academia, the approach is one of scaling up annotation efforts, with recent resources such as the close analysis of 35,000 words of student essays [29] and 100,000 words from the US2016 presidential debates [32]. But these are not enough on their own for building argument mining systems. They need to be coupled with operationalised theories of argumentation.

Walton's theory of argumentation schemes has delivered here too. In just the same way that students learn to use schemes to get a whiff of some flavour of reasoning, so too can algorithms benefit from what is effectively the expert knowledge engineering that has been baked into scheme structure. The important step is that many argumentation schemes are stereotypically associated with specific types of utterance and specific types of proposition. Sometimes these types are characteristic of conclusions. So, for example, arguments from positive and negative consequences typically involve normative statements in their conclusions. Practical reasoning inevitably concludes in action-oriented claims. In other cases, it is the premises that are recognisable. Arguments from expert opinion and from witness testimony typically include reported speech amongst their premises. Arguments from authority and position to know refer in their premises to characteristics of authority and position to know. The key insight is that these types are often much easier to recognise automatically than are reasoning structures. A normative statement often involves specific lexicalisation — the modalities 'should' and 'must', for example. Reported speech depends upon verbs of speech and in some cases, punctuation devices. Recognising these statement types is thus (at least in some cases) much more tractable than the problem of argument mining in general — and once an algorithm has recognised that there is a statement that is normative or that involves reported speech, it has a whiff of an argumentation scheme and can focus its computational gaze upon proximate sentences to see if that argumentation scheme plays out. Though automatically attempting to identify argumentation schemes has been attempted in various studies (first in [7]), this intuition that statement types can be used as an indicator of reasoning type and thence reasoning structure has been pursued by Lawrence and Reed [17]. They have shown that, at least for some types of argumentation scheme, seeking to identify the component parts of the scheme can be a valuable step in identifying that argumentation is happening — and thence in what way it is structured.

Though scheme-based argument mining is far from a panacea in the search for automated recognition and understanding of argument structure, it is a testament to Walton's theory that it can be so straightforwardly operationalised and brought to bear at the very cutting edge of AI development. And despite the modest abilities of the state of the art in such algorithms for argument mining, the very first public deployment of the techniques has made a significant impact, again in a pedagogical setting.

4 Argumentation Schemes in the Classroom: The Case of Fake News

The BBC has a long-running programme of engaging school pupils in media literacy, current affairs and the journalistic process. A collaboration between the News and Education arms of the corporation, BBC Young Reporter runs on an annual cycle, delivering classroom activities, lesson plans, technology, media, and access to BBC staff and resources to a broad range of school pupils in the UK and beyond. In 2018, the focus of the programme was on Fake News. As a part of the initiative, the Centre for Argument Technology provided a software solution, *The Evidence Toolkit* (`arg.tech/bbcschoolreport`), complemented by teaching plans and sample materials, for supporting pupils aged 16 to 18 in teasing apart the structure of news articles in order to determine their veracity [34].

At the core of *The Evidence Toolkit* lies a simplified set of argumentation schemes, and the analysis process in which students are inculcated is one of identifying scheme usage and then answering the critical questions associated with the schemes. The idea is that the critical questions help to point students towards appropriate critique to evaluate the veracity of the article. After distinguishing claims presented as fact from those presented as opinion, students can identify one of five schemes: expert opinion, popular opinion, example, personal experience, and statistical generalisation (here cast as an argumentation scheme), before being led through two to five critical questions. So, for example, if the student identifies an opinion, and then the scheme from expert opinion, the software presents the questions: Did the source actually make the attributed statement?; Is the source a credible expert on this subject?; Is the source duly impartial and not profiting from lending their support?; and, Do other experts agree with the source? Though requiring some paring down and simplification for a general audience, the lineage to Walton's original presentation is clear.

The structure of the interaction required the software to have a representation of the 'right' analysis, so that it could provide students with feedback at every step. To do this, five articles were manually curated and analysed to provide a gold standard model answer. In addition, however, the brief required the system to work on other news articles unseen by the development team: that is to say, students could run the same process on a news article of their choosing. The challenge is that the software therefore needs to be able to process an article, automatically identifying its claims and evidence, automatically recognising argumentation schemes, and automatically determining whether or not the critical questions are sufficiently answered. This is precisely the task of argument mining. As a result, *The Evidence Toolkit* also incorporates algorithms for argument mining and is the first public deployment of such algorithms. The results from these algorithms are couched as recommendations and hints so that the flexibility introduced as a result of being able to use the system on any news article of choice is safely traded off against errors or failures in automatic processing.

The software was deployed by the BBC into over 3,000 educational institutions, and was used over 25,000 times. Feedback was striking. It earned an overall rating of 4.15 out of 5 (where the average for applications on the BBC platform lies around 3.5). The user feedback moreover showed not only an accessible user experience (78% found it easy to use), but a successful one: 84% said the critical thinking tools explained in *The Evidence Toolkit* help to check the reliability of news, with 75% saying that it made them think more deeply about the topics at issue in the news articles. Putting critical literacy high on the BBC's agenda and applying argument technology to drive it also appears to reflect positively on the organization itself, with 73% stating that *The Evidence Toolkit* changed their view of the BBC for the better.

With Walton's theory of argumentation schemes front and centre in the application, it is a powerful indicator of the pedagogic appeal of the approach that feedback from this demanding user group was so consistently positive, and it is furthermore an indication of the general applicability of his insight that it can be put to work in tackling an issue of such societal importance and timeliness as fake news.

5 Representational Standards

One of the key challenges facing computational work in argumentation, informal logic and critical thinking has been the need to avoid reinventing the wheel that is the underlying computational model of argument structure. Of course that model is inevitably influenced by a conception of argument, but it should not be influenced by the software tools and systems that use it. For a significant period there was a tendency to build new ad hoc models of argument every time there was a need for a new tool; every time a new project involved the development of a software system. The problem is that this stymies any hope of incremental development and data re-use: each time, work starts from scratch and relearns where the challenges and pitfalls lie. Such an approach is inefficient and costly.

Early on in the development of computational models of argument as a field, this challenge started to be mapped out. Influenced by developments in computational theory of argument structure including Dung [5]; Krause et al. [16]; Prakken and Sartor [24]; Reed and Rowe [27]; Garcia and Simari [10] and others, a template for computational interoperability was laid out reflecting the work in philosophy from scholars including Toulmin, Freeman, Hitchcock and Walton in particular. That model was the Argument Interchange Format, AIF, [4] and has provided a blueprint for data and systems development in the field.

The AIF is a lightweight knowledge representation format that captures argument structures as a graph, connecting pieces of information ('I-nodes') via various types of schematic relations ('S-nodes'), where those relations are of various types, including applications of rules of inference ('RA-nodes'), and applications of rules of conflict ('CA-nodes'), amongst others. Though the AIF offers a slim vocabulary, it allows a rich representation of argument structure, including distinctions between linked, convergent, divergent and serial structures, and between rebutting and undercutting attacks [25]. Perhaps more significantly, the AIF also handles argument structure explicated according to different (and sometimes substantially different) accounts of argumentation, including, in philosophy, pragma-dialectics [6, 30, 40], and the box-and-arrow model perhaps best summarised by Freeman [9]. In addition, it is compatible with theories of argument structure that have emerged from linguistics and pragmatics such as Inference Anchoring Theory [3], and from the mathematics of computation, including Dung [5] and ASPIC+ [23]. As a result, AIF can serve as an interlingua bridging not just between conceptualisations of argument but also between disciplines and goals (both goals of argument and goals of research).

Whilst many systems are compatible with the AIF, a smaller number are 'native,' in that they are designed from the start to work with the standard. One example of such a native AIF tool is OVA, designed to support analysts and students in analysing argument structure [14]. OVA streamlines the process of identifying argumentative passages in source text, allowing the analyst to indicate how premises are connected to conclusions, and how larger scale argumentative structures are composed from smaller pieces. Inheriting from its predecessor, Araucaria [27], OVA makes significant use of Walton's theory of argumentation schemes. S-nodes throughout the AIF may be left unadorned as unspecialised RA-nodes or CA-nodes, or they may be marked explicitly as instances of particular types of reasoning — of particular argumentation schemes. OVA allows users to mark RA-nodes as scheme instances and extends the notion to allow in addition the definition and marking of schemes of conflict on CA-nodes. Following the approach in Araucaria, OVA does not tie itself (any more than the AIF does) to a specific set of schemes. One might, for example, use schemes from the set presented in [36], or alternatively opt for the extended set presented in [38]. Alternatively, one might look at sets of schematic reasoning developed by other scholars — Pollock [25], for example, or even further afield, casting the rhetorically-inspired patterns of Perelman and Ohlbrechts-Tyteca [22] as schemes. Furthermore, it might be, as suggested above, that a particular project, team, educator or researcher needs a bespoke set of schemes: again, following in the spirit of Walton's pragmatism, such definition is straightforwardly accommodated.

As *The Evidence Toolkit* demonstrates, once schemes are recorded, their critical questions naturally lend themselves to supporting an analyst in the task of argument evaluation. Furthermore, the definitions of the schemes lay out the implicit premises, assumptions and presumptions which can be used to help identify enthymemes. These implicit components of schemes can also act as 'growth points' for arguments; points upon which attacks can be focused. In one of his collaborations with AI researchers, Walton explored this connection between argument scheme structure and argument attack (particularly undercutting attack) [11] in the context of an AI legal reasoning system called Carneades. Carneades too is AIF compatible, and as a result, arguments analysed using OVA, with scheme usage marked, can be sent to Carneades for automatic computation and evaluation.

OVA is currently used by tens of thousands of users in schools, colleges and universities in over 80 countries, and because of the rapid progress that the tool allows analysts to make, there are now significant datasets involving argumentation schemes, including several corpora that are built solely from instances of schemes. This opens up an intriguing prospect. Argumentation theory as a discipline has to manage the tension between the normative and the descriptive in its modus operandi, with normative models inevitably informed by practice, and an empirical agenda increasingly forming one branch of the discipline as a whole. Walton was an adept at managing this tension, developing robustly normative models driven by insight derived from case studies — and then demonstrating the value of those models by re-applying them to realistic cases. His approach was, however, hampered by a paucity of data: there simply were not the large datasets available to support robust empirical investigation. With such datasets newly available, it may be possible to revise and extend accounts of argumentation schemes through empirical study, testing and validation, to generate accounts that have both greater detail and greater applicability.

6 Dialectics and Argument Technology

Argumentation schemes have a peculiar dual nature. On the one hand, they are, self-evidently, monological. A scheme constitutes a pattern of reasoning that is carried out (or reported, or offered, etc.) by an arguer, a method they use for getting from A to B, intellectually speaking. There need not be an antagonist; the argument can be entirely solipsistic, or delivered as soliloquy. Yet on the other hand they are, self-evidently, dialogical. A key part of the power of argumentation schemes lies in the critical questions by which the reasoning can be tested, probed and critiqued: whether they form a part of an internal dialogue or of one with a real opponent, these critical questions are there to be asked. This duality was explored by Reed and Walton [28] which sought to provide an account of how the asking of critical questions can be couched in terms of formal dialectics, thereby reconciling the monological and dialogical facets of schemes. The Argumentation Schemes Dialogue system, ASD, provided a technical resolution of the dichotomous nature of schemes, but also marked a reunification of two major themes of Walton's work, argumentation schemes and formal dialectic.

Following in the tradition established by Hamblin [12], Walton's work on formal dialectics had a profound influence on research in AI. Building on his early characterisation of commitment dynamics and fallacious manoeuvring in dialogue [35] and his collaboration with Woods [41] exploring properties of argumentative interaction over the course of dialogical interaction, his work with Krabbe [37] was of such insight and breadth that it served as a catalyst that introduced many AI researchers to argumentation theory. The Walton-Krabbe account of dialogical types spawned a rich area of research [21] using formal dialectic as a starting point for defining protocols of interaction between autonomously acting software agents.

More recently, however, formal dialectic has also been attracting sustained interest within the area of argument technology as a way to architect conversations not between autonomous agents, but between agents and humans in what are variously known as human agent collectives or mixed initiative teams — groupings that bring together a heterogeneous mix of humans and AI systems to work collaboratively to solve problems as diverse as disaster management and intelligence analysis.

Various research projects have explored the implementation of formal dialectics in such settings [8], but in each case systems are built from the ground up; newly constructed from scratch each time. An opportunity was being missed. Walton and Krabbe [37] describe the classes of components of formal dialectic — the locution rules that describe what can be said, the structural rules that describe how one locution can follow another, the commitment rules that describe how locutions update a shared information state [31], and win/loss rules that describe termination. For the characterisation of these rules, they use variations of the formal characterisations from Hamblin, Woods and Walton. The specifications in [37] look very much like computer programs — yet of course, no such programming language existed.

The Dialogue Game Description Language, DGDL, [39] aims to fill this gap. By taking the vocabulary of Walton and Krabbe, and combining it with a programmatic interpretation of the language of formal dialectic in Hamblin and Woods-Walton, DGDL provides a domain-specific programming language by which systems of formal dialectic can be rapidly implemented as executable computer programs. Wells and Reed [39] demonstrated how DGDL can be used to express a wide range of formal dialogue games in the philosophical literature, including Hamblin's H, Walton's CB and its variants, Mackenzie's DC, Walton and Krabbe's PPD and RPD, as well as others from Girle, Hintikka, Lorenzen and Rescher. Though these games were primarily designed to explore specific phenomena in philosophical contexts (persuasion and argument stability, cumulativity, silence-means-assent, etc.), DGDL was also then used to capture games designed from an engineering perspective, including those aiming to support information seeking in educational environments [20] and those enabling deliberation between agents [18].

DGDL is an interpreted language: it requires an interpreter to execute. This interpreter is provided by the Dialogue Game Execution Platform [2]. With DGEP deployed as a general-purpose platform, DGDL becomes a flexible tool for rapid development of dialogue games, bringing development times down from years to days. More significantly, DGEP inherits the intimate connection with argumentation theory borne of the roots of formal dialectic theory, but hinging directly upon the model in Walton and Krabbe. Thus as moves are made in a dialogue game executing on DGEP, the status of the dialogue as it unfolds can be recorded in AIF. The intuition that is exploited was hinted at in Freeman [9], in which the distinction between linked and convergent argumentation is explicated by imagining an interlocutor asking two different questions in the two cases: "Why is that relevant?" (for linked argument structures) and "Can you give me another reason?" (for convergent argument structures). But instead of seeing these questions as imaginary ways of testing an extant argument structure, they can instead be operationalised, described as possible moves in a dialogue game (much as they are in Walton and Krabbe), and then when they are executed their effect is to create argument structures that are linked or convergent. In this way, DGDL and DGEP tie together in both computational theory and implemented software the connection between the process constituted by conducting a formal dialogue and the product constituted by the structure that such a dialogue constructs or navigates. That connection was first suggested in Walton and Krabbe's discussion, in the context of presenting their Permissive Persuasion Dialogue, PPD, and Rigorous Persuasion Dialogue, RPD, of internal and external stability adjustments, whereby updates to what from an AI perspective are knowledge structures are in turn governed by the dynamics of the PPD and RPD dialogue systems. This is what is now available as a general purpose programming language and its execution environment.

7 Dialectics in the Classroom: The Automated Polemicist

Using formal dialectic to structure student learning has been explored extensively in the Interloc platform [26] and its benefits clearly demonstrated. Interloc, however, uses a single hard-coded dialogue game. DGDL in contrast facilitates rapid iteration of dialogue game systems. One example is the Polemicist system (www.polemici.st) used in both educational and public communication settings. Polemicist comprises a web interface (one of a number that can provide human interaction with DGEP), a DGDL specification of a PPD-like game running on DGEP, and a series of agents, each with access to a set of arguments represented in AIF. The dialogue game implemented by the application supports up to nine players of whom exactly one must be human. Neither DGDL nor DGEP distinguishes between human and software players. The arguments that populate the knowledge bases of the software players are drawn from manual analysis (using OVA) of transcripts of a BBC Radio 4 debate programme, The Moral Maze. Each software agent has access to the arguments of one of the eight human participants in the radio programme. The human player adopts the role of the chair of the debate and can ask questions, probe positions, accept interruptions, ask for alternative points of view and contribute their own opinions. The software players can respond to questions, offer opinions, and indicate when they want to interrupt. Though the discourse material available to the software players in Polemicist is limited to what was said in the original programme, the flow of the dialogue and the navigation of the debate is under the control of the human player. Polemicist was made available to the public by the BBC as part of a suite of argument technology to accompany an episode of the Moral Maze broadcast in 2017.

Though only a prototype, Polemicist demonstrates the connection between, on the one hand, dialogical interaction amongst members of a mixed initiative team combining humans and software agents, with, on the other, AIF resources that are navigated and updated on the fly. The software represents the first example of using a system of formal dialectic to help the public navigate a complex, nuanced and emotional debate (the topic of the programme was abortion) and the first time that a system for mixed initiative argumentation has been deployed to the general public.

8 Building on the Foundations

Walton's insights in two distinct areas, argumentation schemes and formal dialectics, have had a lasting impact on research in AI, and on the shape of the field of argument technology in particular. His focus on empirical grounding combined with practical applicability of theory has found a receptive audience in the computational sciences and has laid critical foundations for the development of both theory and applications of computational models of argument.

One of the most exciting new developments in argument technology is the development of techniques for argument mining. With much of the low hanging fruit now harvested, the community is faced with deep challenges which frame research questions not only for AI and natural language processing, but also for cognitive, linguistic and philosophical models of argumentation. The promise of such techniques, however, is tantalisingly great. The challenges for automation mirror those of manual annotation: to take discourse and first, to recognise where argumentation is taking place; second, to identify which spans of discourse contribute to the argument; third, to determine what types of reasoning pattern are being employed in premise-conclusion complexes; and fourth, to connect the individual components into a larger graph mapping out a debate at scale. If these tasks can be accomplished (even if imperfectly) automatically and at scale, it will become possible to offer unique insight for experts and the general public alike, and to empower contributors and potential contributors across a broad range of the social, geographical, political and economic spectrum. Wherever argumentation and debate contribute to process — from democracy to corporate decision making, from global geopolitics to neighbourhood mediation — argument technology holds the potential for transforming efficiency, accessibility and engagement with such processes. The scale of the opportunity (and challenge) cannot be understated: political polarisation, social media filter bubbles, fake news, and disengagement, disempowerment and disillusion with democratic process are the defining characteristics of the first quarter of the twenty first century and argument technology has already, even in its most tentative first steps, demonstrated the potential to have major impact on these problems.

The need for applicable, computable, accessible argumentation theory has never been greater. Heavily indebted to the framework and approach that Walton's work has delivered, argument technology is now well equipped to tackle both the computational and societal challenges that lie ahead.

Acknowledgements

The research described in this paper that has originated in the Centre for Argument Technology at the University of Dundee (www.arg.tech) has been supported by various research projects and represents the work of many current and past members of the team. The author would like to acknowledge the financial support of EPSRC (under grants EP/N014871/1, EP/K037293/1, EP/G066019/1 and EP/G060347/1), ESRC (ES/V003909/1), AHRC (AH/V003305/1), JISC (CIINN01), VW Stiftung (92 182) and the Leverhulme Trust (RPG-2013-076 and F/00-143/C), and to recognise in particular the lynchpin role played by Dr John Lawrence in the development of much of the technological infrastructure of the systems mentioned here. Finally, the relationship with the BBC has rested upon the decade-long enthusiasm and commitment of Christine Morgan, Head of Radio, Religion and Ethics at the BBC, whose tireless efforts have helped to prove the value of argument technology in large-scale, public-facing settings.

References

[1] Atkinson, K. and Bench-Capon, T. (2007). Practical reasoning as presumptive argumentation using action based alternating transition systems. *Artificial Intelligence*. 171, 855-874.

[2] Bex, F., Lawrence, J. & Reed, C. (2014) "Generalising argument dialogue with the Dialogue Game Execution Platform" in Parsons, S., Oren, N., Reed, C. & Cerutti, F. (eds) *Proceedings of the Fifth International Conference on Computational Models of Argument (COMMA 2014)*, IOS Press, Pitlochry, pp141-152.

[3] Budzynska, K., & Reed, C. (2011). Whence inference? Technical report, University of Dundee.

[4] Chesnevar, C., McGinnis, J., Modgil, S., Rahwan, I., Reed, C., Simari, G., South, M., Vreeswijk, G., and Willmott, S. (2006). Towards an argument interchange format. *Knowledge Engineering Review*, 21(4), 293—316.

[5] Dung, P. (1995). On the acceptability of arguments and its fundamental role in non-monotonic reasoning, logic programming and n-person games. *Artificial Intelligence*, 77(2), 321—358.

[6] Eemeren, F.H. van, & Grootendorst, R. (1984). *Speech acts in argumentative discussions: A theoretical model for the analysis of discussions directed towards solving conflicts of opinion*. Dordrecht: Floris Publications.

[7] Feng, V.W., & Hirst, G. (2011). Classifying arguments by scheme. In *The 49th annual meeting of the association for computational linguistics* (pp. 987—996)

[8] Ferguson, G. & Allen, J. (2007). Mixed-Initiative Systems for Collaborative Problem Solving. *AI Magazine*. 28. 23-32.

[9] Freeman, J. (1991). *Dialectics and the macrostructure of argument*. Foris.

[10] García, A.J. and Simari, G.R. (2004) Defeasible Logic Programming: An Argumentative Approach. *Theory and Practice of Logic Programming*. 4(1-2): 95-138

[11] Gordon, T.F., Prakken, H. and Walton, D. (2007) The Carneades model of argument and burden of proof, *Artificial Intelligence*, Volume 171, Issues 10—15, 875-896.

[12] Hamblin, C.L. (1971) Mathematical models of dialogue, *Theoria* 37, 130—155.

[13] Hitchcock, D. (1998) Does the Traditional Treatment of Enthymemes Rest on a Mistake?. *Argumentation* 12, 15—37.

[14] Janier, M., Lawrence, J., & Reed, C. (2014). OVA+: an argument analysis interface. In Parsons, S., Oren, N., Reed, C., & Cerutti, F. (Eds.) *Computational models of argument - proceedings of COMMA 2014*, volume 266 of frontiers in artificial intelligence and applications (pp 463—464). IOS Press.

[15] Katzav, J. & Reed, C.A. (2004) "On argumentation schemes and the natural classification of arguments", *Argumentation*, 18 (2), pp239-259.

[16] Krause, P., Ambler, S., Elvang-Goransson, M. and Fox, J. (1995), A logic of argumentation for reasoning under uncertainty. *Computational Intelligence*, 11: 113-131.

[17] Lawrence, J. & Reed, C. (2020) "Argument mining: A survey" *Computational Linguistics* 45 (4), pp765-818.

[18] McBurney, P., & Parsons, S. (2002). Games That Agents Play: A Formal Framework for Dialogues between Autonomous Agents. *Journal of Logic, Language, and Information*, 11(3), 315-334.

[19] Moor, A. & Aakhus, M. (2006). Argumentation Support: From Technologies to Tools. *Communications of the ACM*. 49. 93-98.

[20] Moore, D., Hobbs, D. J. 1996. Computational use of philosophical dialogue theories. *Informal Logic* 18(2), 131—163.

[21] Norman T.J., Carbogim D.V., Krabbe E.C.W., Walton D. (2003) Argument and Multi-Agent Systems. In: Reed C., Norman T.J. (eds) *Argumentation Machines*. Argumentation Library, vol 9. Springer, Dordrecht

[22] Perelman, C., and Olbrechts-Tyteca, L. (1969). *The New Rhetoric: a treatise on argumentation.* University of Notre Dame Press.

[23] Prakken, H. (2010) An abstract framework for argumentation with structured arguments, *Argument & Computation*, 1:2, 93-124

[24] Prakken, H. and Sartor, G. (1997) Argument-based extended logic programming with defeasible priorities, *Journal of Applied Non-Classical Logics*, 7:1-2, 25-75

[25] Pollock, J. (1995). *Cognitive carpentry.* MIT Press.

[26] Ravenscroft, A., Mcalister, S. and Sagar, M. (2010). *Digital Dialogue Games and InterLoc: A Deep Leaning Design for Collaborative Argumentation on the Web.* Bentham Science E-Books.

[27] Reed, C.A. & Rowe, G.W.A. (2008) Argument Diagramming; The Araucaria Project. In Okada, A., Buckinghum Shum, S. & Sherborne, T (eds) *Knowledge Cartography*, Springer, pp164-181.

[28] Reed, C. A. & Walton, D. (2007) "Argumentation Schemes in Dialogue" in Hansen, H. V., Tindale, C. W., Johnson, R. H. & Blair, J. A. (eds). Dissensus and the Search for Common Ground (OSSA 2007), Windsor.

[29] Stab, C. and Gurevych, I. (2014) Annotating Argument Components and Relations in Persuasive Essays. In: *Proceedings of the 25th International Conference on Computational Linguistics (COLING 2014)*, pp. 1501-1510

[30] Toulmin, S. (1958). *The uses of argument.* Cambridge University Press.

[31] Traum D.R. and Larsson S. (2003) The Information State Approach to Dialogue Management. In: van Kuppevelt J., Smith R.W. (eds) Current and New Directions in Discourse and Dialogue. *Text, Speech and Language Technology*, vol 22. Springer, Dordrecht.

[32] Visser, J., Konat, B., Duthie, R., Koszowy, M., Budzynska, K. & Reed, C. (2019). "Argumentation in the 2016 US presidential elections: Annotated corpora of television debates and social media reactions." *Language Resources and Evaluation.*

[33] Visser, J. Lawrence, J. & Reed, C. (2020) Reason-Checking Fake News. *Communications of the ACM* 63 (11) pp38-40

[34] Visser, J., Lawrence, J., Wagemans, J. & Reed, C. (2020b, to appear) "Annotating Argumentation Schemes" *Argumentation.*

[35] Walton, D.N. (1984). *Logical dialogue-games and fallacies.* University Press of America.

[36] Walton, D. (1996). *Argumentation schemes for presumptive reasoning.* Lawrence Erlbaum Associates.

[37] Walton, D., & Krabbe, E. (1995). *Commitment in dialogue.* SUNY Press.

[38] Walton, D., Reed, C., & Macagno, F. (2009). *Argumentation schemes.* Cambridge University Press.

[39] Wells, S., & Reed, C.A. (2012). A domain specific language for describing diverse systems of dialogue. *Journal of Applied Logic*, 10(4), 309—329.

[40] Wigmore, J. (1931). *The principles of judicial proof*, 2nd edn. Little, Brown & Co.

[41] Woods, J. and Walton, D. (1982). Question-begging and cumulativeness in dialectical games. *Noûs* 16 (4):585-605.

Argument Schemes and Dialogue Protocols: Doug Walton's Legacy in Artificial Intelligence

Peter McBurney*
Department of Informatics, King's College London
peter.mcburney@kcl.ac.uk

Simon Parsons[†]
School of Computer Science, University of Lincoln
sparsons@lincoln.ac.uk

Abstract

This paper is intended to honour the memory of Douglas Walton (1942–2020), a Canadian philosopher of argumentation who died in January 2020. Walton's contributions to argumentation theory have had a very strong influence on Artificial Intelligence (AI), particularly in the design of autonomous software agents able to reason and argue with one another, and in the design of protocols to govern such interactions. In this paper, we explore two of these contributions — argumentation schemes and dialogue protocols — by discussing how they may be applied to a pressing current research challenge in AI: the automated assessment of explanations for automated decision-making systems.

Keywords: Dialogue protocol, Argument Scheme, XAI.

The authors would like to thank John Woods for the invitation to contribute to this special issue, and for his forbearance when it became clear that they had been overly optimistic in their estimation of the speed with which they could complete the task of writing this paper.

*PM wishes to thank Dylan Cope and Liz Sonenberg for many interesting conversations on the topic of explanations.

[†]SP acknowledges funding from EPSRC grant EP/R033722/1, and thanks Nadin Kökciyan, Quratual-ain Mahesar, Isabel Sassoon and Elizabeth Sklar for many interesting conversations on the topic of explanations.

1 Introduction

Both of us, along with many researchers in artificial intelligence (AI), especially those working on computational models of argumentation and multiagent systems, have been greatly influenced by the work of Doug Walton. At the end of this paper (see Section 5) we will say a little about this from a personal perspective, but we want to spend the bulk of this paper exploring why we think Doug's work has been so influential. In short, it is because two aspects of his work — the work on dialogue protocols, exemplified by [62], and that on argument schemes, exemplified by [61] — provide a basis[1] for a solution to some of the major problems in artificial intelligence.[2] We will illustrate this by taking one such problem — the need to provide explanations for the reasoning performed by AI systems — and showing how Doug's work provides an underpinning for a possible solution. We start with this problem, and why it has become a prominent problem.

1.1 Why explanations are necessary

The third edition[3] of Russell and Norvig's "Artificial Intelligence: A Modern Approach", published in 2009, includes a history of AI from its birth (which they date to 1956, at the Dartmouth workshop, though acknowledging that work on AI was done before this point) to the time of writing. The period from 2001 is headed "The availability of very large datasets", and points to the ability of systems bootstrap from large collections of data as possibly leading to AI systems that no longer need the careful knowledge engineering that was previously necessary. The subsequent decade has seen this prediction, if not borne out[4], at least extensively tested, with impressive results on a range of applications.

Much of this success has been due to techniques from *deep learning*, that is techniques that make use of neural networks with many layers. These methods were coming into their own while Russell and Norvig were putting the third edition

---

[1]The use of the indefinite article is deliberate here. There are undoubtedly other solutions which would have other bases. However, that does not undermine the importance of that based on Doug's work.

[2]At the end of writing this paper we discovered another tribute to Doug Walton that focuses on the same two of his contributions, this time in the area of AI and law, namely [7].

[3]Though a fourth edition was published in early 2020, it is not yet easily available in the UK at the time of writing.

[4]It is noteworthy that much of the recent cutting-edge work on machine learning has been looking at ways to incorporate engineered knowledge into the learning process, suggesting that researchers in machine learning are beginning to feel that there are limitations to the idea of extracting all that is needed to solve every problem directly from data.

together[5], and have come to dominate work on machine learning and AI. Indeed, for many outside the field of AI, and a good number of those within who have graduated in the last few years, machine learning *is* AI, and the only kind of machine learning worth considering is deep learning. While the performance of deep learning systems is extremely impressive, there are a number of (well-known) issues that widespread use of such systems raises. Chief among these[6] is the fact that it is frequently obscure *why* a deep model gives a specific answer. This is in contrast to earlier AI methods — for example the rule-based methods of expert systems, or the causal probabilistic networks that led to the previous wave of AI applications — where it is straightforward to extract a trace of the reasoning that led to a conclusion and one could pose "what if?" questions about related situations. It is in contrast to other machine learning methods, for example decision trees, where structural information about a domain can be extracted from the model that has been learnt.

The reason that this is significant is because, as AI applications become more widespread, there will be an increasing need to be able to explain not just *what* decisions were reached, but *how* those decisions were reached. In other words, there is a requirement for AI to be *explainable*. This requirement is driven by regulatory pressure. For example, GDPR[7] regulation in the EU, requires that organisations that use AI systems to make decisions

> shall implement suitable measures to safeguard [the subject of those decision]'s rights and freedoms and legitimate interests, at least the right to obtain human intervention on the part of the [organisation making the decision], to express his or her point of view and to contest the decision.

This is widely understood to mean that decisions made by those AI systems must be such that that can be explained to the subject of those decisions, since how

---

[5]In [26], three of the pioneers of deep learning date the breakthrough in such methods to 2009 for speech recognition and 2012 for image processing.

[6]Two others, in passing, are the following. (1) the fact that deep learning not only benefits from huge amounts of data, but *requires* it. As a result, if you work in a domain that does not have tens of thousands of examples that your system can learn from, you will not be able to create robust models. Unfortunately, areas like medicine fall into this category. Another example is the creation of software for control of autonomous ships, where there is a severe lack of publicly-accessible data on collisions. Nowadays there are very few collisions between large ships; there are many more near-collisions, but most of these are not reported outside the companies involved. (2) training deep models uses a large amount of power, and since the methodology for learning the hyper-parameters that determine whether or not a particular model is effective is basically brute-force search, training a good model is very energy inefficient. In a climate emergency, one might question the morality of widespread use of deep learning.

[7]https://gdpr-info.eu/

else would that subject be able to express their views and contest the result in any meaningful way?

Similarly, the European Union's Markets in Financial Instruments Directive II (MiFID II[8]), which came into force in January 2018, requires companies which provide financial information or services in which wholly automated decision have material impacts on individuals or on small and medium-sized enterprises to provide those impacted with human-understandable explanations of how the automated decisions have been made[9]. Indeed, the policy statement from the European Commission to the European Parliament relating to AI (published in April 2018) emphasizes Explainable AI as a key area of research and innovation for the next EU Multiannual Financial Framework (2021–2027), along with the areas of unsupervised machine learning and energy and data efficiency.[10]

These regulatory pressures are also present elsewhere in the world. For example, in January 2019 the Personal Data Protection Commission (PDPC), a Government agency in Singapore, released a draft model framework for the Governance of AI systems in large organizations and enterprises [1]. After public consultation during 2019, a revised version was released in January 2020. The framework is a voluntary collection of ethical principles and governance considerations that are recommended by the PDPC for adoption by organizations; the Framework is not legally binding. The Model Framework proposes two high-level guiding principles for design and deployment of AI applications:

- Organizations using AI in decision-making should ensure that the decision-making process is explainable, transparent and fair; and

- Applications of AI should be human-centric.

The Singapore Model Framework also provides guidance on when and how applications of AI should incorporate human involvement in decision-making processes.

This section has discussed the pressures from Governments and industry regulators on adopters of automated decision-making systems to ensure that these systems explain their decisions. Another pressure will likely come from the legal system. If a

---

[8]https://www.esma.europa.eu/policy-rules/mifid-ii-and-mifir/

[9]Legal or regulatory requirements to provide explanations for decisions reached by automated systems have led some people to propose inserting a dumb human into the decision process so that the process no longer appears completely automated. However, if the human only ever approved the decisions and never rejected them, then it is unlikely that European courts would accept such gaming of these regulations.

[10]https://ec.europa.eu/digital-single-market/en/news/communication-artificial-intelligence-europe

human car driver is faced with an untenable choice, for example driving straight on and thereby hitting an oncoming car or swerving off the road and hitting a pedestrian, and if there is a subsequent legal case, judges and juries may well accept (as they do now) an explanation from the driver along the lines of, *"I was faced with an impossible choice, and in the heat of the moment I chose one way rather than the other."*.

However, if the car in question is an autonomous vehicle, that response will most likely not be acceptable to courts. Instead, courts will want to ask how that trade-off was made by the vehicle in that moment. Was it pre-coded? If so, how did the software developers make that pre-coded decision? If not, how did the software developers allow the machine to decide itself between the two options (e.g., did it make a random choice?). Courts may well also probe what ethical considerations the developers considered before coding the vehicle. What ethical training had they had before considering any ethical issues? What directives or ethical advice, and from whom, had they received beforehand? Etc. Such probative questioning by courts will not stop at the first response as with a human driver. Hence, we expect the legal system's response to cases concerning accidents involving autonomous vehicles to add further pressures on developers of AI systems to provide explanations of the decisions made or recommended by those systems.

1.2 Fairness and explanation

Note that this desire for AI systems to be explainable, is related to concerns about the *fairness* of AI and, more broadly, what is known as *algorithmic decision making*[11].

---

[11]The term "algorithmic decision making" is used to refer to situations in which decisions are made by a system that involves software with no human oversight or involvement. Clearly decisions made by a software system that uses AI and which has no human oversight or involvement are a subset of those reached using algorithmic decision making. In our view, "algorithmic decision making" is a bad piece of terminology, since it is perfectly possible for a human to follow an algorithm as part of making a decision in such a way that they exercise no free will, making the decision determined purely by the process encoded in the algorithm. In other words, the use of the term "algoithmic" does not imply the use of software, or the exclusion of a human from the process. One of us (SP) remembers making decisions in exactly this way when he worked in a temporary position at a Job Centre in the summer of 1988. One of the parts of the job was reviewing the record of people receiving unemployment benefit and, provided that they met some criteria to do with the length of time they had been out of work, inviting them for an interview. ("Inviting", in this case, meaning "threatening them with a loss of benefits if they did not attend".) The process was as mechanical as described — we were not allowed to exercise judgement, and what we did could easily have been carried out by software. We suspect that the reason that such a poor term as "algorithmic decision making" has come into use is a combination of its euphony (much better than "software decision making" or "computerised decision making") and the fact that many people do not know the difference between an algorithm (the process itself) and its implementation.

The concern is that whenever software is used with no human intervention, there is the possibility for it to produce results that are biased, in the sense of discriminating against individuals. Of course decisions involving humans can also be biased if the humans are biased, but part of the concern with software decisions is that they can be unscrutable (and so be hard to identify and rectify) and that they can exist even when the software designers and deployers have no intention of being biased[12].

Two well-known cases are the admissions process for students at St George's Hospital Medical School in London, and the COMPAS recidivism risk calculator. In the case of St George's [28], the medical school created a piece of software to screen applications for places to train to be a doctor. There were two aims. First, they wanted to ensure that all applicants were treated the same, something that can clearly not be the case when decisions are reached by humans (especially when the decisions are distributed across a group). Second, they wanted to reduce the load on their staff. The medical school was heavily over-subscribed (with 12 applicants for each place in 1988), and the idea was to have the software screen out some applicants so that the admissions team had less applications to consider. The system was carefully designed and then tuned until it had close agreement with the manual process. Unfortunately, the manual process was itself flawed, and the software system was found to be discriminatory, with an investigation by the Commission for Racial Equality finding that:

> as many as 60 applicants each year among 2000 may have been refused an interview purely because of their sex or racial origin. [28]

COMPAS, is a software system developed by Northpointe Inc. to help assess the risk that, on the basis of their history, an individual would reoffend. The performance of the system was analysed by ProPublica [4, 25] and found to exhibit racial bias. The analysis considered more than 10,000 real cases from one county in Florida, and compared the rate of recidivism predicted by the COMPAS software against what the individual actually did in the next two years. The headline finding was that:

> Black defendants were often predicted to be at a higher risk of recidivism than they actually were. Our analysis found that black defendants who did not recidivate over a two-year period were nearly twice as likely to

---

[12]Note that it is possible for bias to exist not only in the data used for training purposes or as inputs to some software analysis procedure, but even in the underlying conceptual abstractions that allow the data to be recognized as *data* and thus enable its collection; for an example, see [19]. Econometricians analyzing national accounts data face similar issues, for example, when the definition of employment ignores unpaid work done by family members within households or on farms.

be misclassified as higher risk compared to their white counterparts (45 percent vs. 23 percent). [25]

In both these cases, the software system making decisions does so in a way that is biased. In the case of the admissions system, the software was designed to replicate an existing decision process that was already biased. In the case of the recidivism system, the designers apparently [3] tuned the system to ensure that its accuracy was the same for both black and white individuals — they assumed that doing this would make its decision fair. However, as above this turned out not to be the case for some reasonable definitions of "fair", in particular the one alluded to in the quotation above, that the rate at which defendants were wrongly classified as higher risk should be the same regardless of whether the defendant was black or white. Subsequent analysis [3] has shown that it is impossible for both these notions of fairness — that the accuracy of predictions do not vary by race, and that there is no disparity in incorrect misclassification as higher risk — to be simultaneously satisfied. Indeed, aiming for equal accuracy of predictions leads directly to a disparity in misclassification to a higher risk category. Such concerns about the fairness of AI lead back to the desire for AI to be explainable because if one can check the reasoning that an AI system uses, then it will be possible to check that reasoning for bias [64].[13]

1.3 Explainable AI

The last few years has seen a surge in work on explainable AI, or XAI. Much of this work has centred around creating explanations for machine learning models, especially those that look to many users like "black boxes", in other words inscrutable oracles that are inherently impossible for people to understand. A typical approach is to take a black box model and train another model that is easier to understand on the same data, and use that second model to explain the decisions made by the first. This is the thrust of [16], which creates an ensemble of decision trees as an explanation of a, more complex, deep neural network model. Another take on the same issue is to explain a decision by plotting out the local area around the point where the decision needs to be made and creating a model of that [43]. The intuition here is that the inscrutability of models — that they consist of complex multi-dimensional surfaces separating different outcomes — will often not exist at a local level, allowing simple, and hence easy to understand, rules to be identified that explain the decision. A criticism of this work, and much of the other efforts in XAI is

---

[13] Of course, this is not the only way to ensure fairness, and much of the work on the fairness of AI systems does not attempt to do this through explainability.

that they are developed by the same people who build the black box models in the first place, start from the thing to be explained, and create a solution by simplifying it. This is a process that takes very little account of what the people who want the explanations would find helpful [39].

Miller's [39] examination of the literature on explanation, follows [23], among others, in suggesting that many explanations presented by people focus on describing the underlying causal mechanisms, and, further [24], that these explanations are presented in the form of a conversation. As [39] discusses, [5] goes further in suggesting that explanations are presented not just as conversation, but as *arguments*, in the sense of the provision of justifications for the assertions that are made. The research in [5] is drawn from the analysis of a number of explanations from human conversations — that is where one person explains something to another. Given this, admittedly rather limited[14], evidence, it seems plausible that an argumentation-based approach to explanation will be a promising approach for adoption by AI systems. Below, we sketch some requirements for computer-based explanations, giving a first-principles analysis to complement the discussion above, and the point to ways in which Doug Walton's work can be used to underpin these requirements.

2 Asking for and assessing explanations

2.1 What do we need for computer-based explanations?

What would we require in order to have automated explanations? A first requirement — and challenge — would be to generated explanations automatically for AI decision systems. As mentioned above, for some types of AI systems, such rule-based expert systems and causal probabilistic networks, automatically generating explanations is straightforward, by generating a trace of the reasoning undertaken by the system in reaching a conclusion.

For other types of AI systems, especially those which operate at a low level of granularity, such as image classification programs analyzing individual pixels and their neighbourhoods, this is not necessarily at all straightforward. Why automated generation of explanations for such systems is difficult is because the level of operation of the AI system is at lower level of the objects being classified than any level containing human meaning. For a human being recognizing images of faces for example, parts of the face are arguably very important to recognition and classification, for example, colour of hair, shape of hairline, size of ears, presence or absence

---

[14]As [5] explain, their analysis is based on 30 examples of explanation, but they are from a single conversation, itself taken from [51]

of a beard, etc.[15] Such parts have human meaning and can be readily described to other humans as the reason for a particular classification. If, instead, an AI system uses lower-level elements of images, such as individual pixels, or the relationships between nearby pixels (eg, identifying edges by means of observed differences in pixel colours) for facial recognition or classification, then these lower levels will typically have no human-understandable meaning. It is generally not obvious how the use of such lower-level elements could be aggregated or assembled automatically into a higher-level explanation able to be understood by a human. Thus, automated generation of explanations is difficult challenge for these types of AI systems.

In this paper, however, we will ignore the challenges involved in the generation of explanations. Our focus will be on assessment of an explanation that has somehow been produced, by automated means or manually. Given that an explanation has been created, what is needed for its automated assessment by some entity seeking to obtain an explanation for a decision of an AI system? Based on human-to-human explanations, we might expect any machine assessment to have several features.

The first feature is a means for the formal representation of explanations, where by *"formal"*, we mean machine-readable. This is necessary for automated parsing of the explanation, as the first stage in a process of automated analysis and assessment, and possibly also automated comparison with alternative arguments. As mentioned above, this process is well-known and straightforward in cases where explanations may be constructed from sequences of syllogistic or mathematical deduction (as in rule-based Expert Systems) or from sequences of causal influences between time-ordered events (as in Bayesian Belief Networks). Automated parsing and reasoning over such explanations is routine in AI and in Computer Science[16]. However, there are many other types of inference besides logical deduction and other types of explanation besides sequences of causes and effects. It behooves us therefore to seek more general formalisms for representing explanations.

2.2 The role of argument schemes

One such generalization are argumentation schemes with critical questions. Doug Walton was a pioneer in the study of argumentation schemes, both individual

---

[15]This account of how humans recognize faces differs from that given in Oliver Sack's book, *"The Man Who Mistook His Wife for a Hat"* [46]. When one of us (PM) wrote to Sacks in 1987 to contest his account and to propose an alternative, Sacks replied with a suggestion for an experiment to decide between the two alternative explanations. Only decades later did PM learn that Sacks suffered from prosopagnosia.

[16]For instance, every version of Microsoft's Windows Operating System since the release in 1995 of *Windows95*, has included a Bayesian Belief Network for the diagnosis of the likely causes of printer faults.

schemes and collectively.[17] He told one of us (PM) that he had been led to consider these schemes for pedagogical reasons — to make it easier for his students to recognize and critically analyze informal arguments. Only later did he realize that their study could have theoretical and practical implications. His 1996 book [54] appears to have been his first work looking at multiple schemes, but he had written earlier books on particular types of informal argument, for example, on *Ad Hominem* arguments [52] and Slippery Slope Arguments [53].

Arguments schemes are a form of default reasoning where a claim is posited as presumptively true or to be endorsed by default. A rational reaction to the claim may investigate the assumptions being made, implicitly or explicitly, in endorsing the claim and assess whether or not these assumptions hold in any particular case. We could consider consideration of the assumptions to be an assessment of the validity of application of the scheme in a particular case. Endorsing a claim (especially a claim proposing that an action be executed) may entail commitments to endorsements of other claims or to other actions. Arguably, a rational decision–maker (one making decisions based on reasoned grounds) would therefore only endorse the default claim both knowing these commitments and *taking any decision under advisement*, i.e., informed by that knowledge. Thus, a rational decision–maker would also assess the commitments that endorsement of a default claim would entail. As well as eliciting the assumptions behind a presumptive conclusion, critical questions can explore the existence and nature of such entailments.

Argumentation schemes with their associated critical questions have found application in AI, for example in the development of automated argument in practical reasoning [8], in automated dialogues over commands [9], and in automated selection of statistical models for data analysis [48]. Many argument schemes involve default conclusions which are logically fallacious, and so their study has been undertaken in that branch of argumentation theory known as Informal Logic. Even though logically fallacious they may plan an important role in society, particularly in situations where information is incomplete, inconsistent or uncertain. As an example, *Ad Hominem* arguments are criticized by most scientists, since they appear draw conclusions about the content of an argument from personal attributes of the proponent of that argument. Science, it is often argued, should be an objective activity, and so *Ad Hominem* arguments are typically disparaged by scientists. Yet, these arguments play a great role in legal proceedings, because they allow the court to assess the testimony of witnesses and of experts. Over time, in most legal jurisdictions, rules have developed as to when and how such arguments may be made in considering testimony.

---

[17]See [30] for a history of argumentation schemes and related forms of reasoning.

Another example are *epideictic* arguments, which involve drawing conclusions about the substance of claim from the form of its presentation. Although clearly logically fallacious, there are circumstances where this form of reasoning is rational, as William Rehg has argued [42]. Indeed, there are circumstances where epideictic reasoning is also commonplace, as in assessments made by venture capitalists of potential investment proposals from start-ups. In this situation, potential investors may have little past experience on which to base an investment decision, and the start-up may face an uncertain and fast-changing business environment. The marketing plans and financial forecasts of the start-up management team will almost certainly not prove accurate, and so the team's ability to modify their plans in the light of operational experience becomes a better indicator of their potential success than the contents of the current plans themselves. Such abilities may best be assessed, not by the written plans and forecasts, but by the management team's ability to respond to probative questioning from the venture capitalist.

Not only is the use of such logically fallacious informal arguments widespread, there is a strong argument that modern society could not function without their use. Philosopher Charles Willard, for instance, has argued [65] that in a society which depends on complex technology that is too vast and changing too quickly for any one person, or even a small group of people, to ever master completely, then we all need to rely on arguments from authority and on assertions made by experts. The COVID-19 pandemic[18] that so occupies our current attentions illustrates our society's reliance on such arguments with great immediacy. The point is not to avoid such a reliance, because that is infeasible, but rather to make our reliance as rationally justified as possible (within the time available in each case) by means of rational interrogation of the claims of authorities and of experts, and of their supporting arguments and sub-claims.

Thus, for instance, in the case of COVID-19, many governments have relied on advice from expert epidemiologists. Given a particular claim from a particular expert epidemiologist, we could interrogate it according to the Argument Scheme from Expert Opinion that Walton articulated and studied in [55, page 210]. This scheme was presented as an argument with two premises and a default claim, along with six critical questions. Using the notation of a later presentation[19] of the argument

---

[18] For the benefit of any readers who were born after the pandemic, particularly if it has largely disappeared from the historical record in the interim, we note that this paper was written in the throes of the second wave, and that the pandemic as a whole greatly disrupted all aspects of life across the world, including the writing of this paper.

[19] As a trivial example of the wide-ranging nature of the impact of the pandemic is the fact that one author (PM) owns a copy of [55] but has not been to his university office, where the book is located, for 8 months.

scheme [60], E is an expert in some field of knowledge F comprising a finite collection of propositions. The argument scheme consists of:

1. **Major Premise:** Source E is an expert in field F containing proposition A.

2. **Minor Premise:** E asserts that proposition A in field F is true (false).

3. **Conclusion:** Proposition A may plausibly be taken to be true (false).

Walton [55, page 223] proposes six critical questions for this scheme, labelled **CQ1** through **CQ6**, as follows:

CQ1. Expertise Question: How knowledgeable is E as an expert source?

CQ2. Field Question: Is E an expert in the field F that A is in?

CQ3. Opinion Question: What did E assert that implies A?

CQ4. Trustworthiness Question: Is E personally reliable as a source?

CQ5. Consistency Question: Is A consistent with what other experts assert?

CQ6. Evidence Question: Is E's assertion based on evidence?

To these six questions, we would add another:

CQ7. Self-interest Question: Is it the case that E does not stand to gain by our endorsement of proposition A?

As a simple example of the use of this scheme, consider that Anthony asserts that wearing a mask is an effective way to limit the spread of the Severe acute respiratory syndrome coronavirus 2 (SARS-CoV-2) that causes COVID-19. Considered through the lens of the Argument Scheme from Expert Opinion, we might want to check that we can provide positive answers to the critical questions before we are prepared to accept Anthony's argument. In this case we can accept the argument, since (CQ1) Anthony is a extremely knowledgeable, having been extensively cited; (CQ2) Anthony is an expert in a relevant field, that of infectious diseases; (CQ3) Anthony made assertions in [27] implying that wearing a mask was an effective way to limit the spread of SARS-CoV-2; (CQ4) we have no knowledge of Anthony lying, so can consider him a trustworthy source; (CQ5) his advice is consistent with what other experts, for example those in the World Health Organisation[20]; (CQ6) Anthony's

---

[20]https://apps.who.int/iris/handle/10665/332293

assertion is backed by evidence, listed in [27]; and (CQ7) there is nothing to suggest that Anthony has anything to gain by our endorsement of his claim that wearing a mask is an effective way to limit the spread SARS-COV-2.

Note that in order to accept Anthony's claim, we need to examine *all* of the critical questions. If we cannot give a positive answer to any one of the questions, the conclusion should not be accepted. For example, consider Donald, who makes the opposing claim to Anthony, that wearing masks is not helpful in the context of the COVID-19 pandemic. Even if one accepts that Donald is an expert in the field, the fact (CQ4) that he is known to have repeatedly lied on the matters related to the pandemic[21] means that we cannot answer the "trustworthiness question" in the affirmative, and hence Donald's argument claim cannot be accepted.

In the above example, there is only level of analysis. We took the argument, and applied critical questions to that argument. However, a multilevel analysis may sometimes be appropriate. Consider Neil, who claims, for example, that during the pandemic, no more than six people should gather together indoors to limit the spread of the disease. As an epidemiologist, Neil is an expert on disease transmission, and when asked for evidence to support his argument (CQ6), would point to the computational diffusion model that generated the results. In other words, the claim about the "rule of six" rests on the output of a computational model.

Why should we accept that output? Well, the epidemiologists who developed the model would claim, in effect, that it is an oracle which, much like an expert, considers a range of factors that are outside the grasp of most humans. The oracle weighs these factors and produces a summary that the non-experts can use to guide their behaviour. Since the computer model is treated as an expert, we might consider the evidence that it produces for Neil's claim in the same way that we consider the claim itself, that is as an instantiation of an Argument from Expert Opinion. If we do this, then we might want to subject it to a second level of analysis, to check whether neil is justified in relying on it. If we do so, then, in order to answer the "trustworthiness question" (CQ4) it might be wise to ask the opinion of professional programmers, another group of experts part of whose expertise is the ability to establish whether software is reliable, that is whether the outputs of that software are trustworthy.

When experienced programmers look at models like Neil's, they usually find they were built incrementally and with very poor or no software engineering practices. That is, there is no or little documentation, no standard good development models, no agreed statement of specifications, no formal design, no rigorous testing of the

---

[21]See [11] for a list of Donald's many lies on the subject between the start of the pandemic and November 2nd 2020, and [63] for a record of his lies as President. As of September 3rd 2020, the number of lies that Donald had told since taking office was more than 22,000 over the course of 1320 days.

components, and no independent testing by professionals other than the programmers who built the model.[22] This might lead us to question the relaibility of the model, and hence whether Neil's original claim holds.

In contrast to the situation in computational epidemiology, some disciplines which regularly use simulation models, such as economics, specialist expertise increasingly exists on how to evaluate such models, for example [32, 38]. Thus we see an instance of Willard's argument on the inter-connected complexity of contemporary life: evaluation of a statement about the best public policy to reduce the risk of infection during the pandemic may require, for its resolution, evaluation of claims about particular computational simulation models in epidemiology, which, in turn, may require evaluation of claims about software engineering best practice and their application to the particular epidemiological model; few people if any have the necessary skills to evaluate all these claims across the different disciplines involved, from public policy to epidemiology to simulation modeling to software engineering.

2.3 The role of dialogue

A second feature is that evaluation and assessment of explanations might best be undertaken within the context of a dialogue, between an *explainer*, either the entity which generated the explanation or an entity able and willing to answer questions about the explanation, and an *explainee*, an entity seeking to assess the explanation, For human interactions, if one person seeks from another person an explanation of something, and the two have an appropriate social relationship allowing them to engage in a conversation of equals,[23] then our contemporary western cultural experience would lead most of us to expect the two entities to engage in a dialogue involving questions and responses about the explanation. We do not call these responses *"answers"* because they may not be intended by the responder to be answers to a prior question and because, even when so intended, they may not satisfy the questioner.

The questions may serve a number of purposes: they may seek further clarifica-

---

[22] As an example of such analysis, see the anonymous critique a software developer of the code of the Imperial College COVID-19 epidemiological model by published on the Web in May 2020 at: https://lockdownsceptics.org/code-review-of-fergusons-model/

[23] Habermas [20] discusses such social relationships. In this sub-section, we are ignoring interactions which are normally adversarial, such as criminal and military interrogations or courtroom cross-examinations. One of the dialogue types which Walton and Krabbe include in [62] is Eristic dialogues, which are adversarial interactions where one or both parties give vent to anger or frustration. Even these dialogues have been studied by argumentation theorists, e.g., by Dov Gabbay and John Woods [17, 18]; this work has potential applications, for instance, in customer service centre operations.

tion of the explanation; they may seek clarification regarding a response to a prior question; they may seek to identify or make explicit any underlying assumptions in the explanation or in the responses; they may seek to identify consequences of the explanation or of the response (for example commitments to particular beliefs or actions entailed by endorsing the explanation or a response); they may seek to expose internal inconsistencies in the explanation, or in the responses, or in both explanation and responses when considered together; they may seek to contest or argue with the explanation, or its assumptions or consequences, or those of the responses; and, as with any linguistic interaction between two or more parties, the questions posed may seek to clarify previous utterances or concern the operation of the dialogue itself, for example, if there is sufficient time to ask further questions. In other words, this conversation between some person or machine seeking an explanation generated for an AI system and a person or machine who has proposed such an explanation could easily take the form of a dialogues involving questions and responses. For convenience in this paper, let us call these Explanation-Question-Response (EQR) dialogues.

To enable machines to automatically engage in such EQR dialogues, we need to define the rules of the dialogue — their formal syntax, their semantics, and their pragmatics. Although these terms are taken from linguistics, over time they have come to have subtly different connotations in disciplines other than linguistics, firstly in mathematical logic, and then in computer science and AI. In particular, as we discuss in [35], for autonomous computational agents engaged in dialogic interactions, a formal semantics is needed for the agents (and their human or machine designers) to be able to verify, as best they can, that different agents engaged in a dialogue share the same understanding of each other's utterances and of the dialogue itself.[24] Moreover, having a formal semantics and pragmatics for utterances and dialogues can greatly facilitate (or hinder) the computational implementation of interactions. In [35] we discuss these issues at length; here we will briefly mention each element with respect to EQR dialogues.

Syntax

The rules of syntax for a computational dialogue typically govern the permitted forms of utterances and the rules applying to their use. An agent communications language such as ACL developed by the Foundation for Intelligent Physical Agents FIPA (now IEEE FIPA) [15], for example, specifies very strictly the form of each of the 22 permitted utterances, although it has no rules or protocols regarding their

---

[24] Michael Wooldridge showed in [66] that a sufficiently clever software agent can always present to an external observer an insincere representation of its own internal state.

combination. Computer scientists attempting to use this language for agent communications quickly realized that more structure was needed, and so developed specific interaction protocols, for example, for running Dutch auctions [14]. Such protocols, although well and good for their particular intended purposes, lack generality. What was needed was a general theory of dialogue which allowed for different types and purposes of dialogues.

This was found in Doug Walton's 1995 book with Erik Krabbe, *"Commitment in Dialogue: Basic Concepts of Interpersonal Reasoning"* [62]. This work presented a classification of human dialogues in terms of three dimensions:

- What the participants each knew before the dialogue commenced;

- What each participant intended to achieve by participating in the the dialogue (i.e., the goals of each participant); and

- What the goals of the dialogue are.

With these dimensions, Walton and Krabbe identified and analyzed six types of dialogue: Information-Seeking, Inquiry, Persuasion, Negotiation, Deliberation and Eristic. This classification and these dialogue types have been quite influential within AI with computational models being proposed for each of these types (see [35] for a review of applications). Walton and Krabbe do not claim their list is comprehensive, and indeed other types have been studied by researchers in AI. In earlier work [34, 35], we presented a list of the key elements needed for specifying the syntax rules of a dialogue between computational agents, drawing both on Speech Act theory from the Philosophy of Argumentation (as does the ACL language of IEEE FIPA) and on Walton and Krabbe's classification in [62].

Although very influential in AI, the Walton and Krabbe classification is not without some challenges. In a context of autonomous agents, one would have to ask how a dialogue type, an entity without agency, could have goals. At best, *"the goals of the dialogue"* might be understood as the maximal subset of shared goals of the participants for their participation in the interaction, but that would assume they share any goals. In a multi-agent context, that assumption may not apply. In other work, one of us identified *"the goals of the dialogue"* with the set of possible outcomes of interactions conducted under the rules of that dialogue and used these sets of outcomes to design an efficient means of storage of dialogue types [40].

Moreover, in any computational system where participating agents may be designed by independent teams of software developers, there is no guarantee that the stated goals of each participating agent are in fact their real goals. Even without any insincerity on the part of the participants of their design teams, software agents may have buggy code, and so may act contrary to their stated goals.[25] For example, a participant in a dialogue may wish never to a reach a conclusion, or may wish to delay reaching a conclusion until after some other event has occurred, or may join an interaction in order to delay or distract another participant, or just to cause confusion.[26]

For EQR dialogues, we could conceive the appropriate dialogue type to be a Persuasion dialogue, where the Explainer is trying to persuade the Explainee to accept or endorse the explanation provided by the Explainer. The incorporation of critical questions, however, may lead us to consider these interactions as Information-Seeking dialogues (where Explainee is seeking an explanation from Explainer) or Information-Giving dialogues (where Explainer is providing information in the form of an explanation about the operations of some AI system to Explainee). Information-Giving dialogues are not analyzed in the Walton and Krabbe typology [62]. However, in may applications of EQR dialogues, Explainee may wish to see how well and by what means Explainer is able to convey an understanding of the operations of the AI system in this particular case, for this particular decision, and so the dialogue may be closer in form to the Query dialogues of [12], where questioner wants to hear and understand, not just a claim itself, but the arguments for the claim.

Semantics

As far as we are aware, Charles Hamblin was the first person to present a semantics for question–response interactions, in his 1957 PhD thesis [21]. Hamblin's semantics was based on alternative possible worlds, with different responses corresponding to certain propositions being true in different possible worlds.[27] Hamblin later expanded these ideas in a paper that became well-known in linguistics [22]. The subject of the semantics of questions and of question–response interactions has since become a topic of great interest in theoretical linguistics, and there are now several alternative theories; see Floris Roelofsen's linguistics encyclopedia entry [45] for a recent review.

---

[25] For the same reason, the consoling assumption of mainstream economists that agents always act in their own self-interest cannot be made by computer scientists.

[26] Some of these disruptive behaviours have been observed in industry-wide discussions over new computer standards [37].

[27] Hamblin's PhD, which was submitted in 1956, included one of the earliest instances of possible worlds semantics, alongside those of Richard Montague (initially in 1955), Carew Meredith and Arthur Prior (1956), Stig Kanger (1957), A. Bayart (1958, 1959), Saul Kripke (1959, 1962) and Jaako Hintikka (1962). See [13] for a partial history of possible worlds semantics. Hamblin had been a student of Karl Popper and Hamblin's own student Jim MacKenzie argues in [31] that Hamblin was strongly influenced by the ideas of both Popper and Wittgenstein.

In this paper we are proposing the use of argumentation schemes and critical questions for modeling arguments and questions in EQR dialogues. The various semantics for question–response interactions explored in linguistics do not formally incorporate the structure of argumentation schemes and critical questions. The critical questions are not randomly asked, but are specific to the presumptive claim of a specific argumentation scheme, and to its specific (albeit possible implicit) assumptions and its specific potential consequences. We believe this argumentation theoretic structure is important for understanding (and thus for modeling and automatically generating) the reasons why particular questions are asked and for the overall structure of the EQR dialogue in which the questions sit. The semantic frameworks found in linguistics, because they are not based on an explicit argumentation theory, seem too coarse for this purpose.

As an example of how a computational semantic structure can incorporate an explicit philosophy of argument based on argumentation schemes, we mention the work of Katie Atkinson and Trevor Bench-Capon in [6]. Their approach used a framework based on the Alternating-Time Temporal Logic of [2] to create a formal semantics for the syntax for multi-party practical reasoning presented earlier in [8]. The last-cited work articulated a framework for dialogues over what actions to take in some situation (i.e., practical reasoning dialogues) building on Walton's Argumentation Scheme for Practical Reasoning in [54]. We believe that a similar approach would be fruitful for EQR dialogues.

Pragmatics

The pragmatics of utterances and dialogues concerns not their form (the syntax), nor their relationship to truth or reality (their semantics), but other aspects of their meaning unrelated to truth. The most common aspect of meaning unrelated to truth concerns how and when utterances are used, for example: what pre-conditions apply to their use, and what consequences usually follow from their use. In the English language, for instance, asking *"Do you have the time?"* normally results not in an affirmative *"Yes"* response if the responder has the time, but in the provision of the time itself. So part of the meaning of this question is the fact that responders to the question usually answer another question, *"What is the time?"*

It makes sense to talk about the pragmatics of dialogues as well as of utterances, particularly when dialogues are nested, concatenated or interleaved. For example when participants in a Negotiation dialogue start to enact an Information-Seeking dialogue, one may ask if this diversion is somehow necessarily pre-determined by the first dialogue or its contents, or whether it is an appropriate diversion at this point or elsewhere in the first dialogue, etc; see [34] for a discussion of these issues.

Although Speech Act theory from the Philosophy of Language, which is focused on the pragmatics of utterances, has been very influential in the branch of AI devoted to agent communications, the computational study of pragmatics of utterances and of dialogues is still only its infancy in AI.[28] As an example of such work, our paper [36] presents a formal game-theoretic semantics for dialogues over actions, in which the semantics provides a framework for two pragmatic features of speech acts over actions: firstly, the fact that in modern western cultures, such speech often require acceptance by the intended recipient (so-called *"uptake"*) before such utterances create any action commitments; and secondly, that once a commitment is incurred, the rights of revocation of that commitment may no longer lie with with the person who made the utterance.

At first glance, uptake and revocation may be considered unimportant for EQR dialogues because these dialogues do not appear to be concerned with actions. However, insofar that explanations for decisions made or recommended by AI systems do involve actions, whether these actions are before, alongside, or subsequent to the operation of the AI system, these two considerations will be important. For instance, if future regulations or laws governing AI systems require that any implementation of an automated AI decision-system includes both an explanation of how the decision was reached for the intended subject of the decision and also an endorsement (i.e., uptake) of that explanation by the subject (acting as an explainee) before any execution of the decision, then these two pragmatic aspects will be crucially important. In the developed world we now have several decades of experience asking medical patients for their informed consent before implementing medical procedures and treatments, so modeling and implementing these aspects may well be relatively straightforward.[29]

3 The Nosenko Case

The case of Yuri Nosenko, a Soviet citizen who defected to the USA on 4 February 1964, is instructive. Nosenko arrived claiming be employed by the USSR Komitet Gosudarstvennoy Bezopasnosti (KGB) and to have first-hand knowledge of the period in which Lee Harvey Oswald, President Kennedy's assassin, spent as a defector in the USSR, including having seen his KGB files. Based on this knowledge, Nosenko claimed that the USSR had not used Oswald to assassinate Kennedy and indeed that the KGB had played no role in his death.

Opinion was strongly divided within the US Central Intelligence Agency (CIA)

---

[28] Arguably, it may only be in its infancy in Linguistics also.
[29] Although it is not clear that it will be; see [49] for a critique of these practices in medicine.

as to whether Nosenko was a genuine defector or a Soviet plant, intending with his defection to deceive the CIA in some way or simply to cause confusion.[30] He apparently had detailed knowledge of some aspects of KGB operations, but lacked knowledge of others (such as KGB office and human resource procedures). Over the course of the seven years following his defection, management at CIA went through periods of apparent strong belief in Nosenko's sincerity, and periods of apparent strong disbelief. In the former periods, Nosenko was treated well, given free accommodation and even given money. In the latter periods, he was held in solitary confinement and interrogated with ferocity. Among the strongest sceptics of Nosenko's sincerity was the long-term CIA Chief of Counterintelligence, James J. Angleton.[31]

Eventually, CIA leadership in 1969 officially accepted Nosenko as genuine, and he was put on the payroll as a consultant, helping to train CIA officers, for example. As late as 2007, however, Tennent Bagley, a CIA officer who had been involved in the case from the start, published a detailed account arguing for the case that Nosenko was indeed a plant [10]. Nosenko died in 2008.

A key first question for the CIA was thus whether or not to believe Nosenko was genuine. If he was genuine, then so too presumably were his claims about the files he had seen on Oswald, and the denial of Soviet involvement in the Kennedy assassination. But this first question was not the only important question. A second key question, independent of the first, was what should CIA let the Soviets believe was their (the CIA's) answer to the first question. In other words, even if CIA believed (or did not believe) Nosenko, what should they allow the KGB to know — that they did believe him or that they did not?

These two questions arise in any case of a defector, and indeed the KGB would have faced the same two questions in reverse when Oswald had defected to the USSR in 1959; likewise, the CIA would have faced them again when Oswald returned to

---

[30]That intelligence agencies on both sides of the Cold War sought to create confusion in their opponents is well-attested, e.g., see [44]. As an example in the reverse direction to the Nosenko case, Lukes has argued [29] that the show trial and execution of Deputy Prime Minister and former Communist Party General-Secretary Rudolf Slánský and other leading Government officials in Czechoslovakia in 1952 was facilitated by a western intelligence operation which sent false compromising letters to leading party members as part of an operation to sow confusion in Czechoslovakia. A book by journalist Stewart Steven [50] claimed that all the show trials across the region in the late 1940s and early 1950s were the result of a sophisticated western intelligence effort, called *Operation Splinter*, to cause division between the ruling communist parties in the the USSR and those in its Eastern European satellites; however, the claims of the book may be false, and the publication of the book in 1974 may itself have been a disinformation effort intended to cause confusion.

[31]A story based on the Nosenko case features in a 2006 film by Robert de Niro about the life of Angleton, *The Good Shepherd*.

the USA in 1962.[32] The answers to these questions had a special resonance in this case because of the Kennedy assassination aspect. For the CIA to lead the KGB to believe that the CIA doubted the sincerity of Nosenko would have then led the KGB to believe that the CIA doubted Nosenko's claims of no Soviet involvement in Kennedy's assassination. Even if the CIA did doubt those claims, was it in the interests of the CIA (or the USA) for the KGB to think that the CIA may consider the Soviets responsible for the assassination? While enquiries were still ongoing — the Warren Commission into the assassination only reported in September 1964 — it would have behooved the CIA to not allow a clear indication of its conclusion to the first question to be communicated to the USSR, even if a determination had been reached.

Two further complications arise here. One is that the evidence in this case, both that from questioning Nosenko and that from other information, was not clear cut.[33] If the KGB intended to sow confusion with a false defector, then these inconsistencies may well have been deliberate. On the other hand, even if not deliberate and Nosenko sincere, the KGB may also have known about the inconsistencies. Hence, if CIA wanted to convince KGB that their determination about Nosenko's sincerity was itself sincere, then they could not reach that determination (or pretend to reach that determination) too quickly or readily. In other words, the seven-year back-and-forth CIA effort to decide what to think about Nosenko may itself have been a feint, to convince the KGB that the final conclusion was reached with difficulty, and was thus itself sincere.[34] Why that would be necessary is because of the second complication: In any military conflict, it is usually very difficult to communicate a message to your enemy and have them believe it straight away; they will naturally be suspicious of any message you send them directly. For this reason, intelligence agencies may not initially reveal or expel agents of foreign powers they learn are working inside them, because such agents can be useful for the communication of messages to the enemy which the enemy are more likely to believe than direct communications.[35]

In the Nosenko case then, we have a Nosenko-explainer answering questions from

---

[32]The fact that the USSR accepted Oswald as a defector but sent him to the relative isolation of factory work in Minsk, may have been an indication of a lack of trust by the KGB in his sincerity. Similarly, the fact that Oswald does not appear to have faced any impediment to his return to the USA, with the US Embassy in Moscow even lending him money for the fare, despite his earlier renunciation of his US citizenship and public defection to the USSR, would have led some in the KGB to conclude that his first defection had not been genuine, i.e., that he had been a US plant (although not necessarily working for the CIA).

[33]As Bagley shows in [10].

[34]The difficulty of computational modeling of feints in human interactions is discussed in [33].

[35]Some people believe this is one reason why the UK intelligence agencies were slow to expose the Cambridge spies in the 1940s and 1950s.

a CIA-explainee. The explainer may have been seeking to deceive the explainee, and the explainee would have tried to detect such deception. Even if deception by the explainer existed and was discovered by the explainee, the explainee may not have wished to inform explainer Nosenko of this. The CIA-explainee may also have wished to deceive the USSR (specifically the KGB) about whether or not they believed the explanation given by explainer Nosenko. Hence, the explainee's actions, including the environment of the interrogations (e.g., the use of solitary confinement), the lines of questioning adopted, and the order and content of specific questions, may have been part of a larger deception effort aimed at the KGB. Even Bagley's late book [10] may have been part of some greater deception effort.

The purpose of this example is to show the difficulty of accounting for *all* relevant factors and considerations in any computational modeling of explanation dialogues. Both the explainer and explainee may have multiple objectives or agendas in which the Explanation-Question-Response dialogue plays only a small part. These objectives may be in conflict with one another, and may change in the course of the interaction. To achieve particular objectives, either or both the parties may seek to deceive the other, and to deceive external entities who are not parties to the EQR dialogue.

4 Conclusions

Alongside the recent rise to prominence of Machine Learning and Deep Learning within AI has arisen the associated challenge of automatically generating explanations for how automated decision-systems reach the conclusions they do. This challenge is driven by strong pressure from governments and industry regulators in many sectors of the economy to make automated decision-systems and recommendation-systems transparent and fair. For most model-driven AI systems, such as rule-based expert systems, generating explanations for automated decisions is relatively straightforward. For many machine learning and deep learning systems, this task is not. In either case, creating automated explanations leads to a subsequent research challenge: How may we analyze and assess these explanations, and how may we undertake this task automatically?

In this paper, we have outlined an approach to the challenge of automated assessment of explanations drawing on two areas of the philosophy of argumentation to which Doug Walton made important contributions: the study of argument schemes and their associated critical questions, and the classification of types of dialogue he developed with Erik Krabbe. Both these areas have had strong influence in Artificial Intelligence over the last quarter century, particularly in the area known as Agent

Communications. This area seeks to enable automated communications between autonomous intelligent software agents, in other words automated machine-to-machine communications.

In presenting the approach in this paper we have not considered other work of Walton's which is relevant, in particular his study of explanation dialogues, e.g., [56, 57, 58, 59]. We have also not yet considered aspects highlighted by the Nosenko case in Section 3, such as the broader intentions of the participants and the possibility of deception by either or both of Explainer and Explainee. Despite the study of lies, deception and equivocation having a long history in philosophy and theology, computational models of these phenomena are only just emerging, e.g., [47]. Applying these various elements to this challenging problem domain remains future work.

5 Memories of DW

SP: Before I sat down to write this, I thought my first memory of Doug was from the 1996 Formal and Applied Practical Reasoning (FAPR) conference in Bonn which (and this I *am* sure about) was when I first came across the Informal Logic school of work on argumentation. I, like a number of the other attendees at the conference, came to work on logic and argumentation through the AI tradition, and were wholly, and embarrassingly, unaware of this other tradition. When I started to write this, I thought I would check that Doug was there and revisit what he presented. However, I can find no record of his presence — neither in the proceedings nor in any of the material that is now online[36]. As a result, I am no longer sure whether I know Doug from FAPR, or that I became aware of his work around this time though the work of people like Chris Reed, who was quick to connect work on argumentation in AI with that from philosophy. Ultimately, though, it doesn't matter where I first met Doug. What is important, is that he became a near ubiquitous presence in my academic life (and I mean that in a good way). Very quickly Doug's work — initially that on dialogue, subsequently that on argument schemes — became pretty central to a lot of what I work on, and Doug himself turned out to attend many of the events that I went to. He was always interesting to listen to, and though I know some folk who found some of his examples to be a little, shall we say "traditional", I always found him to be both courteous and respectful of everyone I saw him interact with. He was always generous with his time, and in that, and his astonishing productivity, I have long thought of him as a role model, and will continue to do so.

---

[36]Of course, this was still in the dark ages pre Web-2.0, and, as a result, very little of the conference was ever published online.

PM: I first met Doug in 2000 at Pitlochry in Scotland, at the week-long *Argumentation and Computation Symposium* which Chris Reed and Tim Norman organized for philosophers of argumentation to meet computer scientists, held at Bonskeid House.[37] Doug was friendly and courteous, and – I say this as an Australian and intending it as a compliment – very Canadian. I subsequently met him frequently at various conferences and workshops and he was always the same. He was also very helpful to me in providing memories of Charles Hamblin, an Australian philosopher of argumentation and pioneer of computer science, whom he had met and had worked with.

I recall an incident at a workshop held in Bologna, Italy at the 10th International Conference on AI and Law, held at Alma Mater Studiorum – Università di Bologna, in 2005. A student gave a presentation to the workshop which included a discussion of the model of dialogues in Walton and Krabbe [62]. Unknown to the presenter, Doug was sitting in the front row of the audience. Someone asked a question about the dialogue model, to which the presenter responded with a statement that he did not know the answer, and that only the authors of the book would know the answer. Members of the audience who knew Doug laughed, and someone said, "Well, let's ask the author himself!" Doug responded quite humbly, and to the great surprise of the presenter.

References

[1] Model Artificial Intelligence Governance Framework: Second Edition. Personal Data Protection Commission, Government of Singapore, January 21st, 2020.

[2] R. Alur, T.A. Henzinger, and O. Kupferman. Alternating-time temporal logic.

[3] Julia Angwin and Jeff Larson. Bias in criminal risk scores is mathematically inevitable, researchers say. https://www.propublica.org/article/bias-in-criminal-risk-scores-is-mathematically-inevitable-researchers-say, December 30th 2016.

[4] Julia Angwin, Jeff Larson, Surya Mattu, and Lauren Kirchner. Machine bias. https://www.propublica.org/article/machine-bias-risk-assessments-in-criminal-sentencing, May 23rd 2016.

[5] Charles Antaki and Ivan Leudar. Explaining in conversation: Towards an argument model. *European Journal of Social Psychology*, 22(2):181–194, 1992.

[6] Katie Atkinson and Trevor Bench-Capon. Practical reasoning as presumptive argumentation using action based alternating transition systems. *Artificial Intelligence*, 171(10–15):855–874, 2007.

---

[37]The main outputs of the Pitlochry symposium are published in [41].

[7] Katie Atkinson, Trevor Bench-Capon, Floris Bex, Thomas F. Gordon, Henry Prakken, Giovanni Sartor, and Bart Verheij. In memoriam Douglas N. Walton: The influence of Doug Walton on AI and law. *Artificial Intelligence and Law*, 28(3):281–326, 2020.

[8] Katie Atkinson, Trevor Bench-Capon, and Peter McBurney. Computational representation of practical argument. *Synthese*, 152(2):157–206, 2006. Section on Knowledge, Rationality and Action.

[9] Katie Atkinson, Rod Girle, Peter McBurney, and Simon Parsons. Command dialogues. In Iyad Rahwan and Pavlos Moraitis, editors, *Proceedings of the Fifth International Workshop on Argumentation in Multi-Agent Systems (ArgMAS 2008)*, Lisbon, Portugal, 2008. AAMAS 2008.

[10] Tennent H. Bagley. *Spy Wars*. Yale University Press, New Haven, CT, USA, 2007.

[11] Christian Paz. All the President's Lies About the Coronavirus: An unfinished compendium of Trump's overwhelming dishonesty during a national emergency. The Atlantic, https://www.theatlantic.com/politics/archive/2020/11/trumps-lies-about-coronavirus/608647/, November 2nd, 2020. Accessed: 2020-11-14.

[12] Eva Cogan, Simon Parsons, and Peter McBurney. New types of inter-agent dialogues. In Simon Parsons, Nicolas Maudet, Pavlos Moraitis, and Iyad Rahwan, editors, *Argumentation in Multi-Agent Systems*, Lecture Notes in Artificial Intelligence 4049, pages 154–168. Springer, 2006.

[13] B. Jack Copeland. Notes toward a history of possible worlds semantics. In *The Goldblatt Variations: Eight Papers in Honour of Rob*, Uppsala Prints and Preprints in Philosophy, pages 1–14. Department of Philosophy, Uppsala University, Uppsala, Sweden, 1999.

[14] FIPA. Dutch Auction Interaction Protocol Specification. Technical Report XC00032F, Foundation for Intelligent Physical Agents, 15 August 2001.

[15] FIPA. Communicative Act Library Specification. Standard SC00037J, Foundation for Intelligent Physical Agents, 3 December 2002.

[16] Nicholas Frosst and Geoffrey Hinton. Distilling a neural network into a soft decision tree. *arXiv preprint arXiv:1711.09784*, 2017.

[17] Dov M. Gabbay and John Woods. More on non-cooperation in Dialogue Logic. *Logic Journal of the IGPL*, 9(2):321–339, 2001.

[18] Dov M. Gabbay and John Woods. Non-cooperation in Dialogue Logic. *Synthese*, 127(1-2):161–186, 2001.

[19] Jeremy Green. *The Social Construction of the XYY Syndrome*. PhD thesis, University of Manchester, Manchester, UK, 1983.

[20] Jürgen Habermas. *The Theory of Communicative Action: Volume 1: Reason and the Rationalization of Society*. Heinemann, London, UK, 1984. Translation by T. McCarthy of: *Theorie des Kommunikativen Handelns, Band I, Handlungsrationalitat und gesellschaftliche Rationalisierung*. Suhrkamp, Frankfurt, Germany. 1981.

[21] Charles L. Hamblin. *Language and the Theory of Information*. PhD thesis, University of London, London, UK, 1957.

[22] Charles L. Hamblin. Questions in Montague English. *Foundations of Language*,

10(1):41–53, 1973.

[23] Denis J Hilton. Logic and causal attribution. In D. J. Hilton, editor, *Contemporary science and natural explanation: Commonsense conceptions of causality*, pages 33–65. 1988.

[24] Denis J Hilton. Conversational processes and causal explanation. *Psychological Bulletin*, 107(1):65–81, 1990.

[25] Jeff Larson, Surya Mattu, Lauren Kirchner, and Julia Angwin. How we analyzed the COMPAS recidivism algorithm. https://www.propublica.org/article/how-we-analyzed-the-compas-recidivism-algorithm, May 23rd 2016.

[26] Yann LeCun, Yoshua Bengio, and Geoffrey Hinton. Deep learning. *Nature*, 521(7553):436–444, 2015.

[27] Andrea M. Lerner, Gregory K. Folkers, and Anthony S. Fauci. Preventing the Spread of SARS-CoV-2 With Masks and Other "Low-tech" Interventions. *JAMA*, 10 2020.

[28] Stella Lowry and Gordon Macpherson. A blot on the profession. *British Medical Journal*, 296(6623):657–658, 1988.

[29] Igor Lukes. The Rudolf Slánský affair: new evidence. *Slavic Review*, 58(1):160–187, 1999.

[30] Fabrizio Macagno, Douglas Walton, and Chris Reed. Argumentation schemes. History, classifcations, and computational applications. *Journal of Logics and their Applications*, 4(8):2493–2556, 2017.

[31] Jim MacKenzie. What Hamblin's book *Fallacies* was about. *Informal Logic*, 31(4):262–278, 2011.

[32] Robert E. Marks. Validating simulation models: A general framework and four applied examples. *Computational Economics*, 30(3):265–290, 2007.

[33] Peter McBurney, David Hitchcock, and Simon Parsons. The eightfold way of deliberation dialogue. *International Journal of Intelligent Systems*, 22(1):95–132, 2007.

[34] Peter McBurney and Simon Parsons. Games that agents play: A formal framework for dialogues between autonomous agents. *Journal of Logic, Language and Information*, 11(3):315–334, 2002. Special Issue on Logic and Games.

[35] Peter McBurney and Simon Parsons. Dialogue games for agent argumentation. In Iyad Rahwan and Guillermo Simari, editors, *Argumentation in Artificial Intelligence*, pages 261–280. Springer, Berlin, Germany, 2009.

[36] Peter McBurney and Simon Parsons. Talking about doing. In Katie Atkinson, Henry Prakken, and Adam Wyner, editors, *From Knowledge Representation to Argumentation in AI, Law and Policy Making: A Festschrift in Honour of Trevor Bench-Capon on the Occasion of his 60th Birthday*, pages 151–166. College Publications, London, UK, 2013.

[37] Jeremy McKean, Hayden Shorter, Michael Luck, Peter McBurney, and Steven Willmott. Technology diffusion: analysing the diffusion of agent technologies. *Autonomous Agents and Multi-Agent Systems*, 17(3):372–396, 2008.

[38] David F. Midgley, Robert E. Marks, and D. Kunchamwar. The building and assurance of agent-based models: An example and challenge to the field. *Journal of Business*

Research, 60(8):884–893, 2007. Special Issue on Complexities in Markets.

[39] Tim Miller. Explanation in artificial intelligence: Insights from the social sciences. *Artificial Intelligence*, 267:1–38, 2019.

[40] Timothy Miller and Peter McBurney. Efficient storage and retrieval in agent protocol libraries using subsumption hierarchies. *Multiagent and Grid Systems*, 9(2):101–134, 2013.

[41] Chris Reed and Tim Norman (Editors). *Argumentation Machines: New Frontiers in Argument and Computation*. Kluwer Academic, Dordrecht, The Netherlands, 2003.

[42] William Rehg. Reason and rhetoric in Habermas's Theory of Argumentation. In W. Jost and M. J. Hyde, editors, *Rhetoric and Hermeneutics in Our Time: A Reader*, pages 358–377. Yale University Press, New Haven, CN, USA, 1997.

[43] Marco Tulio Ribeiro, Sameer Singh, and Carlos Guestrin. Anchors: High-precision model-agnostic explanations. In *Proceedings of the 32nd AAAI Conference on Artificial Intelligence*, pages 1527–1535, 2018.

[44] Thomas Rid. *Active Measures: The Secret History of Disinformation and Political Warfare*. Macmillan, USA, 2020.

[45] Floris Roelofsen. Semantic theories of questions. In *Oxford Research Encyclopedia of Linguistics*. Oxford University Press, 2019.

[46] Oliver Sacks. *The Man who Mistook his Wife for a Hat*. Summit Books, New York, NY, USA, 1985.

[47] Stefan Şarkadi. *Deception*. PhD thesis, King's College, University of London, London, UK, 2020.

[48] Isabel Sassoon, Sebastian Zillessen, Jeroen Keppens, and Peter McBurney. A formalisation and prototype implementation of argumentation for statistical model selection. *Argument and Computation*, 10(1):83–103, 2019.

[49] Carl E. Schneider. *The Practice of Autonomy: Patients, Doctors and Medical Decisions*. Oxford University Press, Oxford, UK, 1998.

[50] Stewart Steven. *Operation Splinter Factor*. Granada, London, UK, 1974.

[51] Jan Svartvik and Randolph Quirk. *A corpus of English conversation*. Gleerup, Lund, Sweden, 1980.

[52] Douglas Walton. *Arguer's Position: A Pragmatic Study of Ad Hominem Attack, Criticism, Refutation, and Fallacy*. Greenwood Press, Westport, CT, USA, 1985.

[53] Douglas Walton. *Slippery Slope Arguments*. Clarendon Press, Oxford, UK, 1992.

[54] Douglas Walton. *Argumentation Schemes for Presumptive Reasoning*. Lawrence Erlbaum Associates, Mahwah, NJ, USA, 1996.

[55] Douglas Walton. *Appeal to Expert Opinion: Arguments from Authority*. Pennsylvania State University Press, University Park, PA, USA, 1997.

[56] Douglas Walton. A new dialectical theory of explanation. *Philosophical Explorations*, 7(1):71–89, 2004.

[57] Douglas Walton. Examination dialogue: An argumentation framework for critically

questioning an expert opinion. *Journal of Pragmatics*, 38(5):745–777, 2006.

[58] Douglas Walton. Dialogical models of explanation. In *Proceedings of the International Explanation Aware Computing (ExaCt) workshop*, pages 1–9, 2007.

[59] Douglas Walton. A dialogue system specification for explanation. *Synthese*, 182(3):349–374, 2011.

[60] Douglas Walton and Thomas Gordon. Modeling critical questions as additional premises. In F. Zenker, editor, *Argument Cultures: Proceedings of the 8th International Conference of the Ontario Society for the Study of Argumentation (OSSA), May 18-21, 2011.*, pages 1–13, Windsor, ON, Canada, 2011. OSSA.

[61] Douglas Walton, Chris Reed, and Fabrizio Macagno. *Argumentation Schemes*. Cambridge University Press, Cambridge, UK, 2008.

[62] Douglas N. Walton and Erik C. W. Krabbe. *Commitment in Dialogue: Basic Concepts of Interpersonal Reasoning*. SUNY Series in Logic and Language. State University of New York Press, Albany, NY, USA, 1995.

[63] In 1,323 days, President Trump has made 22,510 false or misleading claims. The Washington Post, https://www.washingtonpost.com/graphics/politics/trump-claims-database/, September 3rd, 2020. Accessed: 2020-11-14.

[64] Anne L Washington. How to argue with an algorithm: Lessons from the COMPAS-ProPublica debate. *Colorado Technology Law Journal*, 17:131, 2018.

[65] Charles Willard. Authority. *Informal Logic*, 12(1):11–22, 1990.

[66] Michael J. Wooldridge. Semantic issues in the verification of agent communication languages. *Journal of Autonomous Agents and Multi-Agent Systems*, 3(1):9–31, 2000.

Critical Questions to Argumentation Schemes in Statutory Interpretation

Michał Araszkiewicz
Department of Legal Theory, Jagiellonian University, Kraków, Poland.
michal.araszkiewicz@uj.edu.pl

1 Introduction

Professor Douglas N. Walton, with his enormous legacy of 50 books and over 400 papers published, made outstanding contributions to different spheres of argumentation research, such as theory of argument structure [45, 54], dialogue theories and systems [52], and the issue of burden of proof [49]. Walton's ideas have been used extensively in the field of computational argumentation [15, 53, 16]. Doug Walton's work considerably influenced the research on the computational models of legal reasoning in the area of artificial intelligence (AI) and law. Walton also participated intensively in the research in the field (this area of his contribution was recently summarized in [13]; see [47, 48, 49]). In this paper, we focus on the subject to which he turned his attention during the last decade — the modeling of statutory interpretation with the use of argumentation schemes — developed in a series of publications coauthored by Doug Walton [55, 35, 57, 58]; in the following part of the paper, we will refer to this proposal as the WSM model).[1] Legal interpretation has only recently became the topic of intensive interest in AI and law, and the WSM model stands out as one of the most important contributions that combines insights from computational modeling of legal reasoning, philosophy of argument, and the knowledge of language.

This paper discusses the main features of the model and applies it to the classical set of canons of legal interpretation identified as commonly used in ten different

---

The writing of this paper has been supported by the Polish National Science Centre (Research project agreement UMO-2018/29/B/HS5/01433).

[1]I am aware of the book by Walton, Sartor and Macagno [59] in which the WSM model is extended and developed. However, I have purposely not consulted the content of this paper with Giovanni Sartor or Fabrizio Macagno, as in my opinion it will be interesting to compare the directions of the development taken by the authors of the model with the proposal presented here. This paper was finished before the publication of the said monograph [60].

jurisdictions in the comparative project *Interpreting Statutes* [21], which served as one of the points of departure for the development of the WSM model. We focus on the sets of critical questions attached to particular interpretive argumentation schemes, as opposed to general critical questions that may be used in connection with any interpretive argument, which have been already presented in [57]. We develop the sets of critical questions assigned to the first four (out of eleven) argument types listed by MacCormick and Summers [22], that is, the argument based on the standard ordinary meaning, the argument based on standard technical meaning, the canon of contextual harmonization, and the argument based on precedent.

The order of investigations is as follows. In Section 2, we provide a general theoretical and terminological framework related to the issue of legal interpretation. The next Section recalls the basic features of the theory of argumentation schemes and the WSM model application to the domain of legal interpretation. Section 4 offers a systematic elaboration of the four abovementioned argument types identified in MacCormick and Summers [21], focusing on the critical questions assigned to them. In the last Section, we develop directions for future research concerning theorizing about legal interpretation with the use of argumentation schemes.

2 Modeling Statutory Interpretation

A significant part of legal-theoretical work is devoted, directly or indirectly, to the problems of legal interpretation. The systematic research on legal interpretation traces its roots back to the 19th century [36], and since that time, it has generated a multitude of theories, models, and conceptions. For obvious reasons, we cannot characterize the development of this research in this paper; however, it is possible to indicate important methodological criteria that are useful in systematizing the approaches to theorizing about legal interpretation.

First, the theoretical elaborations of legal interpretation may be classified with regard to their primary purpose into three basic categories: descriptive, analytical, and normative (see [2, 61, 33] for a broader elaboration of the methodological problems of legal reasoning modeling). Descriptive accounts intend to represent actual features of legal interpretation. The scope of such theories depends on the aim of a particular research — what types of objects will be considered as the subject of description; for instance, the psychological processes related to the process of interpretation, the behavior of relevant actors manifested, for instance, in the courtroom in the course of the dispute on the content of law, or the linguistic features of arguments expressed in some sources, for instance, judicial opinions. Analytical approaches aim to elucidate the important structural features of the analyzed phenomenon through the lens of an assumed conceptual scheme or a formal model. The important task of an analytical theory is to explicate the object of interpretation, the structure of interpretive activity, and how the result of this activity should be understood. The distinction between descriptive and analytical models is not rigid, and sometimes, the latter are presented as a subcategory of the former. Lastly, the normative accounts' purpose is to formulate a set of guidelines or more decisive directives concerning the process of interpretation, the criteria of acceptability of interpretive hypotheses, etc. Obviously, in any general theory, or a more concrete model of legal interpretation, all three aspects are present to some extent; however, the classification is useful with regard to indication of the prevailing purpose of a given theoretical account (cf. the methodological discussion in [4]).

Second, theoretical work on legal interpretation may be systematized on the continuum delineated by the distinction universal-local. Universal theories of interpretation intend to cover any case of interpretation — they may be referred to as theories of interpretation in the widest sense [61] and they do not limit themselves to the subject of legal interpretation — but they are concerned with the interpretation of any object of culture. In legal philosophy, we may distinguish, for instance, between theoretical accounts of any legal interpretation and of particular subcategories of interpretation. These categories may still be relatively broad (e.g., as in general theories of statutory interpretation) or much narrower (as in models of interpretation performed by the Court of Justice of the European Union). Statutory interpretation performed by the highest courts is a particularly extensively investigated category (as in [21]). It is subject to dispute whether understanding of a legal provision requires interpretation in any case or whether there exists a class of such easy cases that interpretation thereof does not involve interpretation. Anticipating future investigations, let us state here that the modeling of legal interpretation based on argumentation schemes offers fruitful tools to discuss this issue.

Third, theoretical accounts of legal interpretation may make more or less extensive use of philosophical concepts and theories. This distinction becomes apparent in the context of understanding the notion of meaning in a model of legal interpretation. A standard understanding of the process of interpretation consists of the assignment or meaning of statutory terms. However, the concept of meaning belongs to the class of notoriously debated philosophical notions [38]. Some legal-theoretical works make a direct reference to the debate concerning meaning, while others adopt a "philosophically neutral" attitude. This is particularly common as far as the descriptive models of legal interpretation are concerned: as judges typically do not refer to any specific theory of meaning, the theoretical modeling also remains on the intuitive level and does not explicate the adopted understanding of meaning. In such accounts, the assumptions concerning meaning remain implicit, and it is difficult to subject them to debate. Analogous observations may be made in connection with the use of other philosophically engaging concepts, such as rationality, or, on the level of politico-philosophical background of the legal interpretation, concepts like freedom or equality. A particular theoretical account of legal interpretation may also be only partly explicit with regard to the adopted philosophical assumptions.

Fourth, a theoretical elaboration of interpretation may focus on the process of inventing interpretive statements or on justification thereof. This distinction is (distantly) analogous to the philosophical-scientific distinction concerning the context of discovery and the context of justification, discussed in the area of philosophy of science, for instance, by K. R. Popper [30]. Generally speaking, the former approach emphasizes the process of forming legal interpretation heuresis (or, depending on the particular approach, invention, discovery, or construction), while the latter investigates whether interpretive statements are properly justified. The process of heuresis may be accounted for on different layers: it may concern low- or high-level cognitive and decision processes. As far as the latter layer is concerned, it is often pointed out that the actual reasons that motivate the judges to decide in a particular manner (moral concerns, political preferences, self-interest, etc.) may differ from the articulated reasons based on legal sources. However, it is not plausible to assume that articulated normative reasons do not guide the behavior of judges in any case. Moreover, in the context of legal discourse, only the acceptable reasons may be used for the sake of argumentation; judicial bias or impartiality may be invoked as premises in legal argumentation only if the use of arguments based on such premises is acceptable in a given community. As far as the justification of legal interpretation is concerned, the theory thereof is typically built as a subdomain of a broader theory of justification in legal reasoning. A properly justified conclusion should be justified both internally and externally [61]. A statement is justified internally if it is a conclusion of a properly structured reasoning scheme. A statement is justified externally if the premises of the said reasoning scheme satisfy the required criteria (depending on the nature of the premises and the adopted assumptions, we may require that they are true, more than less probable, proven beyond reasonable doubt, acceptable, reasonable, etc.). A specific case of internal justification is logical validity: a conclusion is justified if it is a logical consequence of an accepted set of premises. However, we may adopt a broader notion of internal justification concerning any structure of reasoning pattern accepted as "appropriate". The criteria of "appropriateness" may be subject to debate.

Fifth, the models of legal interpretation focusing on the issues of justification may be systematized on a continuum from the strictly formalistic ones (which account for the notion of interpretation in terms brought by formal semantics) to entirely informal ones, emphasizing the role of weak criteria such as reasonableness of argument. Traditionally, logically oriented models of legal reasoning were juxtaposed against informal approaches, such as the topical-rhetorical approach [27]. During the last five decades, a major part of the work in the field of the theory of legal reasoning was devoted to the integration of the advantageous features of both approaches: the structural emphasis of the logical models and the apparent realism and richness of argumentative models. The notion of transformations developed by Alexander Peczenik [26] is worth mentioning in this respect. According to this scholar, the inference patterns used in legal reasoning have the nature of transformations (or jumps), which means that they have a non-deductive character, and if the set of their premises is extended for a pattern to conform to a deductive scheme, then at least one of the premises is controversial. The work of Robert Alexy [3]), Neil MacCormick [20]), and Aulis Aarnio [1]) are other notable examples of attempts to reconcile the formal and informal aspects of legal justification. These contributions increased the awareness of the problems related to the modeling of legal reasoning with the use of deductive logic and motivated the extensive application of nonmonotonic logics in the field of law.

To a certain extent independent of the developments of the (methodology of) theories of legal interpretation in scholarly works, a general framework of legal interpretation has been developed in the judiciary. According to the widespread view, the interpretation of statutory provisions is achieved through the application of canons of legal interpretation. There exist a few different systematizations thereof, among which the one presented by MacCormick and Summers [22] is particularly worth mentioning, because it was based on the comparative reconstructive work concerning the practice of justificatory statutory interpretation in ten different countries. Notably, the authors openly assumed that the canons provide the basis for arguments [22, pp. 511–515].

1. Arguments from a standard ordinary meaning

2. Arguments from a standard technical meaning

3. Contextual-harmonization arguments

4. Arguments based on precedents

5. Arguments based on statutory analogies

6. Conceptual-logical arguments

7. Arguments based on legal principles

8. Historical arguments

9. Arguments based on statutory purpose

10. Arguments based on substantive reasons

11. Arguments based on legislative intent

Of course, one may criticize the abovementioned catalogue as non-exhaustive or too simplistic. However, the listed arguments were identified as common in each of the investigated jurisdictions; thus, the list conveys important information concerning the conceptualization of statutory reasoning as expressed in the rationales of opinions of the highest courts.

More importantly, in the *Interpreting Statutes* project, the interpretive canons were explicitly accounted for, as providing a basis for arguments (as opposed to strict rules). The view according to which interpretive reasoning consists in construction and use of arguments enables considering the recurring structures of reasoning, at the same time not restricting the scope of investigations to the deductive patterns. Moreover, it enables a focus on the characteristic content of the premises of particular argument types.

As the purpose of the *Interpreting Statutes* project was not to develop any formal model of legal interpretation, it is not surprising that the involved authors did not discuss the general logical features of the reconstructed arguments. However, due to the analysis of the different types of relations between arguments, including conflicts, the project provided grounds for analysis of statutory interpretation in terms of nonmonotonic inference and argumentation formalisms.

3 Argumentation Schemes and WSM Model

The theory of argumentation schemes [54] is one among many approaches to the systematization of arguments. Its distinctive feature, which considerably contributed to its successful reception in various domains, concerns the careful focus on the content of characteristic premises of non-deductive reasoning patterns and the elaboration of the sets of critical questions, that is, the typical ways to attack an argument attached to each argument scheme. This approach enables accounting for the dialogical character of argumentation, where parties exchange positions in the situation of difference of opinions.

An argumentation scheme may be defined as a prototypical pattern of defeasible inference. An inference pattern is defeasible if it is possible that all its premises satisfy the required status (depending on the context, it may be the status of "true", "justified", "proven", "reasonable", etc.), but their conclusions may still not be accepted for some reason.

The arguments that are used in actual discourse are classified as instances of particular argumentation schemes. Alternative variations exist in each scheme's formulation. In particular, the schemes may be represented in a more descriptive form (informal and often enthymematic), or they can be reconstructed into a more complete form by making the hidden premises explicit and presented in a variety of semi-formal, and eventually formal, accounts.

Reasoning with argumentation schemes is naturally modeled in formal argumentation systems, such as ASPIC+ [23] or Carneades [15]. The family of formalisms applicable to defeasible reasoning is based on nonmonotonic logics, a family of logics initiated by the research of Reiter [32] and Pollock [29] and then developed extensively in the 1990s and later through the works of Vreeswijk [44], Hage and Verheij [18], Verheij [41], Prakken [31], and Hage [17], among others. Nonmonotonic consequence operators allow for a situation where a formula logically follows from a set of formulas F, but does not follow from a set of formulas G, where G is a superset of F. This feature of nonmonotonic consequence operators enables them to elegantly model reasoning with imperfect knowledge, in particular the knowledge that expresses relations that hold generally, normally, in typical situations. The general scheme for defeasible arguments is defeasible Modus Ponendo Ponens:

Premise 1. Generally, if P then Q.
Premise 2. P.
Therefore, Q, if there are no reasons not to accept Q.[2]

The structure of defeasible MPP is not readily visible in every formulation of argumentation schemes, but each argumentation scheme may be easily reconstructed as an instantiation of this general pattern [43, 46]. It also enables clearly defining the types of attack on an argument based on a scheme.

Each argument may generally be attacked in three manners. The terminology used to refer to these attacks was diversified and unstable in the literature of the subject; however, during the last decade the following tripartite division has been widely adopted in the state of the art. First, the undermining attacks question one or more of the premises of an argument. Some undermining attacks are simple in the sense that they indicate that the argument does not have any grounds in the first place. However, others may have a more sophisticated nature; for instance, they may point out that the premises of an argument are not properly justified if the standard of their acceptance has not been met. Second, rebutting attacks offer justification for a conclusion incompatible with the conclusion argued for. A

---

[2]This version of a defeasible MPP is a generalization of the account presented in [54, p. 366] that restricts the set of reasons for not accepting Q to the existence of exceptions to the rule expressed in Premise 1.

rebutting attack neither questions any premise of an argument, nor the relation between the premises and the conclusion, but provides a reason for an alternative conclusion. Third, undercutting attacks focus on the relation between the premises and the conclusion of an argument, to the effect that the premises no longer provide adequate justification to accept the conclusion. The account of undercutting attacks we propose here is a generalized one, and it encompasses the following more specific types of attack: (i) those that show that the conclusion does not follow from the premises; (ii) those that introduce an information that casts doubt on whether the conclusion should be accepted; and (iii) those that demonstrate that the premises do not support the conclusion to a sufficient degree. A classification of a particular attack as an undercutter or an underminer may depend on the manner in which a particular argument is reconstructed. If an argument is presented in its natural, often enthymematic form, then the undercutter is naturally modeled as an attack on the relation between the premises and the conclusion. However, if an argument is reconstructed in a more complete form, and its initially enthymematic premises have been revealed, then a particular undercutting attack may be transformed into an undermining attack, directed against one of the added premises. This added explicit premise will often express a defeasible generalization. The important contribution of the argumentation schemes theory consists in the fact that it offered a specific set of attacks attached to each discussed pattern, and these questions often concern the typically assumed generalizations that provide justification for a conclusion in typical or default situations.

We may refer these types of attacks to the general defeasible MPP scheme in the following manner. The reasons expressed in the antecedent of Premise 1 provide ground for tentative acceptance of the conclusion Q if the antecedent P holds (as per Premise 2). The acceptance is tentative in the sense that it is subject to attacks on the argument supporting conclusion Q, which may either (i) concern the premises of the argument (undermining attack), (ii) attack the relation between the premises and the conclusion, which may also be reconstructed as an attack on an implicit premise (undercutting attack), or (iii) provide reasons concerning a different conclusion (rebutting attack). Therefore, whether a conclusion of a defeasible argument will eventually be accepted depends strongly on the (i) status of its premises and, consequently, the possibilities of questioning them; (ii) scope and character of its implicit premises, indicating the vulnerability of an argument to questions following from the information not taken into account explicitly in the argument's original formulation; and (iii) availability of competing reasons that may lead to adopting conclusions different from the one originally argued for. If an argument is already presented as an instantiation of a defeasible MPP reasoning pattern, then the undercutting attacks generally concern the existing exceptions from the rule expressed in Premise 1. We should note that this list of exceptions does not have to be known at the outset of the argumentative discourse; however, the lists of critical questions assigned to a particular scheme will indicate the most typical ways of attacking an argument, and some of them may, and often will, concern the typical exceptions from the defeasible generalization expressed in the first premise.

Let us now turn to the brief characterization of the WSM model, which is an application of the argumentation schemes theory to the domain of legal interpretation. It is worth noting that the rudiments of this approach were already visible in the perspective taken in the *Interpreting Statutes* project, where statutory interpretive canons were explicitly accounted for as arguments having the characteristic of defeasible inference patterns (although the notion of defeasibility has not been invoked openly, which is understandable on historical grounds). However, the research program concerning the application of argumentation schemes approach to modeling of legal interpretation was explicitly announced in a conference paper by Walton, Sartor, and Macagno [55] and then expanded in subsequent papers [56, 57], receiving an elaborated form in the journal paper [57] and published in a slightly modified version as a chapter of a handbook [58]. Importantly, the systematization of interpretive arguments as presented in Interpreting Statutes is among the important points of departure for the WSM model, although the authors also referred to the classification developed by Tarello (see [57, p. 54]; [39]).

In the WSM model, the interpretive arguments are represented in a general template that consists of three parts:

- The major premise is a general canon: if interpreting an expression (word, phrase, sentence) in a legal document (source, text, statute) in a certain way satisfies the condition of the canon issue, then the expression should/should not be interpreted (depending on whether the canon is a negative or positive one) in that way.

- The minor premise is a specific assertion: interpreting a particular expression in a particular document in a certain way satisfies the condition of the canon.

- The conclusion is a specific claim: the particular expression in that document indeed should/should not be interpreted in that way [58, p. 58].

The structured version of this template may be represented as follows. Big letters represent sets of particular objects, and small letters in italics represent their instantiations.

Major Premise: C (canon): If the interpretation of an E (expression) in a D (document) as M (meaning) satisfies the C's condition, then E in M should (not) be interpreted as M.

Minor Premise: The interpretation of e in d as m satisfies the C's condition.

Conclusion: e in d should (not) be interpreted as m.

This structure encompasses both positive and negative interpretive arguments, that is, arguments that provide reasons pro or contra a particular conclusion. It is worth noting that the model not only represents complete interpretations (which establish the relation of equivalence between an expression and its meaning), but also partial interpretations (representable by means of concept inclusion relation). In addition to the informal approach, the WSM model also proposes to use description logics (DL) to develop a formalized account, which offers the possibility of implementing the framework in computational models of legal reasoning. The proposal concerning the use of DL is on point, because the conclusions of interpretive arguments may be accounted for as expressing the relations between the scopes of concepts [7]. Obviously, the template proposed in the WSM model conforms to the defeasible MPP inference pattern.

It is worth noting that the general template represents only a top level of the interpretive arguments. In judicial practice, the scheme typically needs to be completed by supporting arguments justifying the top-level premises. Precisely on the level(s) of supporting arguments, the nuanced character of interpretive argumentation based on individual arguments becomes apparent. This phenomenon is illustrated by the discussion of actual examples in Walton, Sartor, and Macagno [57, p. 7, 74].

Let us now characterize the WSM model in terms of the methodological distinctions discussed above. The adoption of a common defeasible MPP template and partial formalization using description logics lead to the conclusion that locates the model in the sphere of analytical, rather than dominantly descriptive accounts. Moreover, if the proposed structure of the arguments serves as the pattern for the evaluation of particular arguments, it may also perform a normative function. However, in the commented version, the model serves predominantly analytical purposes, enabling the rational reconstruction of argument patterns. It is worth noting, however, that the descriptive aspect is not absent from this proposal, because the investigated patterns are used to reconstruct actual cases and are, to a large extent, based on the set of arguments developed in *Interpreting Statutes*, eventually based on the broad analysis of actual interpretive arguments.

The scope of the model is broad, definitively more universal than local. The general template provided in the WSM model may be fruitfully applied to the representation of any arguments that consist of ascribing meaning to expressions found in different sources. Although the authors focus on statutory interpretation, the template may also be applied to interpret the rationes found in precedents. Moreover, the template is general enough to be applied also in the sphere of interpretation outside the context of law, wherever it is possible to indicate defeasible general rules providing grounds for justificatory argumentation. However, on the layer of analyzed argumentation schemes, the presented model remains rather conservative, focused on classical interpretive arguments and contexts. The WSM model as presented in the referred sources may be described as moderately philosophical. The concept of meaning is explicitly discussed as related to the intention of the sender of a communication [57, p. 58]. Therefore, the pragmatic layer of the process of interpretation is emphasized [58, pp. 519–520]. However, as "meaning" (M) is one of the variables used in the general argumentation scheme template and is substituted by linguistic expressions, the WSM model directly represents an account of interpretation as transition from linguistic expressions of the statutory text to other linguistic expressions denoted as "meanings". The interpretive arguments are eventually represented as defeasible patterns that make use of description logic formalism; thus, "meanings" are eventually represented as elements of language of the applied logic. The authors also note the philosophical problems related to the concept of "meaning" and particularly of "ordinary meaning" in legal interpretation [57, p. 71], and generally do not investigate the influence of a particular conception of meaning on the set of acceptable interpretations (cf. [38]). This fact enhances the descriptive layer of the WSM model, for real-life judicial argumentation is not an area for discussing philosophical conceptions of meaning. As far as the point of focus on intensional or extensional aspects of meaning is concerned, the choice of DL for the modelling of relations between concepts implies the preponderance of extensional perspective (cf. [8, 9]).

As is common with models representing legal interpretation as argumentation, the WSM model focuses primarily on the process of justificatory reasoning. The model encompasses both internal and external aspects of interpretive justification. The former is encompassed by the logical structure of the general template, that is, the defeasible MPP pattern. The model does not focus extensively on the problems of external justification of legal interpretation, that is, the eventual justification of the premises of interpretive arguments; in particular, the authors do not investigate extensively the procedural aspects of legal argumentation (as in [3]) or general theories of justification, such as coherentism [5, 26, 10, 6]. Although focused on the aspects of justification, the argumentation schemes theory also has a strong heuristic potential because of its focus on dialogical exchange of arguments supporting or demoting different interpretations in the discourse. The situation of dialogue (or a more extensive polylogue) based on the available catalogue of arguments may naturally foster the generation of new interpretations or modifications of the existing ones.[3]

Lastly, the WSM model may be characterized as moderately formalistic. The proposed account for interpretive arguments is semi-formal, and it may be further formalized with different tools. The important feature of the model is that it is not too remote from the source form of interpretive arguments, expressed in natural language, with the reservation that the defeasible MPP pattern structure has become explicit in the model through the general template.

The representation of all arguments based on interpretive canons through a common template has important consequences: the model does not formulate a specific set of critical questions assigned to all argument types, but instead three general critical questions that may be used in connection with the use of any canon:

(CQ1) What alternative interpretations of E in D should be considered?

(CQ2) What reasons are there for rejecting alternative explanations?

(CQ3) What reasons are there for accepting alternative explanations as better than (or equally good as) the one selected? [57, p. 63].

These critical questions may be asked with regard to any argument based on the interpretive canon. However, CQ1 and CQ3 represent only one type of attack, namely, the rebuttal, because the alternative interpretations are naturally the conclusions of arguments other than the considered one. CQ2 has a more general character, but it concerns only attacks on arguments that aim to rebut the original argument. The three general questions encompass an important feature of the process of legal interpretation: the ability of different canons to generate at least a few alternative interpretations of a particular statutory expression. Moreover, sometimes arguments based on one canon may have support for some different conclusions. However, two important points are in order here. First, the presented general approach does not encompass the critical questions that may be raised in connection with a specific type of interpretive argument only. Second, the actual availability of alternative interpretations following from different interpretive canons depends on the possibility of using these other canons in particular interpretive situations. In particular interpretive contexts, the process of legal interpretation may be constrained by the so-called interpretive meta-directives. They may, for instance, preclude the use of a

---

[3]The organization of the environment of lawyers in the manner that they can share their work and observe it and comment on it in real time naturally fosters reconciliation of the presented opinions (contrary to the situation where a lawyer works in isolation from other lawyers). This insight was presented to me by Tomasz Grzegory, Legal Director in Google. I think that the work of lawyers in such environments may be very naturally modelled with the argumentation schemes-based approach.

particular type of legal argument with regard to a category of norms or cases (for instance, the use of statutory analogy is not accepted with regard to criminal law provisions that define the types of prohibited acts), allow the use of particular types of canons only if certain conditions are met, or assign a default preference relation between the arguments based on different canons. Therefore, the available answers to the CQ1 outlined above may be, in many jurisdictions, subject to restrictions of different types. Of course, interpretive meta-directives provide grounds for defeasible arguments that are in general subject to attack. However, in actual contexts, the interpretive meta-directives are often not invoked explicitly, but their content may rather be inferred from the overall structure of the interpretive discourse. Therefore, they often play the role of implicit premises in interpretive reasoning, in particular influencing the distribution of the burden of argumentation in a debate on the meaning of statutory expressions. Moreover, the content of applicable interpretive meta-directives may itself be subject to debate.

4 Critical Questions to Argumentation Schemes Based on Interpretive Canons

In this section, we present sets of critical questions assigned to the four types of arguments discussed in Interpreting Statutes as argumentation schemes in the style of the WSM model. Hence, we extend the perspective brought by the three general critical questions proposed in the WSM model by introducing sets of questions that may be used in connection with particular types of interpretive arguments.

It is not necessary to present each scheme separately, as they all follow the basic defeasible MPP scheme-based template. We only present the first scheme, based on the standard ordinary meaning canon, as reconstructed in Walton, Sartor, and Macagno [57, p. 59].

In the process of developing the sets of critical questions, we deliberately omitted the questions that simply inquire whether the premises of the arguments actually hold, because such questions may be rightly assessed as redundant [42, p. 182]. We have also omitted the questions that may provide a basis for a rebutting attack, namely, the questions concerning alternative interpretations generated by different arguments, as they are encompassed by the general question CQ1 provided in the WSM model. Consequently, we focus first on the questions referring to the types of information that provide a basis for undercutting attacks on arguments based on canons. In other words, the questions at least introduce doubts as to whether the premises of argument provide adequate support for the conclusion. As the arguments are presented instantiations of the defeasible MPP scheme, these questions are naturally interpreted as pointing out the exceptions from the defeasible generalization expressed in Premise 1 of any argument, or concerning the scope of application of a particular canon. This manner of representation of undercutting attacks is natural if a defeasible argument is already represented as an instantiation of a defeasible MPP rule. We also indicate some questions that attack Premise 2 of each argument, although not directly, but rather by referring to types of information that may weaken its justification. However, we do not intend to provide definitive classifications of questions into these two categories. It is important to note that both the possible exceptions to the defeasible generalization expressed in the canon — restrictions on the scope of its application — and the information casting doubt on Premise 2 are types of information that are revealed during the discussion of an interpretive argument and reconstruction of its supportive arguments.

1. Argumentation Scheme Based on the Canon of Standard Ordinary Meaning

Major Premise. If the interpretation of an E (expression) in a D (document) as M (meaning) satisfies the *standard ordinary meaning*, then E in M should be interpreted as M.

Minor Premise: The interpretation of e in d as m satisfies the standard ordinary meaning.

Conclusion: e in d should be interpreted as m.

Standard ordinary meaning is one of the most commonly used, and at the same time, one of the most elusive concepts in the theory of legal interpretation. For obvious reasons, we cannot enter into the discussion of the controversies related to this notion here. It is sufficient to recall the opinion that "standard ordinary meaning," "plain meaning," and similar expressions should be regarded as juridical constructs in the same sense as, for instance, "legal principle" [21, p. 517]. Therefore, there exists no "standard ordinary meaning" of statutory expressions except for what eventually is regarded as such meaning by the relevant communicational community of lawyers. The juridical understanding of "ordinary meaning" should, therefore, not be too easily identified with any account of meaning conceptualized in the field of linguistics. Moreover, it should not be identified with colloquial meaning due to the fact that statutory language, in general, belongs to the official register of an ethnic language. Consequently, the standard ordinary meaning of statutory terms should be regarded as the ordinary meaning of an expression recognized as such in legal discourse, rather than in different areas of discourse. Ordinary meaning should be distinguished from technical meaning (both legal or juridical, or developed in other areas of expertise), and standard ordinary meaning should be distinguished from any type of "non-standard" or "special" meaning, although these distinctions are rather fuzzy, and the criteria providing grounds for these distinctions are controversial and disputable. The legal culture has developed, however, certain sets of circumstances that are generally supportive of assigning "standard ordinary meaning" to particular expressions, and the criteria that demote the strength of arguments based on this canon. The latter circumstances are pointed out in the following set of critical questions.

(CQ1) Does the expression e have a legal definition, or a juridical definition commonly adopted in the literature?

(CQ2) Is the expression e actually used in the contexts of communication outside of a specific area of expertise?

(CQ3) Is the document d a regulation directed toward the general auditorium, or rather toward a specific, professional auditorium?

(CQ4) Is the expression e unambiguous, both syntactically and semantically?

(CQ5) Is the expression e vague or does its scope depend on value judgments?

(CQ6) What is the quality of the drafting of the document d?

(CQ7) Can the document d be described as old regulation?

(CQ8) Is the ascription of meaning m to expression e noted in any reliable sources, including dictionaries?

The following supplementary questions may be applicable if the answer to CQ2 is positive.

(CQ8.1) Is the interpretation of e as m supported by the majority of reliable dictionaries?

(CQ8.2) Are the dictionaries supporting the interpretation of e as m recently published, or rather outdated?

(CQ8.3) Are the dictionaries supporting the interpretation of e as m authored or edited by trustworthy experts?

The above list of critical questions may be briefly commented on as follows. CQ1 restricts the scope of application of the canon based on ordinary meaning, because if there exists a legal definition of a term (especially if the definition is to be found in the document d) or if there is a commonly held juridical definition of the expression e, then, defeasibly, the ordinary meaning canon should not be applicable. CQ2 indicates that the expression e may have a purely technical character; hence, the alleged linguistic convention concerning its ordinary meaning may not exist at all. CQ3 points out that due to the technical character of a particular regulation, the ordinary meaning canon may at least have limited application to the expressions contained therein. CQ4 and CQ5 point out the well-known phenomena of natural language, which may cause the argument from plain meaning to yield equivocal results (in case of ambiguity) or that it will be too weak to determine a concrete meaning of the expression e (in cases of vagueness or sensitivity to value judgments). Potentially poor quality of legislative drafting (CQ6) may also diminish the persuasive force of the argument based on ordinary meaning. If the interpreted document is relatively old (CQ7), then the linguistic conventions associated with its terms may be called into doubt, especially if the interpreted expression is obsolete.

CQ8 and its sub-questions are related to the minor premise of an argument based on ordinary meaning, which states that the interpretation of e in d as m is in accordance with the standard ordinary meaning. The questions ask whether this contention is backed by any reliable sources, including dictionaries. Certain features concerning the referred dictionaries may cast doubt on the advocated meaning of the interpreted expression. Generally, arguments based on lexical dictionaries should not be regarded as providing specifically strong reasons concerning justification of ascription of meaning; they constitute only one type among many reasons that may support an interpretation based on ordinary meaning canon.

The abovementioned observations obviously do not exhaust the scope of problems related to the category of "ordinary meaning" in statutory interpretation, which remains one of the most philosophically engaging notions widely used in legal practice. For instance, we have not investigated the issues concerning the history and evolution of the use of particular expressions: some of them evolve in the extra-legal context and are subsequently assimilated by the law, while others are developed in the domain of legal discourse in the first place, although they are also used in different contexts. Therefore, we may generally distinguish between ordinary meaning of expressions that evolved primarily in general language or in legal language. Non-standard meanings of such expressions may also be considered, which leads to the generation of four categories. Moreover, there may be no clear lines drawn between these categories. These complex issues still require extensive, preferably interdisciplinary, elaboration.

2. Argument Scheme Based on the Standard Technical Meaning.

Some expressions found in the statutory text should be interpreted with regard to their technical meaning, specific to some particular area of knowledge, as opposed to their plain, ordinary meaning (if there exists one). Generally speaking, we distinguish between legal and extra-legal technical meaning, where the former may follow from a statutory definition or a doctrinal theory, and the latter may be determined in the specific area of expertise, such as medicine, engineering, or finances, depending on the regulated domain. Very often, the statutory regulation would take an expression used in a particular area of knowledge and modify it to a certain extent in a legal definition. On certain occasions, legal regulations make use of neologisms. The abovementioned circumstances provide a general basis for considering assigning technical, rather than plain or ordinary, meaning to the interpreted expression. The argument based on the technical meaning canon may be attacked on various grounds, which, to a certain extent, overlap with those concerning the argument based on ordinary meaning.

(CQ1) If a legal definition of the expression e does exist, does it apply to the expression e in the document d?

(CQ2) Is the juridical meaning of the expression e in d settled in a legal doctrine, or rather subject to dispute?

In case there is a doctrinal dispute over the technical meaning of the expression e, particular critical questions may be raised in connection with particular opinions. These questions form a variation of the set critical questions attached to the argument scheme from expert opinion [54, p. 15].

(CQ2.1) Is the author of the cited opinion a reputable member of a legal doctrine?

(CQ2.2) Is the author's opinion justified with appropriate reasons?

(CQ2.3) Is the quoted opinion prevalent in the doctrine, or is it rather a separate and unusual opinion?

(CQ2.4) Is the quoted opinion free of any bias?

(CQ3) Is the document d a regulation directed toward the specific, professional auditorium, or rather toward a general auditorium?

(CQ4) Is the expression e actually used dominantly in the contexts of a specific area of expertise?

(CQ5) Is the meaning of expression e adopted in the specific area of expertise commonly accepted or rather subject to a dispute?

CQ1 assumes that the interpreted expression is the subject of a legal definition, but it investigates whether the interpretation of the expression is actually constrained by this definition. The most evident situation takes place where the legal definition of e is expressed in the same statute (document d) as the provision in which the interpreted expression e occurs. However, this may not be the case. The legal definition of an expression e may occur in a document f different from the document d. In such a situation, the relevance of this definition for the interpretation of the expression e in document d will depend on various factors. For instance, if the legal definition is to be found in a code, then it will be, by default, applicable to the whole branch of law for which the code is relevant. However, specific regulations may provide for a different, more specific definition of an expression e, which may take preference before the definition expressed in a code. A more complicated situation occurs where a legal definition is to be found in a regulation the subject matter of which is distant from the subject matter of the document d. In such a situation, the distance of the subject matter may provide a strong reason against interpreting e in accordance with the legal definition. CQ2 concerns the situation where the expression e is not defined in a statutory provision, but is subject to doctrinal definitions. The existence of such doctrinal understanding of the expression basically precludes the ascription of ordinary meaning; however, as is often the case, the doctrinal debate concerning the meaning of e may be inconclusive. In this connection, other critical questions may also be raised, as doctrinal determination of the meaning of a term is a specific case of argument from expert opinion. CQ3 and CQ4 are analogous to the questions asked in connection with the argument based on standard ordinary meaning, but they naturally have a different direction, while they intend to undercut the assumption that the expression in question does not have a meaning determined outside the specific, "technical" context. Lastly, CQ5 points out the potential semantic disputes in domains of expertise other than law. It should be noted that the critical questions concerning ambiguity, vagueness, evaluative openness or obsolete character of expressions, listed above in connection with the argumentation scheme based on standard ordinary meaning, may also be used in connection with the canon based on technical meaning. Let us note that we may distinguish between standard and non-standard technical meanings.

3. Argumentation Scheme Based on Contextual Harmonization. The contextual harmonization directive states that the meaning of an interpreted expression should not be determined in isolation, but it should take into account the context of the neighboring expressions and provisions, the context of the whole document d in which the interpreted expression occurs, and the context brought by the texts of other relevant normative acts. There may exist a tension between a standard meaning of an expression (ordinary or technical) and its meaning determined by taking into account different contexts. Walton, Sartor, and Macagno [57, p. 68] discuss an example in which, precisely, such conflict is considered.

The list of critical questions assigned to the argumentation scheme based on the canon of contextual harmonization may be reconstructed as follows.

(CQ1) Is the text invoked as an appropriate context for the interpretation of an expression e in document d relevant for the same sphere of regulation or a branch of law?

(CQ2) Is the influence of the context on the interpretation of the expression e in document d reasonably explainable?

(CQ3) Is the regulation involved as the appropriate context for interpretation of the expression e in document d clear and precise?

(CQ4) Are the purposes and values protected by the provisions invoked as an appropriate context for interpretation of an expression e in document d similar or identical to the ones protected by the provision containing the interpreted expression?

The first critical question points to the fact that different branches of law and spheres of regulation may develop their autonomous terminology; hence, the usages of similar or even identical expressions there may not have significance for the interpretation of the expression e. CQ2 requires the proponent of an interpretation based on contextual harmonization to explain precisely and reasonably how the context contributes to the interpretation, which may lead to discovery of fallacious reasoning. CQ3 is a variation of a critical question based on poor quality of legislative drafting or on vague character of terms that were used in the application of the canon. CQ4 introduces a teleological aspect and investigates whether the use of a particular context is acceptable in light of appropriate goals and values.

4. Argumentation Scheme Based on Precedent. The term "precedent" is ambiguous in juridical language, and the concepts associated with it are subject to theoretical debates [37, 19]. However, it is possible to indicate the three widely used senses of "precedent": (1) a judicial decision (or more precisely, a part thereof) that constitutes a formally binding legal norm; (2) a judicial decision that introduces an element of the so-called normative novelty, that is, it introduces new elements to the preexisting system of law; and (3) a judicial opinion that is referred as a (part of) legal basis in the following decisions of application of law, even if the cited decision is neither formally binding nor introduces any normative novelty. It should be noted that each of the three abovementioned criteria is subject to debate. Of course, a detailed investigation of these issues is beyond the scope of this paper; it is sufficient to note that in the context of continental legal culture, it is difficult to draw the line between the interpretation of legal provisions and the development of the legal system through case law (cf. the German concept of *Rechtsfortbildung* [40]), and the introduction of actual new elements of legal norms. However, it is unquestionable that the role of case law is significant in the interpretive discourse and also in jurisdictions where, in principle, judicial opinions do not constitute the source of valid law. Perhaps the naming of this argumentation scheme as based on earlier decisions of application of law would be less controversial than using the term "precedent".

The essence of the canon in question is as follows: if the meaning m has been ascribed to expression e in the document d in earlier judicial opinions, especially if this meaning has been applied uniformly in the existing case law, then it should be interpreted as such unless there are some reasons to depart from it. These reasons may be different in connection with formally biding precedents on the one hand and persuasive precedents on the other hand.

The list of the critical questions to this argumentation scheme may be represented as follows. For the sake of brevity, we will use the expression "precedent-based interpretation" to refer to "an interpretation of expression e in document d as having the meaning m presented and applied in an earlier judicial decision p".

(CQ1) Is the earlier decision, containing the precedent-based interpretation, binding on the court deciding on the interpretation of the expression e in the current fact situation?

  (CQ1.1) CQ1.1. If the precedent-based interpretation is formally biding on the court in the current fact situation, are there any procedural means to overturn it?

(CQ2) Is the precedent-based interpretation universally accepted in the case law, or at least does it amount to a dominating opinion or *jurisprudence constante*?

(CQ3) Are there any judicial decisions that generally argue for rejection of m as the appropriate interpretation of e in d?

(CQ4) Is the precedent-based interpretation correct and properly argued with regard to the preexisting law?

(CQ5) Is the case in the context of which the decision p was issued relevantly similar to the current fact situation?

  (CQ5.1) Are there any other judicial decisions that are at least similar to the current fact situation, but which argue for a different interpretation of the expression e in document d?

(CQ5.2) What are the differences between the case that provided grounds for decision p and the current fact situation?

(CQ6) Are there any amendments to the document d or any other relevant normative acts made since the decision p was issued, or have any other significant changes to the law taken place?

(CQ7) Are there any changes to the social, political, economic, or other context since the decision p was issued?

(CQ8) Is the precedent-based interpretation clear and precise?

(CQ9) Is the precedent-based interpretation well-received in a legal doctrine?

(CQ10) If the procedural regulation allows for it, are there any dissenting opinions, how many of them are presented and how does the quality of argument of the majority opinion relate to the dissenting opinions?

(CQ11) Are there any doubts concerning the composition of the panel of judges that enacted the decision p or other foundational legal issues related to this decision?

If an argument based on precedent is invoked in the process of statutory interpretation, it first has to be checked whether the earlier decision may be formally binding on the court deciding the current fact situation. A positive answer to this question, which may be true not only in the context of the common law culture, but also in many procedural settings in continental legal systems, renders some critical questions less relevant or simply ineffective. For instance, if a court is formally bound by the earlier decision because that decision is prior, typically it is inconsequential to contest the precedent-based interpretation on the basis of its substantial incorrectness (CQ4) or its low degree of precision (CQ5). However, even if an earlier decision is formally binding, in some procedural contexts, there exists a possibility of formally overturning the decision (CQ1.1), even if this possibility may be rarely used in practice.

Especially, if an earlier decision is not formally binding on the court deciding the current fact situation, it is important to check whether an interpretive opinion expressed therein is widely accepted in the judiciary, or if it is rather an isolated statement. In particular, if the precedent-based interpretation expresses a dominating legal opinion, or if it is a part of the line of decisions referred to as *jurisprudence constante*, the persuasive force of an argument based on it will be significant. On the contrary, if the judicial opinions concerning interpretation of an expression e

are disputable and polarized, citing a particular decision may initiate the process of investigations of the details of the developed streams of case law, eventually leading to the rejection of the interpretation assigning meaning m to expression e.

CQ3 encompasses the concept of a general counterexample: a judicial decision that rejects the interpretation of e as incorrect. The generality of such a counterexample should be emphasized: in such a case, an interpretation is rejected as incorrect as such, and not only as inapplicable to a particular class of cases.

CQ4 focuses on the substantial correctness (correctness with regard to the pre-existing law) and the quality of reasoning presented in the quoted earlier decision. A wrongly decided or poorly argued judicial decision quoted as a basis for a particular interpretation of an expression e may at least cast doubt on the appropriateness of this meaning ascription. As we have pointed out earlier, these features may provide a basis for an attack on an argument basically if it is not formally binding in the current fact situation.

CQ5 concerns the classical issues of similarity and dissimilarity between legal cases, extensively discussed and elaborated in the theories and models of case-based reasoning and in the research on analogy [11, 12, 50]. Some interpretations of statutory terms may not have a universal value (as indicated in CQ3) but may be relative to a specific class of states of affairs. This concerns, first and foremost, partial interpretations that do not establish equivalence relations, but indicate what expressions are included in, or overlap with, the scope of other expressions. The characterization of cases is typically modeled with factors: generalizations from the factual descriptions of a case that serve as reasons to decide the case in a particular manner (a summary of this and cognate approaches in AI and law may be found in [14]). However, similarities and differences between cases may also be found on different layers, such as the set of values that may be supported or demoted through particular decisions concerning the cases or the nature of procedural issues related to the case. The more different the decision p is from the current fact situation, the less persuasive the argument for the precedent-based interpretation. The considerations discussed in this context are, to some extent, analogous to the problems concerning the practice of distinguishing in the common law culture.

CQ6 is related to the important, and relatively under-investigated, issue of influence of legislative change on the case law concerning the amended or repealed law. Judicial decisions concerning statutes that were modified typically will lose a degree of relevance in the discussion of legal interpretation here and now, but this effect is subject to many qualifications. For instance, let us assume that judicial decision p fixed the interpretation of an expression e as having the meaning m. Later, the document d is repealed, but the provisions containing the expression e are introduced to a novel statute that regulates similar issues. The interpretation fixed in the decision p apparently may be relevant in connection with the understanding of the new regulation, but this conclusion is not obvious and may be effectively questioned.

CQ7 offers the possibility of attacking a precedent-based interpretation in connection with the changes in the environment of legal regulation. The most spectacular situation of this type concerns the change of political system in a particular jurisdiction (for instance, consider the political transition of the Central and Eastern European countries from socialist regimes to democratic systems during the years 1988–1990) or another major political change, such as the access of a state to the European Union. However, it is also possible to point out other sources that may increase or decrease the significance of an interpretive opinion formed in a prior judicial decision. Such sources include the general change concerning social perspective on certain problems (including the legal position of certain social groups) or the changes in the economic circumstances after the occurrence of natural disasters. These factors may lead to the conclusion that the interpretation of a statutory expression determined in a significantly different context is less relevant nowadays. The eighth critical question concerns the quality of the draft of the quoted judicial opinion and of the interpretation of the expression e eventually advocated by it. It is not unusual that the interpretive conclusion to be found in a judicial opinion generates interpretive doubts itself, especially if it contains a partial, as opposed to complete, interpretation. It is also possible that one judicial opinion will formulate at least two incompatible interpretations of an expression e.

CQ9 concerns the assessment of a precedent-based interpretation in legal doctrine. Legal scholars intend to enter into a dialogue with the most important judiciary bodies and would refer to them in their publications, either affirmatively or critically. If a decision p is subject to intensive and well-argued criticism, it is an important factor that decreases the persuasive force of an argument based on it. On the contrary, if a decision is well-received and praised as a landmark case in commentaries, monographs, and textbooks, an argument based on it is relatively stronger.

The tenth critical question invokes the topic of dissenting opinions. If an interpretive issue decided in the decision p was controversial, and if the procedural law allows for it, it is probable that the issue provoked dissenting opinions. If the decision p is not formally binding in the current fact situation, the argumentation to be found in dissenting opinions may be fruitfully used to question the interpretation advanced by the decision p.

The final critical question concerns serious problems that may arise if the quoted decision was enacted in violation of the principles of the rule of law, but for some reason it has not been formally overturned. For instance, if the composition of the panel of judges was determined to be in violation of the constitutional principles or if the panel obviously acted under the undue influence of some political forces, the impact of the interpretation advanced therein should be minimized in the system following the principles of rule of law.

5. Conclusions and Directions for Further Research on the Application of Argumentation Schemes to Statutory Interpretation. In this paper, we have characterized the WSM model of legal interpretation based on the Waltonian conception of argumentation schemes, and we have reconstructed the sets of critical questions assigned to the four argumentation schemes based on canons. Our analysis confirmed the fruitfulness of the dialectical approach to the modeling of statutory interpretation. Agreeing with the WSM model authors in the opinion that there exist generic critical questions that may be used to attack any argument based on an interpretive canon, we have focused on the reconstruction of critical questions that concern particular argument types.

The analysis has revealed the complicated structure of implicit premises assumed in connection with the use of canons. The source of this complexity is the fact that the concepts that define particular canons, such as "standard ordinary meaning, "standard technical meaning," "context," or "precedent," are juridical constructs that are subject to continuous dispute. In fact, each of such concepts is a part of a broader net of concepts and propositions that constitute a set of background knowledge to any particular situation of statutory interpretation. The process of asking appropriate critical questions reveals the content of these background assumptions. Additional information that is generated in that way serves as the basis of undercutting attacks by weakening the connection between the premises and the conclusion of an interpretive argument. We have purposefully focused on this type of attack because the problem of forming and justifying alternative interpretations has been explicitly considered in the referred version of the WSM model.

We do not claim that the sets of critical questions developed in this paper are exhaustive. On the contrary, it is perfectly possible to extend these sets by indicating new types of critical questions. In the sets developed above, we have also indicated that some critical questions are interrelated: a particular answer to one question may justify or invalidate other questions. The relation between the top-level formulation of an argument based on interpretive canon and its supportive arguments is mirrored by the relation between the main and auxiliary critical questions. The problem of logical relations between particular questions and answers thereto is a research problem in itself, and it can be investigated in the formal models of argumentation by means of erotetic logical systems [60].

In this paper, we have not investigated the particular results of asking critical questions with regard to the distribution of burden of argumentation in the procedural discourse. As indicated above and according to the distinction proposed by Gordon and Walton [15], if a critical question points out the exception from the rule expressed in the major premise of a given argument, the burden of argumentation should rest in the person formulating the critical question. However, if an attack on an argument consists of questioning an assumption, the burden of argumentation should be switched to the proponent of the argument. However, in the practice of interpretive argumentation, whether a given invoked circumstance should count as an exception from a defeasible rule or as an implicit assumption may be subject to dispute.

These observations lead to the acknowledgement of the role of interpretive meta-directives in the process of statutory interpretation. The interpretive meta-directives provide grounds for arguments concerning the sequence of application of interpretive arguments and their default preference relations. In particular, they indicate what reasons should be invoked in considering an interpretation alternative with regard to the one already proposed. Depending on the content of interpretive meta-directives, the application of all interpretive canons may be allowed at the outset or, to the contrary, the application of some arguments may require the satisfaction of certain conditions, especially if the application would lead to the refutation of already obtained interpretive conclusions. However, the interpretive meta-directives also play an important role in the evaluation of a particular argument through the set of critical questions attached thereto. They may, in particular, determine whether a particular circumstance referred to in a critical question requires further justification (to meet the conditions indicated in an interpretive meta-directive) or whether it should be regarded as an effective attack on the argument. Importantly, the sets of interpretive meta-directives are rarely reconstructed to the full extent in legal science; on the contrary, they typically rather manifest themselves though the interpretive practice of law-applying authorities. Moreover, their content and structure may also be subject to debate, as they may be naturally represented as defeasible inference patterns. In particular, the process of determining the content of interpretive meta-directives may be represented as the process of weighing legal principles and values ([28]; see [34] for a general model of balancing reasoning in the law).

The theory of statutory interpretation as defeasible argumentation based on schemes is particularly fit for the application in a research program concerning the interplay between interpretive arguments and interpretive meta-directives. Through the application of a common template, the WSM model and its ancestors may be used not only to develop analytical and normative accounts of statutory interpretation, but also to systematize the results of a descriptive comparative research, similar to the *Interpreting Statutes* project. Moreover, the possibility of implementing the schemes in formal models in computer programs multiplies the potential impact of such research projects. It is highly probable, then, that the argumentation schemes-based approach will become the dominant paradigm in the research on statutory interpretation.

A Personal Note

The first conference I had the pleasure to co-attend with Doug Walton was JURIX 2010: The Twenty-Third Annual Conference on Legal Knowledge and Information Systems, which took place in Liverpool, UK. Doug was a coauthor of a paper with Floris Bex entitled "Burdens and Standards of Proof for Inference to the Best Explanation". As a post-doc just having defended my thesis, I have become very strongly interested in Doug's work. The first occasion to discuss the problems of argumentation with Doug happened shortly after, during the QAJF 2011: Proportionality and Justice conference, which was organized by the European University Institute in Fiesole, Italy (February 24–25). I was extremely honored to have the opportunity to have my talk scheduled just after Doug's insightful presentation entitled "Teleological Argumentation to and from Motives". Doug was kind enough to approach me during the coffee break and ask some questions about my presentation and about my research in general. I was extremely flattered. Doug indicated the strong points of my proposal, but in addition, in a very delicate and elegant manner, he pointed out some features that needed further development. Later, we were able to continue the discussion during dinner, together with Giovanni Sartor and Burkhard Schäfer. One of the topics we had the opportunity to discuss was the systematization of inference patterns present in legal reasoning. What an opportunity to learn, and what a pleasant evening that was!

Since then, I have followed the research of Doug Walton and his collaborators. Even though we have never cooperated formally (for instance, we have never worked on any joint paper or project), we were in contact from time to time on both research-related and organizational issues. For instance, in 2012, I had a pleasure to be the Program Chair of the ARGUMENTATION 2012 conference in Brno, Czech Republic, organized by the groups led by Radim Polčak and Martin Škop, and including my close friend Jaromír Šavelka. Doug submitted a paper coauthored with Giovanni Sartor and Fabrizio Macagno, entitled "Argumentation Schemes for Statutory Interpretation," where they, for the first time, proposed the approach that provided grounds for the WSM model, which is the main topic of this contribution. I have become very interested in the application of the argumentation schemes theory in the field of (computational modeling) of legal interpretation. I also included a lecture about the Carneades system in a course on AI and law, which I have had the pleasure of delivering at Jagiellonian University since 2011.

I have particularly enjoyed the opportunity of long talks with Doug during the MET-ARG 2 workshop, which we co-organized with my friend Tomasz Żurek as one of the events that formed the Warsaw Argumentation Week 2018. Doug was among the speakers at the workshop, and he delivered a fascinating talk entitled "Using Distance in Argument Maps to Model Conditional Probative Relevance". This paper will soon be published as a part of a special issue of this journal, devoted to the problems of evidentiary reasoning in law, one of Doug's favorite research subjects.

Doug Walton was among the brightest and most hard-working people I have had an opportunity to meet in the scientific community. His work speaks for itself, but the people who met him in person will remember him not only as an influential scholar, but also as an outstandingly warm and kind person. Besides, it was lovely to have observed the relationship between him and his wife Karen, who accompanied him at scientific events, and for whose support he was so grateful. I feel privileged to have met Doug. He is truly and will continue to be missed.

References

[1] Aarnio A (1987) *The rational as reasonable: A treatise on legal justification.* D. Reidel.

[2] Aarnio A, Alexy R, Peczenik A, (1981) The foundation of legal reasoning. *Rechtstheorie* 12, pp. 133–158, 257–279, 423–448.

[3] Alexy R (1989) *A theory of legal argumentation*, transl. R. Adler, N. MacCormick, Oxford University Press.

[4] Alexy R (2002) *A theory of constitutional rights*, transl. by J. Rivers, Oxford University Press.

[5] Alexy R, Peczenik A (1991) The Concept of Coherence and Its Significance for Discursive Rationality, *Ratio Juris 3/1*, 130–147

[6] Amaya A (2015) *The Tapestry of Reason, An Inquiry into the Nature of Coherence and its Role in Legal Argument*, Bloomsbury

[7] Araszkiewicz M (2013) Towards Systematic Research on Statutory Interpretation in AI and Law, JURIX 2013, Ashley K. (ed.), *Frontiers in Artificial Intelligence and Applications*, Vol. 235, IOS Press, 15 – 24

[8] Araszkiewicz M (2014) Legal Interpretation: Intensional and Extensional Dimensions of Statutory Terms, in: E. Schweighofer, M. Handstanger, H. Hoffmann, F. Kummer, E. Primosch, G. Schefbeck, G. Withalm (eds.), *Zeichen und Zauber des Rechts. Festschrift für Friedrich Lachmayer*, Weblaw, 469-492

[9] Araszkiewicz M, Żurek T (2015) Comprehensive Framework Embracing the Complexity of Statutory Interpretation, in: A. Rotolo (ed.), *Legal Knowledge and Information Systems - JURIX 2015: The Twenty-Eighth Annual Conference, Braga, Portugal, De-*

cember 10-11, 2015. Frontiers in Artificial Intelligence and Applications 279, IOS Press 145-148

[10] Araszkiewicz M, Šavelka J (eds) (2013) *Coherence. Insights from Philosophy, Jurisprudence and Artificial Intelligence,* Springer

[11] Ashley K (1990) *Modeling legal argument: Reasoning with cases and hypotheticals.* MIT Press, Cambridge MA.

[12] Ashley K (2018) Precedent and legal analogy. In: Bongiovanni G, Postema G, Rotolo A, Sartor G, Valentini C, Walton D (eds) *Handbook of legal reasoning and argumentation,* pp 673–710. Springer.

[13] Atkinson K, Bench-Capon T, Bex F, Gordon TF, Prakken H, Sartor G, Verheij B (2020) In memoriam Douglas N. Walton: The influence of Doug Walton on AI and law. *Artificial Intelligence and Law* 28, 281–326.

[14] Bench-Capon T (2017) HYPO'S legacy: Introduction to the virtual special issue. *Artificial Intelligence and Law* 25: 205–250.

[15] Gordon TF, Walton D (2006) The Carneades argumentation framework — using presumptions and exceptions to model critical questions. In: Dunne PE, Bench-Capon TJM (eds) Computational models of argument. *Proceedings of COMMA 2006,* pp 195–207. IOS Press.

[16] Gordon TF, Friedrich H, Walton D (2018) Representing argumentation schemes with constraint handling rules (CHR). *Journal of Argument and Computation.* 9(2): 91–119.

[17] Hage JC (1997) *Reasoning with rules.* Kluwer Academic Publishers.

[18] Hage JC, Verheij B (1994). Reason-based logic: A logic for reasoning with rules and reasons. *Law, Computers & Artificial Intelligence* 3 (2-3), 171–20.

[19] Lamond G (2016) Precedent and Analogy in Legal Reasoning. *The Stanford Encyclopedia of Philosophy* https://plato.stanford.edu/archives/spr2016/entries/legal-reas-prec/

[20] Neil MacCormick (1978) *Legal Reasoning and Legal Theory,* Oxford University Press

[21] MacCormick DN, Summers RS (eds) (1991) *Interpreting statutes: a comparative study.* Dartmouth.

[22] MacCormick DN, Summers RS (1991a) Interpretation and justification. In: MacCormick DN, Summers RS (eds) 511–44.

[23] Modgil S, Prakken H (2014) The ASPIC+ framework for structured argumentation: A tutorial. *Argument and Computation* 5, 31–62.

[24] Peczenik A (1979) Non-equivalent transformations and the Law, Rechtstheorie–Beiheft 1 In: Krawietz W, Opałek K, Peczenik A, Schramm A (eds.) *Argumentation und Hermeneutik in der Jurisprudenz.* pp. 163–176. Duncker & Humblot.

[25] Peczenik A (1996) Jumps and logic in the law. What can one expect form logical models of legal argumentation? *Artificial Intelligence and Law* 4, 297–329.

[26] Peczenik A (2008), *On Law and Reason* (2nd ed.) Springer, (1st ed. 1989; Kluwer Academic Publishers).

[27] Perelman C, Olbrechts-Tyteca L (1969) *The new rhetoric: A treatise on argumentation,*

trans. University of Notre Dame Press.

[28] Płeszka K (2003) *Reguły preferencji w prawniczych rozumowaniach interpretacyjnych* (The Rules of Preference in Juridical Interpretive Reasoning), *Studia z filozofii prawa (Studies in Legal Philosophy)* 2, Wydawnictwo Uniwersytetu Jagiellońskiego, 77–91.

[29] Pollock J (1987) Defeasible reasoning. *Cognitive Science* 11, 481–518.

[30] Popper KR (1959) *The logic of scientific discovery*. Routledge.

[31] Prakken H (1997) *Logical tools for modelling legal Argument. A study of defeasible reasoning in law.* Kluwer Law and Philosophy Library.

[32] Reiter R (1980) A logic for default reasoning. *Artificial Intelligence*, 13, 81–132.

[33] Sartor G (2005), *Legal reasoning*. Springer.

[34] Sartor G (2010), Doing Justice to Rights and Values: Teleological Reasoning and Proportionality, *Artificial Intelligence and Law* 18, 175–215.

[35] Sartor G, Walton D, Macagno F, Rotolo A (2014). Argumentation schemes for statutory interpretation: A logical analysis. In: Hoekstra R (ed.), *Legal knowledge and information systems (JURIX 2014)*, pp. 11-20. IOS Press.

[36] Savigny von KF (1814) *Vom Beruf unserer Zeit f§r Gesetzgebung und Rechtswissenschaft.* Mohr und Zimmer.

[37] Schauer F (1987) Precedent. *Stanford Law Review* 39, 571–605.

[38] Speaks J (2019) Theories of Meaning. *The Stanford Encyclopedia of Philosophy* https://plato.stanford.edu/archives/win2019/entries/meaning

[39] Tarello G (1980) *L'interpretazione della legge*. Giuffre.

[40] Torggler U (ed.) (2019) *Richterliche Rechtsfortbildung und ihre Grenzen*. Linde

[41] Bart Verheij (1996) Rules, Reasons, Arguments. Formal Studies of Argumentation and Defeat. (1960) Maastricht. Dissertation. http://www.ai.rug.nl/~verheij/publications/proefschrift/

[42] Verheij B (2003) Dialectical Argumentation with Argumentation Schemes: An Approach to Legal Logic. *Artificial Intelligence and Law* 11 (1-2), 167–195.

[43] Verheij B (2003a) DefLog: On the logical interpretation of Prima Facie Justified Assumptions. *Journal of Logic and Computation* 13 (3), 319–346.

[44] Vreeswijk G (1992) Reasoning with defeasible arguments: Examples and applications. *JELIA 1992*: 189–211.

[45] Walton D (1996) *Argumentation schemes for presumptive reasoning*. Lawrence Erlbaum Associates.

[46] Walton D (2004) *Abductive reasoning*. University of Alabama Press.

[47] Walton D (2005) *Argumentation methods for artificial intelligence in law*. Springer.

[48] Walton D (2008) *Informal Logic: A Pragmatic Approach*, 2nd ed. New York: Cambridge University Press.

[49] Walton D (2014) *Burden of proof, presumption and argumentation*. Cambridge University Press.

[50] Walton D (2014a) Argumentation schemes for argument from analogy. In: H.J. Ribeiro

(ed.), *Systematic Approaches to Argument by Analogy.* Cham.

[51] Walton D (2016) *Argument evaluation and evidence.* Springer.

[52] Walton D, Krabbe E (1995) *Commitment in dialogue: Basic concepts of interpersonal reasoning.* SUNY Press.

[53] Walton D, Koszowy M (2017) Arguments from authority and expert opinion in computational argumentation systems. *AI & Society* 32(4), 483–496.

[54] Walton D, Reed C, Macagno F (2008) *Argumentation schemes.* Cambridge University Press.

[55] Walton D, Sartor G, Macagno F (2012) Argumentation schemes for statutory interpretation. In: Araszkiewicz M, Myska M, Smejkalova T, Savelka J, and Skop M (eds.) *International Conference on Alternative Methods of Argumentation in Law*, pp. 63–75.

[56] Walton D, Macagno F, Sartor G (2014). Interpretive Argumentation Schemes. In Hoekstra, R. (ed.), *Legal knowledge and information systems (JURIX 2014)*, pp. 21–22. IOS Press.

[57] Walton D, Sartor G, Macagno F (2016) An argumentation framework for contested cases of statutory interpretation. *Artificial Intelligence and Law* 24:51–91.

[58] Walton D, Sartor G, Macagno F (2018) Statutory interpretation as argumentation. In: Bongiovanni G, Postema G, Rotolo A, Sartor G, Valentini C, Walton D (eds) *Handbook of legal reasoning and argumentation*, pp 519–560. Springer.

[59] Walton D, Sartor G, Macagno F (2020) *Statutory Interpretation: Pragmatics and argumentation.* Cambridge University Press.

[60] Wiśniewski A (1995) *The posing of questions. Logical foundations of erotetic inferences.* Kluwer.

[61] Wróblewski J (1994) *The judicial application of law.* Translated and edited by Bankowski Z, Macormick DN (ed.). Kluwer.

www.ingramcontent.com/pod-product-compliance
Lightning Source LLC
Chambersburg PA
CBHW080238170426
43192CB00014BA/2486